Understanding Behavior in the Context of Time

Theory, Research, and Application

Understanding Behavior in the Context of Time

Theory, Research, and Application

Edited by

Alan Strathman
University of Missouri

Jeff Joireman
Washington State University

LEA LAWRENCE ERLBAUM ASSOCIATES, PUBLISHERS
2005 Mahwah, New Jersey London

Cover art: "Four Stages" by Melissa Montgomery. Used by permission of the artist. Digital photography by Sue Hollingsworth.

Lawrence Erlbaum Associates, Inc., Publishers
10 Industrial Avenue
Mahwah, New Jersey 07430
www.erlbaum.com

Cover design by Kathryn Houghtaling Lacey

Library of Congress Cataloging-in-Publication Data

Understanding behavior in the context of time : theory, research, and application / edited by Alan Strathman, Jeff Joireman.
p. cm.
Includes bibliographical references and index.
ISBN 0-8058-5000-7 (cloth : alk. paper)
1. Time perspective. 2. Time perception. 3. Psychology—Research.
I. Strathman, Alan. II. Joireman, Jeff.
BF468.U53 2005
153.7'53—dc22 2004058488
CIP

Books published by Lawrence Erlbaum Associates are printed on acid-free paper, and their bindings are chosen for strength and durability.

Printed in the United States of America
10 9 8 7 6 5 4 3 2 1

For my mom and dad,
who always wished for me the brightest future imaginable.
A. J. S.

and

For Esther, Trevor, and Joshua
who help me appreciate the past, enjoy the present,
and look forward to the future.
J. A. J.

Contents

IV. Conclusion

Preface

PURPOSE

The goals of this book are twofold. One goal is to present a review of research on the psychological construct known as temporal orientation or time perspective. A second goal is to satisfy the need for a volume that brings together research from the varied domains of behavior that are influenced by time.

OBJECTIVES OF THE CURRENT VOLUME

Rather than focusing on a single aspect of time, such as time perception, time orientation, or temporal construal, we approached the role of time in behavior by first identifying broad domains of behavior in which time is likely to play a role (goal setting, risk taking, interpersonal relations, organizational behavior), and then asking experts in their respective fields to review work in their domain that has either directly or indirectly implicated time as a factor in behavior. Contributors were asked to organize their chapters around theory, research, applications, and directions for future research.

As the reader will see, the chapters differ in their relative emphasis on theory, research, and applications. In some cases, research is only beginning to emerge on the role of time within the given domain, whereas other domains have a relatively long history of research incorporating time. In addition, certain chapters draw on long-established theories, whereas other chapters break new ground by outlining novel theoretical perspectives that help to expand and/or integrate existing domain-specific theories.

This variety provides a thought-provoking mix of ideas that we hope will stimulate future research on the role of time within a variety of domains relevant to people's functioning. Indeed, we believe that in each chapter interested readers are likely to find, in the words of one of our reviewers, a "gold mine of research ideas." Across the various chapters, we believe readers will also come to appreciate the tremendous range of thought, feeling, and behavior that is influenced by time. Our hope is that this book helps scholars learn of the breadth of research already conducted, and stimulates scholars to conduct research that continues to expand on the knowledge already gained.

OVERVIEW OF CHAPTERS

The book is arranged around three classic levels of analysis (intrapersonal-, interpersonal-, and group-level processes). At the intrapersonal level of analysis, Lasane and O'Donnell (chap. 2) begin by reviewing various ways in which time perspective and temporal orientation have been conceptualized and measured. Nurmi (chap. 3) follows by discussing a process model of future orientation and studies addressing how time orientation develops over the life span. Routledge and Arndt (chap. 4) subsequently use terror management theory in an effort to understand the importance people place on time, and why, for instance, people may distance themselves from elderly persons. The next three chapters in this part focus more directly on the outcomes associated with various time orientations. Boyd and Zimbardo (chap. 5), for example, review their research on the Zimbardo Time Perspective Inventory, a measure designed to assess individual differences in five (theoretically distinct) dimensions of time orientation, including past-positive, past-negative, present-hedonistic, present-fatalistic, and future orientations. Through this review Boyd and Zimbardo highlight the benefits of a future orientation, and the disadvantages of past-negative and present orientations, and then forward the more nuanced argument that the best time orientation may, in fact, be an orientation that allows the person to maximize their future outcomes while at the same time taking time to "stop and smell the roses." In a related vein, Finke (chap. 6) reviews research demonstrating that people often discount the value of future rewards, and illustrates how such discounting can have adverse impact on a cluster of behaviors relevant to an individual's physical and financial well-being. Zaleski (chap. 7) subsequently discusses the concept of future anxiety, which he argues will become increasingly relevant in the wake of recent political events such as the September 11th terrorist attacks.

The remaining chapters in Part II focus on how people navigate through time, how they think about events in the past, present, and future, and how temporal factors may impact attitudes and persuasion. Sheldon and Vansteenkiste (chap. 8) discuss how people can use goals to travel through time and how people evaluate whether, in the end, the trip was worth the ride. Sheldon and Vansteenkiste compare and contrast two approaches to evaluating the success of goal setting, the systemic and organismic perspectives, and suggest that well-being will be enhanced when people choose goals that are in accordance with their "true self." Sanna, Carter, and Burkley (chap. 9) next discuss the concept of counterfactual thinking, and integrate counterfactual research within a broader model of mental simulations that outlines how people think about the past, present, and future. Brown and Albarracín (chap. 10) conclude the part with a discussion of the relevance of temporal factors for attitudes and persuasion.

The next part of the book focuses on the role of time within both interpersonal- and group-level processes. At the interpersonal level of analysis, Rumble (chap. 11) begins by discussing the role of time within the theory of social interdependence, and subsequently illustrates how temporal factors have been incorporated into work on both close relationships and negotiation. In the next chapter, Parks and Posey (chap. 12) compare and contrast social interdependence and evolutionary perspectives on cooperation in groups. Anderson and Wood (chap. 13) follow by discussing how temporal concerns may shape aggressive behavior, a discussion couched within the context of the general aggression model. At the group level of analysis, Bluedorn (chap. 14) discusses the idea of future extension, and how future extension relates to the age and productivity of organizations. Joireman (chap. 15) then discusses proenvironmental behavior as a social dilemma, and subsequently reviews research in this area suggesting that a concern with future outcomes can increase people's willingness to sacrifice for the environment. Jones and Brown (chap. 16) round out the discussion of time by reviewing cross-cultural differences in time orientation. Joireman and Strathman provide a concluding chapter in which they summarize the current state of research and offer avenues for future research.

INTENDED AUDIENCE

This book is intended for a professional audience including social, personality, health, environmental, and organizational psychologists. Social psychologists will find numerous chapters addressing a broad range of basic social psychological phenomena including attitudes, counterfactual thinking, terror management, cooperation, and aggression. Personality psychologists will also find much to like, as many of the chapters discuss the role of individual differences in time perspective within various domains. In addition to addressing fundamental issues within social and personality psychology, many of the chapters have an applied emphasis that should appeal to health, environmental, and organizational psychologists. Psychologists interested in basic issues of time perspective, and those interested in the specific domains sampled, will also find many useful ideas in this volume.

ACKNOWLEDGMENTS

We would like to thank all of the authors for their stimulating contributions and those individuals in the following list who provided valuable and timely reviews of the chapters. We hope that, in several years time, all of those involved in the production of this book will be able to reflect back and conclude that their present efforts yielded future benefits.

—Craig A. Anderson
Iowa State University
—Debora Bell
University of Missouri
—Denise Daniels
Seattle Pacific University
—Jessie Fan
University of Utah
—Jeff Greenberg
University of Arizona
—Elizabeth C. Katz
Friends Research Institute, Baltimore, MD
—Laura A. King
University of Missouri
—Neal Roese
University of Illinois
—Amanda Rose
University of Missouri
—P. Wesley Schultz
California State University–San Marcos
—Paul Van Lange
Free University of Amsterdam
—Mark Van Vugt
University of Kent, UK

We would also like to thank four reviewers whose suggestions improved the quality of this work: Jeffrey M. Conte of San Diego State University, Eugene W. Farber of Emory University School of Medicine, Terrell P. Lasane of St. Mary's College of Maryland, and P. Wesley Schultz of California State University–San Marcos.

List of Contributors

Dolores Albarracín, PhD, University of Illinois, 1997, is Associate Professor of Psychology at the University of Florida, PO Box 112250, Gainesville, FL 32611–2250 (e-mail: dalbarra@ufl.edu). She has published multiple papers in the *Journal of Personality and Social Psychology, Personality and Social Psychology Bulletin,* and *Psychological Bulletin,* among others. She has also published two chapters in *Advances in Experimental Social Psychology,* and is completing the edition of the *Handbook of Attitudes and Attitude Change* (with B. T. Johnson and Mark P. Zanna). Her research concerns attitude change and social cognition.

Kathryn B. Anderson, PhD, University of Missouri–Columbia, 1996, is an Associate Professor and Associate Chair of the Psychology Department at Our Lady of the Lake University, 411 S.W. 24th Street, San Antonio, TX 78207 (e-mail: Andek@lake.ollusa.edu). Anderson has published research with colleagues on the effects of uncomfortable situations, personality and gender-related factors on aggressive thoughts, feelings, and behaviors, and the general aggression model. Her current research also tests evolutionary and social structural predictions about the relation between traditional gender role attitudes and perceptions of attractiveness in men and women.

Jamie Arndt, PhD, University of Arizona, 1999, is an assistant professor in the Department of Psychological Sciences, University of Missouri–Columbia, 210 McAlester Hall, Columbia, MO 65211 (e-mail: arndtj@missouri.edu). His research interests include terror management theory; self-esteem and psychological defense; psychodynamic perspectives on motivation; unconscious processes; creativity; psychophysiology.

Allen C. Bluedorn received his PhD in sociology from the University of Iowa in 1976. He is currently the Emma S. Hibbs Distinguished Professor and Chair of the Department of Management at the University of Missouri–Columbia (403D Cornell Hall, University of Missouri–Columbia, Columbia, MO 65211–2600; e-mail: bluedorna@missouri.edu). He is widely published in the organization sciences, and he has served as Representative-at-Large to the Academy of Management's Board of Governors and as Division Chair of the Academy of Management's Organizational Behavior Division. For the last two decades his research has focused on time and organizations, resulting in the following book published in July of 2002: *The Human Organization of Time: Temporal Realities and Experience*, Stanford, CA: Stanford University Press.

John N. Boyd (e-mail: jnboydiii@yahoo.com) is a researcher at Stanford University where he collaborated with Philip Zimbardo on the creation of the Zimbardo Time Perspective Inventory. A decorated teacher, he has authored instructors manuals for a leading introductory psychology text and has published in the areas of human performance, motivation, and time perspective. He received a PhD and MS in psychology from Stanford University and a BA in economics from UCLA.

Rick Brown, PhD, University of Toronto, 2004, is a postdoctoral associate in the Department of Psychology at the University of Florida, PO Box 112250, Gainesville, FL 32611–2250 (e-mail: rickdb@ufl.edu). His interests concern social cognition, attitudes, and persuasion. Much of his current research attempts to shed light on the mental representation of attitudes and attitudinal processes.

William Terrel Brown, PhD in Clinical Psychology, University of Delaware, 2001. Dr. Brown is a licensed psychologist, and currently serves as a research consultant at the Yale Child Study Center in the Yale School of Medicine, 230 South Frontage Road, New Haven, CT 06510, and an Instructor of Psychology at the Norwalk Community College. E-mail: william.brown@yale.edu. His research interests involve investigating the relationships between psychological and social development and educational risk and resilience, and examining the impact of mental health and educational policy on young people's development and educational engagement.

Edward Burkley, MA, is a graduate student in the social psychology program at the University of North Carolina at Chapel Hill. He has conducted research on various topics that involve the relation of the self to social perception.

Seth E. Carter, MS, is a graduate student in the social psychology program at the University of North Carolina at Chapel Hill. His research concerns self-presentation, with a particular emphasis on the interpersonal context in which impression management occurs.

Michael Finke, PhD, Family Resource Management, 1998, The Ohio State University. Assistant Professor, Department of Consumer and Family Economics, 239 Stanley Hall, University of Missouri, Columbia, MO 65211. Email: FinkeM@Missouri.edu. He studies nutrition economics and consumer behavior under risk and uncertainty. His research focuses on intertemporal decision making, including articles establishing a theory of healthy food choice based on consideration of future health consequences, the impact of information and financial resources on food consumption, and time preference and financial resource allocation.

Jeff Joireman (PhD, University of Delaware, 1996) is an Assistant Professor of Psychology at Washington State University and a former Fulbright Scholar. His research interests include decision-making in social dilemmas, pro-environmental behavior, empathy, aggression, social value orientation, and the consideration of future consequences. Email: joireman@wsu.edu

James M. Jones, PhD, Yale University, 1970, is Professor of Psychology at the University of Delaware and Director of the Minority Fellowship Program at the American Psychological Association, Department of Psychology, University of Delaware, Newark, DE 19716. E-mail: jmjones@psych.udel.edu. His research interests include the psychology of temporality and the cultural psychology of African Americans. His TRIOS model (Time; Rhythm; Improvisation Orality and Spirituality) of adaptation and coping encompasses time and extends it to individual differences in cultural processes.

Terell P. Lasane, PhD, received a BS in psychology from Howard University and a MA and PhD in psychology from the University of Delaware in 1993 and 1995, respectively. He investigated punctuality concern as a dimension of time orientation as an undergraduate and examined the time orientation correlates of academic persistence and achievement for his graduate work. He has published several papers on the academic self-concept, examining the role of time orientation, gender role orientation, and ethnic identity on self-regulated academic behavior. He was named Maryland Psychology Teacher of the Year in 2000 by the Maryland Psychological Association.

Jari-Erik Nurmi received his PhD from the University of Helsinki, 1990. Currently, he is working as a Professor of Psychology at University of Jyvaskyla, Finland. His address is University of Jyvaskyla, PO Box 35, 40351 Jyväskylä, Finland (e-mail: nurmi@psyka.jyu.fi). His current research interests include motivation, life-span transitions, adolescent development, and modeling developmental processes.

Deborah A. O'Donnell, PhD, received a doctorate in clinical psychology from Yale University. She is Assistant Professor of Psychology at St. Mary's College of Maryland, Research Consultant to the International Center to Heal Our Children at Children's National Medical Center in Washington, DC, and Clinical Consultant to Child Nurture and Relief, a nonprofit organization working for the psychosocial rehabilitation of orphaned and vulnerable children in conflict areas. Research interests include cross-cultural processes of risk and resilience among violence-exposed youth, the role of self-regulation and future time orientation in overcoming adversity, and prevention and intervention design. She is a licensed clinical psychologist.

Craig D. Parks, PhD, 1991, University of Illinois. Associate Professor, Department of Psychology, Washington State University, PO Box 644820, Pullman, WA 99164. E-mail: parkscd@mail.wsu.edu. Research interests: individual differences in cooperative choice; cognitive influences on cooperative choice; reactions to seemingly uncooperative group members; individual reactions to group failure in social dilemmas and task-oriented groups.

Donelle (Dee) C. Posey, MS, 2002, Washington State University. Current position: doctoral candidate, Washington State University, Department of Psychology, PO Box 644820, Pullman, WA 99164. E-mail: deeposey@wsu.edu. Research interests: structural solutions to resource dilemmas, justice concerns in resource dilemmas.

Clay Routledge, MA, Social Psychology, University of Missouri, 2003. He is currently a doctoral student in the Department of Psychological Sciences at the University of Missouri and is working under the tutelage of Dr. Jamie Arndt. His research interests include existential fear, terror management, psychological defense and growth, intergroup conflict, and social constructivism. Clay can be contacted via e-mail at cdrwbc@mizzou.edu or by post at Department of Psychological Sciences, University of Missouri, Columbia, MO 65211.

Ann C. Rumble, PhD, 2003, Washington State University. Postdoctoral Fellow, Department of Psychology, 238 Townshend Hall, 1885 Neil Avenue Mall, The Ohio State University, Columbus, OH 43210. E-mail: rumble.2@osu.edu. Research interests: social dilemmas, close relationships, interdependence theory, social identity.

Lawrence J. Sanna, PhD is an Associate Professor in the social psychology program at the University of North Carolina at Chapel Hill. He has conducted research on a variety of topics related to the interplay between people's thoughts, feelings, and behaviors over time, with an emphasis on social cognition, mental simulation, and individual and group performance.

Kennon Sheldon, PhD, University of California–Davis, is an Associate Professor at the University of Missouri. He studies goals, motivation, and psychological well-being, using a mixed organismic/evolutionary/cybernetic perspective. His forthcoming book, *Optimal Human Being: An Integrated Multilevel Perspective* (Erlbaum), considers the ideas in this volume's chapter in more detail.

Alan Strathman, PhD, Ohio State University, 1992. Resident Instruction Associate Professor, Department of Psychological Sciences, University of Missouri, 210 McAlester Hall, Columbia, MO 65211 (e-mail: StrathmanA@missouri.edu). Research interests: temporal orientation, particularly the consideration of future consequences; attitudes and persuasion.

Maarten Vansteenkiste is a doctoral student working at the Department of Psychology, Leuven (Belgium). He studies patients' motivation for change from both a theoretical and applied perspective.

Mark D. Wood, PhD, is an Associate Professor of Psychology at the University of Rhode Island, 10 Chafee Road Suite 8, Kingston, RI 02881 (e-mail: mark_wood@uri.edu). He received his Ph.D. in Social Psychology at the University of Missouri–Columbia and completed a 2-year postdoctoral fellowship at Brown University's Center for Alcohol and Addiction Studies. In addition to his interest in alcohol-related aggression, his research focuses on the etiology of alcohol use and misuse and preventive interventions to reduce alcohol-related negative consequences among college students.

Zbigniew Zaleski, Professor in Psychology, dean of the Faculty of Social Sciences at the University of Lublin (Poland). Has teaching and research experience abroad: UCLA, KULeuven (Belgium), Free University of Berlin. Research interests in (cross-cultural approach to) human motivation, time perspective, values, emotion, possession and privacy, interethnic contacts.

Philip G. Zimbardo (e-mail: zim@psych.stanford.edu) is Professor of Psychology at Stanford University. Internationally applauded for his teaching, he is the recipient of distinguished teaching awards from New York University, Stanford University, and the American Psychological Association. He has published a dozen books and more than 100 articles on a wide range of topics, from aggression and shyness to animal and human behavior, individuals, groups, culture, and time perspective. He has been president of the American Psychological Association (2002) and has twice been president of the Western Psychological Association.

I

Introduction

1

A Brief History of Time (Research)

Alan Strathman
University of Missouri

Jeff Joireman
Washington State University

When we say "time is of the essence" we often mean that, either within the confines of a given situation, or within the broader context of our lives, we have a limited amount of time to achieve a valued goal. We should therefore pay close attention to how much time we have left to complete that goal. Time rouses us from our slumber, time structures our days, and time helps to make sense of our lives and the world. And, from the moment we are born, time provides a canvas upon which we paint a past, a present, and a future. This book represents an attempt to understand how people's thoughts, feelings, and behaviors are shaped by time. In this introductory chapter, we briefly review research on time, and give an overview of the chapters contributed to this volume.

TIME IN PSYCHOLOGY

Time has long been of interest to psychologists from a range of disciplines. William James (1890), for example, in his *Principles of Psychology*, included a discussion on the conscious experience of time, particularly as it related to the perception of time duration and the passage of time. Since that time, thousands of studies have examined the issue of time in human behavior (for a summary of this work, see Roeckelein, 2000). Topics of interest include people's awareness (A. J. Edwards, 2002), perceptions (Cottle, 1976), construal (Trope & Liberman, 2003), and experience of time (Gorman & Wessman, 1974); the relevance of time within broad domains of social psychological inquiry (McGrath, 1988); the development and impact of future-oriented thinking on decisions with important personal and social ramifications (e.g., Kirsch, Nijkamp, & Zimmermann, 1988; Strathman, Gleicher, Boninger, & C. S. Edwards, 1994; Zaleski, 1994; Zimbardo & Boyd, 1999); delay of gratification (Metcalfe & Mischel, 1999); and intertemporal choice (e.g., Loewenstein, Read, & Baumeister, 2003).

Reviewing past research in this area is a challenge, given its breadth and the variety of terms that have been used. Searching the literature requires the use of such key terms as time, time perspective, temporal orientation, future orientation, and future time perspective. Even research focused on any one key term can vary dramatically. For instance, research with time as a key word has examined the experience of time, time pressure, time urgency, and time focus (see Lasane & O'Donnell, chap. 2, this volume, for a discussion of this issue).

A detailed review of the literature in this chapter would overlap a great deal with reviews contained in later chapters in this volume. Thus, we provide a brief introduction to research here and encourage interested readers to look to later chapters for more complete coverage of past research.

In the 1950s and 1960s most research conceptualized *time perspective* as a criterion variable. Barndt and Johnson (1955) and Teahan (1958) utilized sentence and story completion tasks to *assess* various aspects of time perspective. Respondents completed the tasks and then were asked to estimate how much time passed during completion of the story or to estimate how far in the future events in the story were likely to occur. In other research (Klineberg, 1968) participants were asked to recall things about which they had recently thought or spoken, or to identify events they were planning for (Lessing, 1972). In each case participants estimated the time frame in which the events would occur. As mentioned, the goal of this research was to find variables (e.g., age) that predicted one's temporal focus.

More recent research has treated time as a *predictor of behavior*. For example, a number of researchers have attempted to identify stable individual differences in some aspect of temporal orientation. Early efforts (Sanders, 1986; Stewart, 1976) resulted in measures without acceptable reliability and validity. More recently, Strathman et al. (1994) developed the *consideration of future* consequences (CFC) scale, whereas Zimbardo and his colleagues (Zimbardo & Boyd, 1999) developed the Zimbardo Time Perspective Inventory (ZTPI). The CFC scale measures the extent to which individuals think about the relatively immediate versus distant consequences of their potential actions (e.g., I am willing to sacrifice my immediate happiness or well-being in order to achieve future outcomes). The CFC scale has excellent reliability and validity (Strathman et al., 1994) and has been found to predict a range of theoretically relevant outcomes including counterfactual reasoning (Boninger, Gleicher, & Strathman, 1994), HIV testing (Dorr, Krueckeberg, Strathman, & Wood, 1999), proenvironmental behavior (Joireman, Lasane, Bennett, Richards, & Solaimani, 2001), and hostility and aggression (Joireman, Anderson, & Strathman, 2003) (for a recent review, see Joireman, Strathman, & Becker, 2004). As Boyd and Zimbardo review in their chapter (chap. 5, this volume), the more recently introduced Zimbardo Time Perspective Inventory measures five dimensions of time perspective, and has been found to predict a number of important outcomes, including risky driving (Zimbardo, Keough, & Boyd, 1997) and drug and alcohol use (Keough, Zimbardo, & Boyd, 1999).

The influence of time has shown up in a variety of other research domains as well. Mischel (1974), for example, examined the concept of delay of gratification and demonstrated that individuals will, in some cases, opt to delay receipt of a reward if they will receive a larger reward at a later point in time (cf. Metcalfe & Mischel, 1999). In his research on cooperation, Axelrod (1984) addressed the importance of the future as it relates to cooperation. Axelrod suggested that cooperation would be enhanced when people expected future interactions, using the term "shadow of the future" to refer to the influence that the future has in present decisions to cooperate. He suggested that stable cooperation requires that the parties involved do not discount the significance of the future to any great extent (see also Parks & Posey, chap. 12, this volume). Most recently, Liberman and Trope (1998; Trope & Liberman, 2003) have demonstrated that time (i.e., whether a behavior is in the near or distant future) can influence the way in which we explain (i.e., construe) our own and others' behavior.

OBJECTIVES OF THE CURRENT VOLUME

The current volume overlaps with the work just outlined, but it is also unique. Rather than focusing on a single aspect of time, such as time perception, time orientation, or temporal construal, we approach the role of time in behavior by first identifying broad domains of behavior in which time is likely to play a role (goal setting, risk taking, interpersonal relations, organizational behavior), and then asking experts in their respective fields to review work in their domain that has either directly or indirectly implicated time as a factor in behavior. Contributors were asked to organize their chapters around theory, research, applications, and directions for future research.

As the reader will see, the chapters differ in their relative emphasis on theory, research, and applications. In some cases, research is only beginning to emerge on the role of time within the given domain, whereas other domains have a relatively long history of research incorporating time. In addition, certain chapters draw on long-established theories, whereas other chapters break new ground by outlining novel theoretical perspectives that help to expand and/or integrate existing domain-specific theories. This variety provides a thought-provoking mix of ideas that we hope will stimulate future research on the role of time within a variety of domains relevant to people's functioning. Indeed, we believe that in each chapter interested readers are likely to find, in the words of one of our reviewers, a "gold mine of research ideas." Across the various chapters, we believe readers will also come to appreciate the tremendous range of thought, feeling, and behavior that is influenced by time. Our hope is that this book helps scholars learn of the breadth of research already conducted, and stimulates scholars to conduct research that continues to expand on the knowledge already gained.

REFERENCES

Axelrod, R. (1984). *The evolution of cooperation*. New York: Basic Books.

Barndt, R. J., & Johnson, D. M. (1955). Time orientation in delinquents. *Journal of Abnormal and Social Psychology, 51*, 343–345.

Boninger, D. S., Gleicher, F., & Strathman, A. (1994). Counterfactual thinking: From what might have been to what may be. *Journal of Personality and Social Psychology, 67*, 297–307.

Cottle, T. J. (1976). *Perceiving time: A psychological investigation with men and women*. New York: Wiley.

Dorr, N., Krueckeberg, S., Strathman, A., & Wood, M. D. (1999). Psychosocial correlates of voluntary HIV antibody testing in college students. *AIDS Education and Prevention, 11*, 14–27.

Edwards, A. J. (2002). *A psychology of orientation: Time awareness across life stages and in dementia*. Westport, CT: Praeger.

Gorman, B. S., & Wessman, A. E. (Eds.). (1997). *The personal experience of time*. New York: Plenum Press.

James, W. (1890). *Principles of psychology*. New York: Holt.

Joireman, J., Anderson, J., & Strathman, A. (2003). The aggression paradox: Understanding links among aggression, sensation seeking, and the consideration of future consequences. *Journal of Personality and Social Psychology, 84*, 1287–1302.

Joireman, J. A., Lasane, T. P., Bennett, J., Richards, D., & Solaimani, S. (2001). Integrating social value orientation and the consideration of future consequences within the extended norm activation model of proenvironmental behavior. *British Journal of Social Psychology, 40*, 133–155.

Joireman, J., Strathman, A., & Becker, C. (2004). *A review of the literature on individual differences in the consideration of future consequences*. Manuscript submitted for publication.

Keough, K. A., Zimbardo, P. G., & Boyd, J. N. (1999). Who's smoking, drinking, and using drugs? Time perspective as a predictor of substance use. *Basic and Applied Social Psychology, 21,* 149–164.

Kirsch, G., Nijkamp, P., & Zimmermann, K. (Eds). (1988). *The formulation of time preferences in a multidisciplinary perspective.* Brookfield, VT: Gower.

Klineberg, S. L. (1968). Future time perspective and the preference for delayed reward. *Journal of Personality and Social Psychology, 8,* 253–257.

Lessing, E. E. (1972). Extension of personal future time perspective, age, and life satisfaction of children and adolescents. *Developmental Psychology, 6,* 457–468.

Liberman, N., & Trope, Y. (1998). The role of feasibility and desirability considerations in near and distant future decisions: A test of temporal construal theory. *Journal of Personality and Social Psychology, 75,* 5–18.

Loewenstein, G. F., Read, D., & Baumeister, R. F. (Eds.). (2003). *Time and decision: Economic and psychological perspectives on intertemporal choice.* New York: Russell Sage.

McGrath, J. E. (1988). *The social psychology of time: New perspectives.* New York: Sage.

Metcalfe, J., & Mischel, W. (1999). A hot/cool-system analysis of delay of gratification: Dynamics of willpower. *Psychological Review, 106,* 3–19.

Mischel, W. (1974). Processes in delay of gratification. In L. Berkowitz (Ed.), *Advances in experimental social psychology* (Vol. 7, pp. 249–292). New York: Academic Press.

Roeckelein, J. E. (2000). *The concept of time in psychology: A resource book and annotated bibliography.* Westport, CT: Greenwood Press.

Sanders, S. (1986). The dimensions of subjective time experience. Unpublished doctoral dissertation. University of Washington, Seattle.

Stewart, R. A. C. (1976). An experimental form of the Stewart Personality Inventory: A simplified format measure of major personality dimensions. *Perceptual and Motor Skills, 43,* 813–814.

Strathman, A., Gleicher, F., Boninger, D. S., & Edwards, C. S. (1994). The consideration of future consequences: Weighing immediate and distant outcomes of behavior. *Journal of Personality and Social Psychology, 66,* 742–752.

Teahan, J. E. (1958). Future time perspective, optimism, and academic achievement. *Journal of Abnormal and Social Psychology, 57,* 379–380.

Trope, Y., & Liberman, N. (2003). Temporal construal. *Psychological Review, 110,* 403–421.

Zaleski, Z. (Ed.). (1994). *Psychology of future orientation.* Lublin, Poland: Towarzystwo Naukowe KUL.

Zimbardo, P. G., & Boyd, J. N. (1999). Putting time in perspective: A valid, reliable individual differences metric. *Journal of Personality and Social Psychology, 77,* 1271–1288.

Zimbardo, P. G., Keough, K. A., & Boyd, J. N. (1997). Present time perspective as a predictor of risky driving. *Personality and Individual Differences, 23,* 1007–1023.

II

Intrapersonal-Level Processes

2

Time Orientation Measurement: A Conceptual Approach

Terell P. Lasane
Deborah A. O'Donnell
St. Mary's College of Maryland

Time, including its experience, definition, and measurement, has interested the human species for centuries. Shadow clocks, sundials, sandglasses, and water clocks represent but a few of the ancient timekeeping devices ingeniously created by our early counterparts. Prisoners, hostages, and others held in dark, windowless dwellings for long periods frequently develop a disrupted and disjointed sense of time perception, robbed of the cues and time measurement strategies available to the rest of us. Becoming agitated, confused, and depressed in this state of time disorientation, many devise makeshift time-tracking systems, cutting marks into their bedposts or belts fashioning a type of time tabulation calendar. The extent to which this time scheme actually reflects the "real" time experienced in the outside world seems less important than the degree to which it provides a framework of consistency, regularity, and predictability for the disoriented individual. We, as human organisms—at least here in Western culture—seem to be uniquely motivated to anchor ourselves in, and orient ourselves in relation to, time.

Time is studied and measured in many disciplines, including history, anthropology, psychology, religion, philosophy, and physics. The linear understanding and experience of time to which most Westerners adhere may in fact be an illusion, at least according to some physicists, who readily admit that they do not understand the construct of time. In most physics theories of time, it is conceptualized as one of the dimensions, as is space. Because time is not a tangible thing physically but can be affected by physical things, physicists assert that it is difficult to explain time beyond witnessing it by the passing of events. As organisms with advanced abilities of abstraction, we construct cognitive, affective, and behavioral structures that allow us to navigate our way through this abstract and intangible yet exceedingly important thing known as time and to conceive of ourselves and others within the context of time.

Social psychology has traditionally understood these cognitive, affective, and behavioral schemata as being formed within, and operating under, two general structural spheres: temporal perspective and temporal orientation. Temporal perspective is seen in most circles as the broader concept of the two and refers to the composite cognitive structures that characterize the way an individual projects, collects, accesses, values, and organizes events that reside in distinct temporal loci. These regions, past, present, and future, are the regions of life space that underlie this time perspective, and within each of these regions, the following dimensions have been assessed (Jones, 1993): extension, density, valence, accessibility, content, and structural organization (see Jones, 1993, for a review).

Extension refers to the length of one's given time span in a given region. Some individuals extend as far as 10 years into the future when cognizing goals or personal wishes, or dealing with anxieties. Others go far into the past in order to examine events that may have occurred in a formative period of development, that have been particularly traumatic and life altering, or that provide the basis for a contingency-reward structure that influences present-day decision making. Still others may have an elongated view of the present frame and may look neither too far behind nor ahead in the cognitions that govern their psychological lives. Extension, as we show, is implicated as a major methodological consideration in measurement of time perspective.

Density is another dimension underlying time perspective. Density, in the physical sciences, is operationalized as mass per unit volume, and this definition provides a useful analog for an understanding of the property of density in the context of time perspective. The density of time perspective can be conceived of as the relative concentration of cognitions that reside in a particular region of temporal space. A person with a dense past perspective, for instance, will possess a large number of thoughts, feelings, images, and memories in a particular time period. Likewise, a person with a dense future perspective may have a concentration of cognitions that are taking place for years after graduation from college, retirement from a job, or when some personal investment has materialized. It is a reasonable postulate of research on time orientation that density in a given temporal space will correlate reliably with various aspects of individual behavior.

Valence refers to the subjective evaluation of the various time regions. For some individuals, sentimental events may produce very positive images of the past. Conversely, for others, the past may be a painful storehouse for memories that produce feelings of hopelessness, despair, and even depression. Others may find satisfaction with present life events and may respond negatively to future events that will result in life events that will appreciably change this current period of contentment and serenity. Still others may find dissatisfaction with present circumstances and may view the future optimistically as a source of escape from the harsh realities that they are currently experiencing. This evaluative component of temporal experiences will likely have a major impact on the way that an individual reacts in day-to-day experiences.

Accessibility is another dimension of time perspective that is related to several of the dimensions we have discussed. The property of accessibility refers to the ease with which an individual can recall and use information from a particular time frame. An individual with a long extension of past memories may access information from the past region with greater facility than an individual who tends to extend far into a dense and positively evaluated future. Individuals with self-regulation difficulties of task completion may find it difficult to accurately estimate the amount of time and attention it takes to complete a particular task. The planning fallacy, which results in overly optimistic and erroneous predictions for how one's plans will proceed, has been linked to poor temporal accessibility. Poor temporal accessibility has been shown to be a predictor accounting for a great deal of the variance in this disruption in effective self-regulation.

Another cognitive structure that underlies time perspective is the actual content of these regions. The content of the past is contained in memories, the content of the present is contained in experience, and the content of the future is contained in expectancies. A great deal of the research on time orientation has shown how life experience largely affects one's ability to cognize, derive motivation, or to be emotionally affected by a particular time frame. Those who come from cultures of relative deprivation, for instance, are often much more present-time oriented than those from relatively economically affluent and thriving societies. Indeed, it is impossible to assess how far one looks into the future or how dense, valued, or accessible the future is without considering the specific feelings, experiences, and thoughts that have occupied the past and present. The future is much more likely to be a source of motivation of present-day activities and decisions if there are no serious threats to mere survival in one's past.

Structural organization is the final dimension that we discuss as a distinct property of time perspective. This property refers to the organizing schemata that connect the past, present, and future. Individuals vary in the way in which they view the linkages between the past, present, and future. For instance, a person who has a strong future perspective may see the past as driving the present and the present as driving the future. More specifically, a college student with a strong dominant future perspective may see hard work and parental guidance as past events that have resulted in present college enrollment. Moreover, these individuals may view the present as a step in a path to the future, which requires completion of a number of subgoals in order for their future goals to be met. On the contrary, a present-oriented college student may view the present as totally distinct from, and unrelated to, a past plagued by disappointment and failure, and such an individual may embrace each moment as intrinsically meaningful in and of itself. In this scenario, the future may be viewed as a broad and unspecific space that is not controllable and is merely the end result of the passing of an infinite collection of random, unrelated "todays." Research exploring the relationship among attributions, locus of control, and self-efficacy bear these relationships out (Lasane & Jones, 1999; Platt & Eisenman, 1968; Wiener, 1986).

DISTINCTIONS BETWEEN TIME PERSPECTIVE AND TIME ORIENTATION

Temporal orientation represents a more circumscribed construct than does time perspective and involves a behavioral predisposition to be more likely influenced by thoughts, emotions, and motivations for a distinct region of time. An individual's time orientation is an individual difference variable that predicts various aspects of an individual's social behavior and the overall self-schema that may reliably drive and influence behavior. Time perspective, then, refers more broadly to the processes utilized in dealing with temporally relevant information, whereas time orientation is the behavioral by-product of the cognitive processes that results in a distinct pattern of responding to objects, events, and situations that implicate a particular temporal space. In the present chapter, we focus on the individual difference variable of time orientation, the nonmutually exclusive counterpart to the broader construct of temporal perspective.

The relation of time to social psychological processes and theoretical traditions has long been undertaken by social scientists; however, it has only been the case in recent years that researchers have made an earnest attempt to provide coherent frameworks that would capture the disparate notions of temporal experience into a unified study (Gjesme, 1983; McGrath, 1988). The preponderance of the research that has addressed the social psychological correlates of temporal experience has examined the predominant tendency for an individual to be oriented toward a distinct region of temporal space. The constructs resulting from this have resulted in the use of interchangeable concepts of time orientation and time perspective. Although most scholars in the field have defined these concepts as the same, some subtle distinctions have been drawn between the two.

The multidimensional aspect of time perspective and its concomitant correlation with time orientation has been cited by several authors and is also levied as a major concern in the valid and reliable measurement of the construct. These concerns converge on the intuitively obvious recognition that it is impossible to measure a construct that has as many conceptualizations as there are measurement tools—both within and beyond the discipline of psychology (Daltrey & Langer, 1984; Gjesme, 1981).

Research with time-related constructs should be regarded with caution because a unidimensional analysis has considerably less explanatory and predictive power than does a program of research that attempts to integrate the cognitive, affective, and behavioral dimensions that are captured in the aspects discussed earlier. Recently, several research programs have endeavored to synthesize these disparate studies and to formulate a theory that incorporates the various dimensions in a unified way (Jones, 1988, 1993; Zimbardo & Boyd, 1999). These endeavors appear warranted in light of the disparate findings that have emerged regarding the uni- versus multidimensional nature of time orientation (Ruiz, Revich, & Krauss, 1967) and the often cited exigency of valid and reliable measurement tools (Daltrey & Langer, 1984; Lessing, 1972; Trommsdorff, 1983). In the present chapter, we give an overview of some of these attempts to measure time orientation.

The bulk of time orientation research has revolved around the cognitive component and has emphasized the future locus of space as the reference point to which behaviors from the past or present are compared. Consistent with this methodological and operational definition bias, many of the social psychology time orientation measurement techniques have focused only on the cognitive dimensions that were first delineated by Wallace (1956).

Kastenbaum (1961) asserted the erroneous tendency of researchers to employ the terms time orientation, time perspective, and time perception interchangeably. He enumerated the following dimensions of future time orientation: extension, how far ahead an individual sees oneself; density, how densely populated an individual views one's future; coherence, the degree of organization with which one sees one's future; and directionality, the degree to which one sees oneself moving ahead in the future. Daltrey and Langer (1984) have used the same dimensions proposed by Wallace (1956) and added the dimension of attitude/affectivity to describe the evaluative manner in which an individual approaches the future. These dimensions overlap considerably with Jones' (1993) conceptualization. These dimensions, with the exception of attitude/affectivity and valence, consider mostly the cognitive aspects of time orientation. Because many scholars have noted that the advancement of a social psychology of time is tantamount to the synthesis of the various conceptualizations, some of these techniques may lack the conceptual breadth needed to capture the content validity of the construct (Nuttin, 1985; Trommsdorff, 1983).

As a psychological construct consisting of cognitive, affective, and behavioral components, time orientation is best conceptualized as a synergistic process involving multiple levels of intra- and interpersonal influence. As such, the existing body of time orientation research has examined the phenomenon from both projective and objective approaches, attempting to operationalize and capture this illusive construct from both subjective intrapsychic and objective direct methods of measurement. Each approach has its own set of strengths and weaknesses.

PROJECTIVE TECHNIQUES FOR MEASURING TIME ORIENTATION

There have been a number of projective techniques that have been employed to measure temporal orientation. Most commonly, the Thematic Apperception Test (TAT; Murray, 1938) is used. The TAT is a 20-item projective measure in which an individual is shown pictures one at a time and asked to make up a story about each picture. This measure has been traditionally used to assess an individual's personality with a special emphasis on dominant drives, emotions, sentiments, complexes, attitudes, and conflicts (Sweetland & Keyser, 1986). Murray postulated that individuals would project their needs, emotions, conflicts, and attitudes onto the ambiguous pictures. Scoring involves subjectively analyzing story content for various themes, with recurrent themes being indicative of various personality dynamics. By focusing TAT administration and coding on motivational, affective, and cognitive components of time orientation, time researchers have attempted to hone in on the intrapsychic processes that drive the mental and psychological structures of time orientation.

Though often a rich source of qualitative data that can provide a window into an individual's drives, thoughts, and feelings, the psychometric properties of the TAT have not been shown to be very impressive (Murphy & Davidshofer,

1998). Some scoring systems have improved reliability, but the validity of the TAT still remains unsubstantiated. Its use has clearly declined over the last 10 to 15 years (Dana, 1996), in both clinical and social psychological research.

An early time orientation study that used the TAT hypothesized that high achievers would have a more predominant and extensive future time perspective than low achievers and that a positive relationship would be found between optimism and extensiveness of future time perspective (Teahan, 1958). Data were obtained from 60 seventh- and eighth-grade males. Thirty of the participants were high achievers (received grades in the upper quartile of their class during the preceding 2 years) and 30 were low achievers (received grades in the lower quartile for same period). Groups were equated for age and socioeconomic status.

The researcher used three instruments to measure time perspective. The first of these was a technique developed by Eson and Greenfield (1962). Participants were asked to record 25 things that they had thought or talked about during the preceding 2-week period. Subjects then rated these items according to whether, at the time they had thought/talked about them, they referred to something in the past, present, or future. The second technique was the story completion technique, first used by Leshan (1952). Subjects were required to write a story starting with a partially completed statement or series of statements. The following sentences were used: (a) "At three o'clock one bright, sunny afternoon in May, two men were out walking near the edge of town ..." and (b) "Joe is having a cup of coffee in a restaurant. He's thinking of the time to come when ...." Finally, participants were presented with three TAT cards: Card 1 (boy with violin), Card 12B (boy sitting alone before a cabin), and Card 14 (silhouette of person in window). Subjects were told to "write a story about this picture." At the end of the testing session, all subjects were asked: "How much time was involved in the action of this story—not in writing it but in the action described? How long would it have all taken if it had really happened?"

Each TAT response was rated by the examiner on 5-point scale according to amount of optimism reflected in outcome. Two other judges rerated the TAT stories of 33 subjects (16 high and 17 low achievers) in a reliability check. Correlations were .85, .91, and .91 for cards 1, 12B, and 14, respectively. The story completion (SC) task was least reliable in terms of correlations between scores based on different stories. Consistent with their hypotheses, these researchers found that high achievers tend to look mostly toward the future and have more extensive future time perspective than low achievers on most of the projective tests. Students high in future extension also appeared to be more optimistic.

Wohlford (1968) undertook a study examining extension of personal time through the use of two projective techniques—the TAT and the SC task—in a test–retest design influenced by three independent variables, including stimulus cue (TAT vs. SC), TAT instructions (structured vs. unstructured), and order of presentation (SC–TAT or TAT–SC). One hundred sixty men and women were tested in two separate studies. Results supported bidirectional (past–future) and cognitive-empirical distinctions. Empirical pretension emerged as a stable variable and as the most important root common to both the SC and TAT methods. Later order and structured TAT administration tended to lengthen retrotension and pretension scores. Variations in method, such as the order of presentation, or the relative structure of the TAT administration, significantly influenced the length of extension scores. When cues were presented later in order, subjects told fantasy stories with longer TAT and SC extension scores. TAT variables attained consistently higher test–retest reliability than the SC variables in spite of the fact that the sets of SC cues were analytically more identical than the sets of TAT cues. The use of structured TAT administration rather than the less structured approach increased interrater reliability for men, but not for women. These results, taken together, suggest that the manner in which projective tests of time orientation are administered seems to impact results.

Other projective techniques, including life graph and drawing exercises, and measures prompting subjects to record events, thoughts, and aspirations in an open-ended fashion, have also been modified and used in time orientation research. As with other projective measures, these techniques have both methodological strengths and shortcomings. Overall, projective measures such as these are less susceptible to faking than are self-report inventories, as the purpose of the measure is usually disguised, and may provide an effective means of establishing rapport. However, their conspicuous deficiencies in standardization practices and normative data make interpretation difficult. Low levels of test–retest and interrater reliability are of particular concern (Anastasi & Urbina, 1997). Validation studies of projective measures have emphasized problems with criterion contamination, failure to cross-validate, and "illusory validation," which arises from our human tendency to notice and recall that information that fits our expectations while ignoring data that is contradictory to expectations (Kinslinger, 1966).

The cognitive extension into the future, or pretension, has been measured via a number of projective techniques. In one representative study, Barndt and Johnson (1955) developed a projective measure of cognitive extension into the future. Their story completion technique provides participants with a story root and requires that they complete the story. According to these researchers, orientation toward the future is acquired by incidental learning during childhood and adolescence. The variety of family atmospheres typical of different social classes would be expected to produce different time orientations. In order to test the construct validity of this measure, they administered this technique to 26 delinquent boys (ages 15–17) who had been committed to a state rehabilitation school by court action, and 26 control boys with no court history from a small high school. Control participants were selected to fall within the same IQ and age range as experimental group participants.

Their technique was similar to the story-telling technique first used by LeShan. Researchers instructed participants to "Tell me a story." The delinquent boys were unable to respond to this ambiguous prompt, so it was revised a bit to the following: "I want to see what kind of a story you can tell. I'll start a story and then let you finish it any way you want to. Let's see how good a story you can tell. I'll start it now. 'About 3 o'clock one bright, sunny afternoon in May two boys were walking along a street near the edge of town.' Now you start there and finish the story any way you want to." The stories were recorded and later transcribed in full. If no time or time interval was included in story, the following prompt was stated "How long was this from the start of the story?" In this way it was possible to score the stories in terms of specific time intervals dating from 3 o'clock. The stories were assigned score values depending on the length of time covered by the action in the story. In support of their hypotheses, the researchers found that delinquent boys produced stories with significantly shorter time spans than control boys. This story completion technique has also been used by a number of researchers in measuring cognitive extension into the future (see Barndt & Johnson, 1955; Kastenbaum, 1961; Klineberg, 1968; Lessing, 1972; Platt & Eisenman, 1968; Ruiz et al., 1967; Teahan, 1958).

Wallace (1956) developed another technique to measure cognitive protension. This method involves instructing participants to list 10 events that may occur to them during the rest of their lives. The extension was measured in this task by obtaining the difference between the subject's age at the time the task was completed and the age at which the respondent anticipated she would be at the most distally located item on the list. This measure of protension has been employed by several other researchers as a method of operationalizing the degree of cognitive protension into the future (Lamm, Schmidt, & Trommsdorff, 1976; Lessing, 1972; Platt, Eisenmann, & DeLisser, 1971; Ruiz et al., 1967).

A variation of the story completion technique was developed by Lessing (1968). In "The Incomplete Sentences Test" participants are provided with nine incomplete sentences that include developmental milestones in the participant's life, including items such as "I don't expect to be going to any kind of school after I am _______ years old." Cognitive future extension is considered the median number of years for the events cited. This technique has not been employed in many subsequent research investigations, and no data regarding its psychometric properties are available.

Another method of personal extension that has not been widely used is the Lines Test developed by Cottle and Pleck (1969). In this technique, the administrator instructs participants to indicate on a 100-mm representation of life's space the portion of the sheet that corresponds to the past, present, and future. The length of each of the corresponding segments is taken as a measure of the orientation that an individual has for each of the distinct temporal loci. Shannon (1975) employed this technique in a cross-cultural investigation examining time orientation in three cultural groups. Anglo-American participants depicted more extended futures and shorter presents than did American Indian and Mexican-American participants. Little age change was observed for the subcultural groups in either present or future extension.

In more recent research, Whitbourne and Dannefer (1986) used projective techniques to examine how individuals' views of personal biographical time change as a function of movement from youth to old age. Hypotheses in the field purport that aging brings with it a change in temporal orientation from a focus on the present to a focus on the past, and an avoidance of extension of oneself into the future. In order to test their hypotheses, these researchers examined 47 men and 47 women (age 24–60; mean age = 41) selected through telephone directory listings in the Rochester, New York area. The initial sample was recontacted by mail 1 year later. They employed a variation of the Life Graph, called The Life Drawings Technique, which is an adaptation of Cottle's nonverbal projective approach to the life graph technique in order to obtain a single measure from which to derive both temporal dominance and future extension. The instructions given to participants were as follows: "I'd like to ask you to take this sheet of paper and 'draw your life' on it, however you would like to, in whatever way you think best represents your life. Then I'd like you to mark it off into segments that you personally consider to represent important eras, epochs, or periods that you have lived through."

Temporal dominance was emphasized in the scoring protocol. Each period marked by a participant was measured and divided by the number of years it was intended to represent. The period with the largest ratio was classified as "past" or "present." Future extension was dichotomously coded, according to whether the future was portrayed on the drawing. Twelve-month test–retest analyses from 55 participants showed the measures to be reliable.

Researchers found that age related to past temporal dominance, particularly for men. Life changes during the 12-month interval appeared to precipitate movement from past to present temporal dominance. Results garnered little support in the life drawings for age-stage theories of adult development. The events depicted as markers had no relationship to age. Tardiness in keeping one's interview appointment was related to present dominance, and among women, having many antiques and knick-knacks in the home was related to past dominance. Sixty-eight percent of subjects over 40 marked no event or demarcated eras whatsoever after the age of 30. Among those under 30, almost every year was used as a marker event, with no particular years being chosen with high frequency. Findings are consistent with Back's (1974) findings of greater acceptance of time's passage among women than men as they age.

Though cognitive extension into the future has been the most widely measured dimension of the broader concept of time orientation, this is by no means representative of the cognitive-affective-behavioral construct. This emphasis reflects our earlier claim that the preponderance of research and measurement on time orientation has focused on cognitive aspects to the exclusion of other dimensions. To that end, coherence, another cognitive aspect of time orientation, has been measured in paradigms similar to those used in the measurement of extension. In fact, the events test used to measure extension into the personal future is often applied to coherence as well. In a study by Lomranz, Shmotkin, and Katznelson (1983), participants were asked to order cards representing various life events based on the sequence that they would be likely to occur and were then asked to state the age of occurrence for each of the events in the events task. The correlation obtained between the event's likelihood of occurrence and the age specified in the later task (i.e., At what age will each event occur?) was used as the operational definition of the participant's temporal coherence.

Cottle's Circles Test (Cottle & Klineberg, 1974) is another projective technique that has been employed in the assessment of temporal coherence. In this method, the degree to which an individual integrates the temporal regions is assessed by giving an individual a blank sheet of paper and reading the following instructions: "Think of the past, present, and future as being in the shape of circles in any way you want that best shows how you feel about the relationship of the past, present, and future. You may use different-size circles. When you have finished, label each circle to show which one is the past, which one is the present, and which one is the future." The representation of the circles are then scored by a trained judge in order to determine how integrated the regions are to the participant.

Density, or the relative concentration of events located in a unit of temporal space, has also been assessed through projective techniques. The protocol presented by Kastenbaum (1961) is used to measure density by requiring that participants predict events that are likely to occur in the future. The probe that is employed in such protocols asks the questions (a) "Who are you?"; (b) "Who were you?"; and (c) "Who will you be?" Density is operationalized, using this method, as the number of personal events that respondents would place in each of the temporal loci represented by the corresponding probe question. Platt and Eisenman (1968) have also employed this measure in their research.

Other measures of density have asked participants to list the number of hopes and fears in certain areas of their lives. Density is usually operationalized in this scheme as the total number of hopes and desires in six areas (family, occupation, personal development, economy, politics, and environment) divided by the total number of hopes and fears (Lamm et al., 1976). A similar formula was employed by Agarawal and Tripathi (1984) to assess future orientation as a density construct.

Directionality appears to be the least measured cognitive component of the temporal orientation construct. The Time Metaphors Test (Knapp & Garbutt, cited in Platt & Eisenman, 1968) is used to measure directionality. With this measure, the administrator instructs participants to select 7 of a list of 14 items that represent the passage of time. Seven of the items represent a "dynamic-hasty" perception of the passage of time whereas the other seven represent a "naturalistic-passive" perception of the passage of time. Directionality is operationalized as the dominant tendency to perceive time in either one of these two different ways.

To date, there have been no projective techniques employed to assess the attitude toward the future that has evolved from the Daltrey and Langer (1984) and Jones (1993) analysis of temporal dimensions. The attitude subdimension does not clearly fall within the cognitive analysis of time orientation and represents more accurately an affective dimension of this construct. Nonetheless, few argue against the theoretical utility of considering valence or time attitude when applying temporal variables to the social psychological study of behavior. Perhaps researchers find it more appropriate to measure these aspects directly.

Direct Methods of Measuring Time Orientation

Numerous attempts have been made to develop inventories that can accurately and objectively assess the time orientation construct. In general, these involve paper-and-pencil measures that instruct participants to respond to a number of statements relevant to time related experiences. In fact, most of the dimensions that have been assessed by projective techniques described earlier have also been measured more directly via self-report measures. It can generally be noted that the projective techniques are used to measure the cognitive dimensions of time orientation and that these methods do not address the affective and behavioral components. Chronologically speaking, these direct methods have followed the use of many of the projective techniques frequently employed in early time orientation research, and many of these direct methods have obviated the need for the use of the projective techniques previously employed. A review of the literature on the measurement of time orientation, for instance, sees paltry attention paid to the methods described previously. This may occur for a number of reasons.

The emphasis on behavioral correlates of time orientation in the direct measurement of time coincided with a shift in focus on these more measurable aspects of time experience instead of a singular emphasis on the internal cognitive correlates of time experience, such as extension, density, and coherence. The greater emphasis on behavior may be linked to the fact that the research that has evolved from work on time orientation is intimately linked to achievement motivation research. In a research program that merged the cognitive elements of time orientation with the behavioral expression of those cognitive processes, Raynor & Entin (1983) expanded the model of achievement motivation to include aspects of future time orientation. In its initial formulation, achievement motivation was conceived of as the constellation of resultant behaviors derived from an individual's motive to achieve success (M_S) and the individual's motive to avoid failure (M_{AF}) (Atkinson, 1964). Achievement motivation was seen as the algebraic sum of the M_S and M_{AF}, and the subjective probability of achieving success in a particular activity. Raynor and Entin (1983) pointed out the limitations of this model and elaborated that future time orientation will ultimately affect the motivation of immediate behavior in one of two possible ways. First, the individual may cognitively evaluate his present actions as having direct bearing upon his future and may act accordingly to the attainment of those goals. Second, the individual may cognitively evaluate that his present actions will not bear upon his future success or failure and may view each act on a path to achievement as an act in and of itself. These distinctions that are drawn are termed contingent and noncontingent paths respectively and will ultimately affect the resultant achievement-related motives.

When one considers the various direct measurements of time orientation, it becomes apparent that the measurement of this construct has been inextricably tied to achievement motivation theory and research. In fact, one of the widely used projective techniques for measuring time orientation, the TAT, was used extensively in its early development to measure the need for achievement. Implicit in this measurement trend is the notion that achievement is invariably in the future. In Western culture, our success as individuals and as larger societies is predicated on the assumption that the future is the location of our most important needs and desires. To ensure economic, environmental, and even personal physical and mental health involves some cognitive representation of the future and some present-day preoccupation with preparing for this period. Recognition of this fact might be observed in statements such as "Tomorrow belongs to those who prepare for it today." The literature on time orientation is replete with instances of the inextricable link between future time orientation and achievement motivation.

One of the early direct self-report methods used to measure time orientation is the Heimberg Future Time Perspective Inventory (cited in Lessing, 1972). According to Lessing, the inventory was developed based on Heimberg's conceptualization of time perspective as "the degree to which the future is perceived as predictable, structured, and controllable." This paper-and-pencil measure contains 25 items on which participants respond on a 1 (completely disagree) to 7 (completely agree) scale. The composite score is a measure of the strength of an individual's cognitive-motivational future time orientation. Items include statements like "I always seem to be doing things at the last moment," "Usually I feel time is going too fast," and "It's really no use worrying about the future, because what will be, will be." Overall, the items on the scale appear to measure the cognitive components of time orientation such as extension, coherence, directionality, and density. In addition, items like the last item cited previously begin to explore the affective components of the construct. The unpublished instrument is reported to have considerable evidence in support of its construct validity. No specific information is available from Lessing's reporting of other psychometric properties of the instrument.

Gjesme (1979) extracted several items from the Heimberg inventory and created a shortened form of the scale. Four factors emerged from the items sampled that accounted for more than 40% of the variance in the measure's items. Involvement, the first factor, measures how focused one's future seems. Anticipation, the second factor, deals with the ability to plan effectively for future events. The third factor, occupation, deals with the way in which an individual perceives the duration and use of time. The final factor, speed, contains items that deal with the way in which an individual perceives the speed at which time is progressing. According to this author, the instrument demonstrated acceptable properties of reliability and validity. Interestingly, the strongest two factors of this scale, involvement and anticipation, measure aspects of temporal experience that are more behavioral in terms of use and structure of time. The clarity of the factor structure of these subscales seems inconsistent with the emphasis of research on time orientation showing more concern with cognitive aspects. Perhaps the actual behavioral use of time is more salient to the functional needs of individuals and larger societies in terms of the consequences of these factors on achievement and goal setting, economic success, environmental concerns, and mental and physical health.

DeVolder and Lens (1982) consider the cognitive-motivational properties of time orientation using work motivation theory, which contends that the general motivation to perform an act is a function of the value of the goals on which the action is dependent and perceived instrumentality between present actions and future outcomes. The latter of these components rests on an individual's cognitive interpretation of the difference between the probability of attaining a goal when not performing some behavior and the probability of attaining a goal when performing the behavior. Each action serves as a direct analog to Raynor's notion of a step in a path of action.

Future expectations, particularly as they relate to environmental variables, is another cognitive-affective component of future time orientation that has been explored in the literature. In a study designed to amass empirical support for a seven-dimension conceptualization of resilience among youth exposed to community violence and to assess the extent to which perceived peer, family, and school support predict various domains of resilience, future expectations was operationalized as one domain of resilience (O'Donnell, Schwab-Stone, & Muyeed, 2002). A longitudinal study was employed that followed twenty-six hundred 6th, 8th, and 10th graders in public schools in New Haven, Connecticut over a 2-year period. Level of Future Expectations (i.e., degree to which youth self-report either high or low levels of expectations regarding outcomes for them in their lives in various domains—work, school, personal life) was measured using a five-item self-report scale derived from Jessor's School Health Study (Jessor, Donovan, & Costa, 1989). Sample Likert-scale items include "What are the chances that you will graduate from high school?" and "What are the chances that you will have a happy family life?"

At baseline, youth victimized by violence had significantly lower levels of future expectations than both nonexposed youth and youth who had witnessed violence. At time 2, high levels of future expectations were predicted by school support among nonexposed youth. Conversely, among victimized youth, high levels of future expectations were predicted by peer support. Additional exploration should be devoted to the relationship between level of future expectations and temporal orientation, particularly in the context of violence and/or other known threats to optimal development.

Strathman, Gleicher, Boninger, and Edwards (1994) have proposed a construct called consideration of future consequences (CFC), hypothesized to be a stable individual difference variable representing the extent to which people consider distant versus immediate consequences of potential behaviors. They suggest that this variable reflects an intrapersonal struggle between present behavior and immediate outcomes versus future outcomes. These researchers developed a measure of this construct using seven samples of college students.

Three of the samples (one composed of college students at the University of Missouri and the other two composed of University of California, Los Angeles students) received a set of 24 statements generated by the researchers. Reliability assessments and factor loadings identified a subset of 12 items that provided a valid and reliable measure of CFC. Subsets of participants from two of the samples returned at a later time to provide data concerning stability of the measure. The CFC scale was found to correlate with Ray and Najman's Deferment of Gratification Scale, with Rotter's Internal–External Locus of Control Scale, and the future orientation items from Zimbardo's Stanford Time Perspective Inventory. These correlations lend support to the hypothesis that consideration of future consequences is related to other similar cognitive-affective components of future time orientation.

In an effort to demonstrate the usefulness of the CFC construct, the researchers aimed to predict response patterns from scores on the CFC scale. Results indicate that CFC may play a regulatory role in the affective responses to negative events (Boninger, Gleicher, & Strathman, 1994). CFC was found to affect the extent to which individuals are influenced by persuasive messages and the extent to which they engage in health behaviors (for a recent review of the CFC literature, see Joireman, Strathman, & Balliet, in press).

The partitioning of psychological time into past, present, and future temporal frames—a dimension referred to as time perspective—has been hypothesized by Zimbardo and Boyd (1999) to be a robust influence on human behavior. Following the conceptual definitions put forth by Lewin (1951), time perspective can be understood as often nonconscious processes functioning to organize the continual flow of personal and social experiences into temporal categories in the service of assigning both order and meaning to life events. Zimbardo and Boyd conceive of time perspective as a situationally determined and relatively stable individual differences process.

The Zimbardo Time Perspective Inventory (ZTPI) was devised to measure personal variations in time perspective and specific time perspective biases. Following repeated scale iterations over a period of years, exploratory and confirmatory factor analyses established a five-factor structure to the inventory, demonstrating acceptable internal and test–retest reliability. The ZTPI instructs respondents to indicate how characteristic a statement is of them on a 5-point Likert scale ranging from very uncharacteristic (1) to very characteristic (5). Exploratory and confirmatory factor analyses revealed five distinct time perspective factors: (a) past-negative, reflecting a generally negative, aversive view of the past; (b) present-hedonistic, reflecting a hedonistic, risk-taking attitude toward time and life; (c) future, reflecting a general future orientation; (d) past-positive, reflecting a warm, sentimental attitude toward the past; and (e) present-fatalistic, representing a fatalistic, helpless, and hopeless attitude toward the future and life. Validity analyses uncovered significant relations among time perspective and a variety of dispositions and behaviors including wearing a watch, sexual experiences, risk taking, grades, stress, and perceived time pressures (see Boyd & Zimbardo, chap. 5, this volume).

FUTURE DIRECTIONS IN TIME ORIENTATION MEASUREMENT

Projective and direct measurement techniques of time orientation, though tapping different constructs and utilizing sometimes discrepant frameworks from which to understand and operationalize the processes underlying time orientation, both appear to support the notion that time orientation is a powerful cognitive, affective, and behavioral compass by which human organisms come to understand and relate to their worlds. Time is the backdrop upon which life plays itself out. Although we, as humans, are incapable of influencing time in an absolute sense—the clock keeps on ticking whether we want it to or not—the measurement of time orientation has shown that we are not only capable of, but likely even driven to, create intrapsychic time orienting structures. As such, time orientation could be argued to represent a process that may underlie and serve as the foundation for many other psychological processes. In its broadest sense, then, time orientation is a diffuse, abstract construct with all of the measurement difficulties inherent in such variables. Yet, as articulated by Jones (1993), as the field becomes increasingly sophisticated and precise in its analytical techniques, finer-grained temporal assumptions are being formed that require more acute temporal measurements. Thus, the field is experiencing an increase in the development of direct measurement techniques aimed at differentiating among multiple components of the time orientation construct and less emphasis on broad, projective measures of time orientation.

Although this trend toward increasingly sophisticated, empirically and psychometrically sound temporal orientation measurement tools is undeniably positive and an advancement in our understanding of this complex and multifaceted construct, the fact that time orientation appears to occupy such a foundational role in the human condition and spans various fields of inquiry indicates that there seems to also be a need for cross-disciplinary collaboration and integration in the conceptual, theoretical, and measurement underpinnings of time orientation.

This cross-disciplinary tenor to time orientation can be seen in the various ways in which time orientation manifests itself operationally both in subfields within psychology and in other disciplines. Measures of reaction time, for instance, are a very commonly used means of quantifying affective and cognitive structures involved in processing information in our world. Cognitive psychologists have examined the role of time orientation in encoding, storage, and retrieval of experienced events. Within the clinical realm, individual differences in time orientation can be seen in the symptomatology of various forms of psychopathology.

Relatedly, the fields of neuropsychology and neuroscience have shed light on some of the brain structures implicated in the regulation of temporal orientation. The frontal lobe, seat of executive-function abilities, and the limbic system, home to the emotion centers of the brain, seem to be most strongly implicated in these processes. Technologically advanced measures of brain activity, such as the electroencephelogram, computed tomography scan, magnetic resonance imaging, and position emission tomography scan, all hold promise in elucidating possible neurological substrates of time orientation. Applied realms such as industrial-organizational psychology, public health, and education all also examine time orientation constructs from varied methodological frameworks.

Last, the field may benefit from culturally sensitive developmental and/or longitudinal studies of time orientation across the life span. Despite largely being conceptualized as a relatively stable individual differences variable, it is still likely that time orientation structures undergo some form of development, change, and evolution across ontogeny, being influenced by cultural and environmental pressures.

REFERENCES

Agarawal, A., & Tripathi, K. K. (1984). Influence of prolonged deprivation, age, and culture on the development of future time orientation. *European Journal of Psychology, 14,* 451–453.

Anastasi, A., & Urbina, S. (1997). *Psychological testing* (7th ed.). Upper Saddle River, NJ: Prentice Hall.

Atkinson, J. W. (1964). *An introduction to motivation.* Oxford, England: Van Nostrand.

Back, K. W. (1974). Metaphors as test of personal philosophy of aging. In E. Palmore (Ed.), *Normal aging II: Reports from the Duke Longitudinal Studies, 1970–1973* (pp. 207–216). Durham, NC: Duke University Press.

Barndt, R. J., & Johnson, D. M. (1955). Time orientation in delinquents. *Journal of Abnormal and Social Psychology, 51,* 343–345.

Boninger, D. S., Gleicher, F., & Strathman, A. (1994). Counterfactual thinking: From what might have been to what may be. *Journal of Personality and Social Psychology, 67,* 297–307.

Cottle, T. J., & Klineberg, S. L. (1974). *The present of things future: Explorations of time in human experience.* New York: The Free Press.

Cottle, T. J., & Pleck, J. (1969). Temporal estimates of linear extension: The effect of age, sex, and social class. *Journal of Projective Techniques and Personality Assessment, 33,* 81–93.

Daltrey, M. H., & Langer, P. (1984). Development and evaluation of a measure of future time perspective. *Perceptual and Motor Skills, 58,* 719–725.

Dana, R. H. (1996). The Thematic Apperception Test (TAT). In C. Newmark (Ed.), *Major psychological assessment instruments* (2nd ed., pp. 166–205). Boston: Allyn & Bacon.

DeVolder, M., & Lens, W. (1982). Academic achievement and future time perspective as a cognitive motivational concept. *Journal of Personality and Social Psychology, 42,* 566–571.

Eson, M. E., & Greenfield, N. (1962). Life space: Its content and temporal dimensions. *Journal of Genetic Psychology, 100,* 113–128.

Gjesme, T. (1979). Future time orientation as a function of achievement motives, ability, delay of gratification, and sex. *Journal of Psychology, 101,* 173–188.

Gjesme, T. (1981). Is there any future in achievement motivation? *Motivation and Emotion, 5,* 115–137.

Gjesme, T. (1983). On the concept of future time orientation: Considerations of some functions' and measurements' implications. *International Journal of Psychology, 18,* 443–461.

Jessor, R., Donovan, J. F., & Costa, F. M. (1989). *School health study.* Boulder: Institute of Behavioral Science, University of Colorado.

Joireman, J., Strathman, A., & Balliet, D. (in press). Considering future consequences: An intergrative model. To appear in L. Sanna & E. Chang (Eds.), *Judgments over time: The interplay of thoughts, feelings, and behaviors.* Oxford: Oxford University Press.

Jones, J. M. (1988). Cultural differences in temporal perspectives: Instrumental and expressive behaviors in time. In J. E. McGrath (Ed.), *The social psychology of time: New perspectives* (pp. 21–38). Newbury Park, CA: Sage.

Jones, J. M. (1993). An exploration of temporality in human behavior. In R. Schank & E. Langer (Eds.), *Beliefs, reasoning, and decision making: Psycho-logic in honor of Bob Abelson.* Hillsdale, NJ: Lawrence Erlbaum Associates.

Kastenbaum, R. (1961). The dimensions of future time perspective, an experimental analysis. *The Journal of General Psychology, 65,* 203–218.

Kinslinger, H. J. (1966). Application of projective techniques in personnel psychology since 1940. *Psychological Bulletin, 66,* 134–149.

Klineberg, S. L. (1968). Future time perspective and the preference for delayed reward. *Journal of Personality and Social Psychology, 8,* 253–257.

Lamm, H., Schmidt, R. W., & Trommsdorf, G. (1976). Sex and social class as determinants of future orientation (time perspective) in adolescents. *Journal of Personality and Social Psychology, 34,* 317–326.

Lasane, T. P., & Jones, J. M. (1999). Temporal orientation and academic goal-setting: The mediating properties of a motivational self. *Journal of Social Behavior and Personality, 14,* 31–44.

Leshan, L. L. (1952). Time orientation and social class. *Journal of Abnormal and Social Psychology, 47,* 589–592.

Lessing, E. E. (1968). Demographic, developmental, and personality correlates of length of future time perspective (FTP). *Journal of Personality, 36,* 183–201.

Lessing, E. E. (1972). Extension of personal future time perspective, age, and life satisfaction of children and adolescents. *Developmental Psychology, 6,* 457–468.

Lewin, K. (1951). *Field theory in the social sciences: Selected theoretical papers.* New York: Harper.

Lomranz, J., Shmotkin, D., & Katznelson, D. B. (1983). Coherence as a measure of future time perspective in children and its relationship to delay of gratification and social class. *International Journal of Psychology, 18,* 407–413.

McGrath, J. E. (1988). Time and social psychology. In J. E. McGrath (Ed.), *The social psychology of time: New perspectives* (pp. 255–267). Thousand Oaks, CA: Sage.

Murphy, K. R., & Davidshofer, C. O. (1998). *Psychological testing: Principles and applications* (4th ed.). Upper Saddle River, NJ: Prentice Hall.

Murray, H. A. (1938). *Explorations in personality.* Cambridge, MA: Harvard University Press.

Nuttin, J. (1985). *Future time perspective and motivation: Theory and research method*. Leuven, Belgium: University Press/Lawrence Erlbaum Associates.

O'Donnell, D. A., Schwab-Stone, M. E., & Muyeed, A. Z. (2002). Multidimensional resilience in urban children exposed to community violence. *Child Development, 73,* 1265–1282.

Platt, J. J., & Eisenman, R. (1968). Internal–external control of reinforcement, time perspective, adjustment, and anxiety. *Journal of Personality and Social Psychology, 79,* 121–128.

Platt, J. J., Eisenman, R., & DeLisser, O. (1971). Temporal perspective as a personality dimension in college students: A re-evaluation. *Perceptual and Motor Skills, 33,* 103–109.

Raynor, J. O., & Entin, E. E. (1983). The function of future orientation as a determinant of human behavior in step-path theory of action. *International Journal of Psychology, 18,* 463–487.

Ruiz, R., Revich, R., & Krauss, H. (1967). Tests of temporal perspective: Do they measure the same construct? *Psychological Reports, 21,* 849–852.

Shannon, L. (1975). Development of time perspective in three cultural groups: A cultural difference or an expectancy interpretation. *Developmental Psychology, 11,* 114–115.

Strathman, A., Gleicher, F., Boninger, D. S., & Edwards, C. S. (1994). The consideration of future consequences: Weighing immediate and distant outcomes of behavior. *Journal of Personality and Social Psychology, 66,* 742–752.

Sweetland, R. C., & Keyser, D. J. (1986). *Tests: A comprehensive reference for assessments in psychology, education, and business* (2nd ed.). Kansas City, MO: Westport.

Teahan, J. E. (1958). Future time perspective, optimism, and academic achievement. *Journal of Abnormal and Social Psychology, 57,* 379–380.

Trommsdorff, G. (1983). Future orientation and socialization. *International Journal of Psychology, 18,* 381–406.

Wallace, M. (1956). Future time perspective in schizophrenia. *Journal of Abnormal and Social Psychology, 52,* 240–245.

Whitebourne, S. K., & Dannefer, W. D. (1986). The "life drawing" as a measure of time perspective in adulthood. *International Journal of Aging and Human Development, 22,* 147–155.

Wiener, B. (1986). *An attributional theory of motivation and emotion*. New York: Springer-Verlag.

Wohlford, P. (1968). Extension of personal time in TAT and story completion stories. *Journal of Projective Techniques and Personality Assessment, 32,* 268–280.

Zimbardo, P. G., & Boyd, J. N. (1999). Putting time in perspective: A valid, reliable individual-differences metric. *Journal of Personality and Social Psychology, 77,* 1271–1288.

3

Thinking About and Acting Upon the Future: Development of Future Orientation Across the Life Span

Jari-Erik Nurmi
University of Jyväskylä, Finland

Thinking about and acting upon the future is a key characteristic of the human mind. People typically think about what is likely to happen in the future, they have future preferences and aims, they make a great deal of effort to realize some of their future options, and, sometimes, they may have regrets that the future did not turn out in accordance with their hopes. This orientation toward the future requires the deployment of a wide range of psychological tools. These include cognitive processes, such as anticipation, planning, and the regulation of behavior, as well as emotions and attitudes, like optimism, pessimism, hope, and despair. Similarly, motivation, expressed as interests, values, and goals, plays an important part in acting upon the future. Future orientation is also closely connected to the development of the individual: Anticipating the future is to predict one's own development; aiming at a particular future is to direct it.

The importance of an individual's *future orientation* differs, however, from one stage of life to another. For example, during childhood cognitive abilities have not yet reached the level that would enable an individual to anticipate and plan his or her future over extended periods of time. Moreover, children are not provided with the amount of autonomy that would allow them to make long-range decisions concerning their future lives. When individuals move into adolescence, a future orientation assumes increasing importance in their thinking. This is due to the fact that they have more *developed* cognitive skills, they are given an increasing amount of autonomy by their parents, and they encounter social pressure to find their way into major adult roles. Thinking about the future remains important during early adulthood, when individuals find themselves facing many transitions. Although one's future orientation remains important during adulthood and old age, it is activated more temporarily in the context of major life decisions, transitions, and crises.

Many conceptualizations have been used to describe thinking about and acting upon the future. One of the first concepts to be used in the field was that of the *future-time perspective* (Lewin, 1942), which refers to the overall importance a person attaches to the future. It is typically operationalized by asking people to assess the importance of immediate versus long-range future outcomes (Shell & Husman, 2001) or the importance of the future compared to the present (Zimbardo, Keough, & Boyd, 1997). A very similar concept is that of *time orientation*, that is, to what extent a person orients to past, present, and future in his or her thinking (Hoornaert, 1973). Another frequently used concept in the research field is that of temporal extension (Lessing, 1972), which refers to how far into the future an individual's thinking and interests are projected. This temporal extension is typically measured by asking individuals to estimate how many years ahead they think their expectations and hopes will be actualized (Nurmi, 1989b; Trommsdorff, Burger, & Fuchsle, 1982). There is also a long history of examining the development and consequences of time perception and the concept of time among younger children (Arlin, 1990).

The conceptualization most widely used to describe people's thinking about and acting upon the future is future orientation (Sundberg, Poole, & Tyler, 1983; Trommsdorff et al., 1982). It is a relatively broad concept that embraces the kinds of expectations, hopes, and fears individuals have (content of future orientation), how far into the future such expectations and hopes are projected (extension), how people think about the factors that influence their future (control beliefs, anticipated causal attributions), how people feel about their future (optimism, pessimism), and the kinds of tools they have developed to attain their goals (Nurmi, 1991). Future orientation is typically *studied* by asking people about their hopes, goals, and fears, after which they are further questioned about the temporal extension, control beliefs, emotions, and planning concerning them. In this chapter the future-orientation concept is used to refer to a wide variety of phenomena related to how people think about and act upon the future.

There are also many other concepts used in psychology that are closely related to the ways in which people think about the future. Examples of such concepts include research on anticipated future life events (Crockett & Bingham, 2000), personal goals (Little, 1983), life planning, problem solving (Strough, Berg, & Sansone, 1996), and identity explorations and commitments (Bosma, 1985). Some research concerning these concepts is also reviewed here.

In this chapter, I first introduce a theoretical framework for future orientation. Both the major psychological processes involved, as well as the importance of life-span development, are discussed. Next, previous research on future orientation from childhood to old age is reviewed. This review is organized according to different stages of human life, that is, childhood, adolescence, and early, middle, and late adulthood. Then, some applications originating from the theory and research on future orientation, such as counseling and interventions, are introduced. Finally, future directions of the research field are discussed.

A PROCESS MODEL OF FUTURE ORIENTATION

Future orientation is a complex phenomenon that consists of many cognitive, motivational, and affective processes (Nurmi, 1991; Trommsdorff et al., 1982). The cognitive processes include, for example, acquiring knowledge about time and the future, anticipating and estimating the likelihood of future events, planning and decision making concerning one's future, and exploring future opportunities. The motivational aspects of future orientation, in turn, include future-oriented goals, interests, values, and commitments, on the one hand, and concerns, doubts, and fears, on the other. The affective aspect of future orientation involves many evaluative emotions and attitudes, such as optimism, pessimism, hope, and despair.

In the effort to understand how these cognitive, motivational, and affective processes function in individuals' thinking, future orientation has been described as a *process* that consists of various successive stages (Fig. 3.1). According to Nurmi (1989a, 1991), individuals' anticipation concerning the future originates from the knowledge, beliefs, and schemata they have formed during the course of their lives concerning the typical life-span development in the society in which they live. A comparison of this knowledge with individual motives and values, originating from a person's previous life history and personal characteristics (e.g., temperament or personality traits), thus lays the founda-

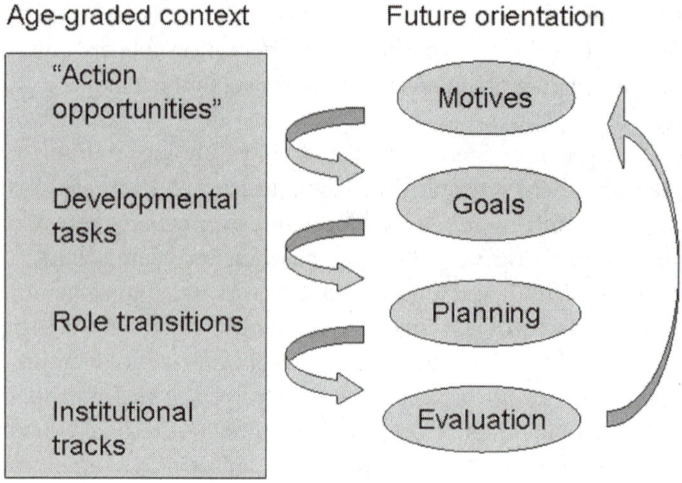

FIG. 3.1. Future orientation in age-graded context.

tion for the construction of future-oriented goals. Comparing one's individual motives with one's knowledge about the future opportunities available assists in the identification of life trajectories that will satisfy one's personal needs. One example of such process is the choice of a career. Young people typically have values and interests that orient their exploration of future career opportunities. Some values lead to an interest in careers in which people expect that they will work in close connection with other people, whereas other kinds of values may lead to emphasizing the importance of money and the level of income in the future profession. However, it is only the process in which adolescents compare their interests and values to the career and educational opportunities available to them that will lead to the construction of realistic career goals.

After setting future-oriented goals, people have to find ways to attain them. This construction of means requires the exploration of future opportunities, invention of subgoals that will lead to goal attainment, planning, strategy construction, investment of effort, and regulation of behavior (Nurmi, 1989a, 1991). As with goal construction, knowledge about opportunities for future action plays an important role in efficient planning and decision making. From a developmental point of view, individuals' abilities to perceive cognitively future events and their distances in time, as well as their planning skills, provide a basis for this aspect of future orientation. After individuals have developed such skills, the planning of the future can be enhanced by providing individuals information about future society and the variety of opportunities it provides for their future development and life decisions. Among adolescents typical sources of such information are parents, peers, school, and media.

Finally, individuals deploy a variety of evaluative tools to deal with the information that they receive concerning the future, and their success in dealing with future challenges (Nurmi, 1989a, 1993). These evaluative mechanisms include, for example, illusory beliefs in a good future, social comparison with those who are doing less well, and the use of self-protecting attributional bias. For example, people typically predict that negative life events are less likely to happen to them compared with their age mates (Blinn & Pike, 1989; Malmberg & Norrgård, 1999). Similarly, they typically believe that they will be able to control many factors that influence their future lives. These evaluative mechanisms in turn influence individuals' anticipation of the future, and the construction of future-oriented goals. These mechanisms can also be used to enhance people's future optimism and future-directed behavior. For example, a typical way to promote unemployed people's job search activity is to enhance their positive self-concept and belief in personal control in finding a job.

Most previous research on people's future orientation has focused on examining the contents of goals and hopes. Some research has also been carried out on the kinds of plans and strategies people have developed to attain their goals. However, only a few studies have examined the evaluative aspect of future orientation.

A LIFE-SPAN APPROACH TO FUTURE ORIENTATION

Future orientation is embedded in an individual's life-span development in two important ways. On the one hand, a particular stage of life, and related challenges, demands, and opportunities, channel the ways in which a person thinks about and acts upon the future (Nurmi, 2004). This channeling is based on the fact that various stages of life are characterized by different age-graded sociocultural structures that create predictable, socially recognized road maps for human lives (Hägestadt & Neugarten, 1985). These structures have been described, for example, as developmental tasks (Havighurst, 1948), age norms (Neugarten, Moore, & Lowe, 1965), role transitions (Elder, 1985), and institutional careers (Mayer, 1986), which channel an individual's future orientation in many ways. For example, age-graded developmental tasks, role transitions, and institutional careers create an "opportunity space" for a particular stage of life, which then provides a basis for the kinds of future-oriented goals people typically construct. Moreover, knowledge of a particular stage of life, and of the societal and institutional opportunities related to it, influence the kinds of plans and strategies people construct: Successful planning requires accurate knowledge of one's future options. Individuals' anticipation concerning the various age-graded demands and developmental standards also provide a basis for the ways in which they evaluate their personal future. There are also cross-cultural and historical differences in age-graded sociocultural structures that are reflected in people's thinking and acting upon the future. For example, the ways in which adolescents perceived their future in the farming society 200 years ago might be assumed to differ substantially from how young people see their future options in the current society. Similarly, differences in normative expectations and opportunities related to gender, social background, and subcultures are reflected in individuals' future orientation.

On the other hand, thinking about and acting upon the future lays the foundation for the ways in which people direct their development and select a variety of life course trajectories (Nurmi, 1993). For example, the prioritization of alternative goals leads to the selection of one specific future option, and the exclusion of the others. Similarly, people's planning skills, together with their knowledge surrounding a certain domain of their future life influence their success in dealing with a particular transition. In addition, optimism and subsequent high level of effort invested in the actualization of a particular goal (Scheier, Carver, & Bridges, 1994) are likely to influence people's life course trajectories. In fact, several successive stages of a future orientation need to be embarked on properly before a particular developmental trajectory is attained.

RESEARCH ON THE DEVELOPMENT OF FUTURE ORIENTATION

The development of future orientation is influenced by many psychological and social factors. For example, future orientation requires a basic understanding of time, knowledge regarding future events, planning skills, and some understanding of one's own abilities to deal with future challenges. Another requirement for future orientation is that a person has developed a whole set of interests and values that span the near and distant future. Finally, as people move from one stage of life to another, changes in age-graded developmental tasks, role transitions, and institutional tracks will affect how they think about the future. Next, I review some of the recent research on people's future orientation at different stages of life.

Childhood

Little research has been carried out on children's future orientation. The reason for this is probably that they are not yet assumed to plan their lives independently from their parents. There are, however, a few studies that have examined children's readiness to think about the future. Friedman (2000), for example, investigated the understanding of future events among children aged 4 to 10. His results showed that, whereas 4-year-olds were unable to differentiate distances into the future, 5-year-olds were already beginning to distinguish events that would occur in the coming weeks and months from those that would not occur for many months. By 8 to 10 years of age, children were able accurately to judge distances into the future by using mental representations of the timing of the events in the annual cycle (Friedman, 2000). These results are in accordance with the findings concerning the perception of time. Arlin (1990), for example, found that a constant sense of objective time develops between 7 and 9 years of age. Kreitler and Kreitler (1987) showed that planning for the near future peaks at the age of 9, whereas the planning of the distant future becomes typical among 11-year-old children.

Little attention has been paid, however, to the particular contents of children's future orientation. In one study, Kreitler and Kreitler (1987) asked 5- to 11-year-old children to describe the kinds of life domains that would require planning. The 5-year-olds considered planning to be applicable primarily to daily activities (eating, going to bed, dressing). The 7-year-olds increasingly reported activities such as manipulating adults, using machines, and doing daily chores (homework, cleaning one's room). In the 9-year-olds, there were a number of children who considered planning as applicable to interpersonal relationships with peers and to achievement-related topics. At the age of 11, planning moves from here-and-now concerns to a greater orientation toward the future. At this age new topics become the objects of planning, such as achievement, one's personal future (studies, career, marriage, leaving the parental home), and those related to society at large (peace, solving social problems, ecology).

In another study, McCallion and Trew (2000) examined education- and occupation-related future hopes and fears among 5- to 9-year-old children. Their results showed that, with age, children expressed a greater number of future-related hopes and fears in both the school and work domains. As children grew older, they also became more aware of the importance of school in relation to their future. Children were also hopeful about succeeding in their future careers. However, they also realized that, if they were not doing well, they would be in danger of losing their jobs.

Oppenheimer and van der Wilk (1984) found clear age differences in the kinds of vocational goals 5- to 12-year-old boys reported: The majority of 5-year-olds preferred white- and blue-collar jobs, whereas the majority of the 6-year-olds mentioned more unrealistic vocations (TV heroes, celebrities, fire-fighters). Thereafter typical white- and blue-collar occupations became popular again. However, at the age of 10 the boys became more uncertain about their professional choice and preferred to postpone making their choice. This may be a sign of increasing realism. A somewhat similar pattern was found by Russell and Smith (1979) among Australian children. Oppenheimer and van der Wilk did not, however, find similar age differences among girls, who, from the age of 5, emphasized vocational goals relating to taking care of others. Russell and Smith also found that 7- to 15-year-old children's career aspirations were strongly own-gender stereotypical.

Some research has also been carried out on children's future concerns and fears. When Slee and Cross (1989) asked 4- to 19-year-olds to identify their worries from a list of fears presented to them, they found that being left alone at night was a typical fear reported by children of 4 to 7 years; war and getting lost appeared in the responses of those 8 to 12 years of age; and not being liked, in addition to war, was mentioned by adolescent participants.

Several studies have investigated children's planning skills. For example, when Hudson and Fivush (1991) examined 3- to 5-year-old children in everyday planning tasks, they found that the children's planning became more complex and flexible with age. Older children also relied less on external support in planning, but when given such support, even the 3-year-olds were able to carry out simple planning tasks. Similarly, Gardner and Rogoff (1990) showed that even 4- to 7-year-olds used both advanced planning and improvisational planning adapted to problem circumstances. Their findings also showed that older children planned more in advance than younger ones. When Oppenheimer (1987) compared the groups of children aged 7, 9, and 11 years, he found that the major differences in planning skills seemed to lie between ages 7 and 9. However, it was not before the age of 11 that children showed competent planning when required to accomplish simple social goals. Kreitler and Kreitler (1987) investigating children's planning across ages 5 to 11 found that, with age, children become increasingly familiar with the manner in which planning occurs, its temporal range, its frequency, some of its difficulties, and its positive and negative results.

Little is known, however, about the consequences of children's future orientation. In one study Wyman, Cowen, Work, and Kerley (1993) examined a group of 9- to 11-year-old urban children exposed to high psychosocial stress. They found that earlier positive future expectations predicted enhanced socioemotional adjustment in school 3 years later and acted as a protective factor in reducing negative effects of high stress.

Adolescence

The vast majority of the research on future orientation has been carried out among adolescents. The reason for this is that the future and future-related decisions play a particularly important role at this stage of life, as young people are expected to make many decisions concerning their adult lives.

When adolescents are asked about their future hopes and interests, they typically report topics that focus on their personal future life, such as education, occupation, family, leisure activities, travel, and self-related issues (Gillies, 1989; Lanz & Rosnati, 2002; Nurmi, 1991; Seginer, 1988). Interestingly, there is only little variation across societies and cultures in such hopes and interests (Nurmi, 1991; Nurmi, Poole, & Kalakoski, 1994). It has also been shown that during their adolescent years individuals become increasingly interested in their future occupation, education, and family (Gillies, 1989; Nurmi, 1989b). By contrast, their interest in leisure activities decreases. These results are in accordance with the life-span theory of future orientation (Nurmi, 1991, 1993): A substantial proportion of adolescents' future hopes and interests focuses on the major developmental tasks of this phase of life.

The research on adolescents´ future orientation has also found some gender differences. Boys are more interested in the material aspects of life, whereas girls are more interested in interpersonal relationships, such as having a family (Nurmi, 1991). However, findings concerning gender differences in interests in occupation and education vary. Some studies have shown that boys are more interested in education and occupation compared with girls (Yowell, 2000); a few studies have found no differences (Malmberg, 1996; McCabe & Barnett, 2000); and others have shown precisely the opposite pattern (Nurmi et al., 1994).

Adolescents also have relatively detailed conceptions of their future lives, that is, the timing of various life events and transitions (Crockett & Bingham, 2000; Nurmi, 1989b). They anticipate their future lives as a sequence of transitions in which completion of their formal education is followed by entry into working life, and then by marriage and parenthood. These anticipations are also in accordance with the statistics on the age at which individuals typically go through these transitions (Crockett & Bingham, 2000; Nurmi et al., 1994). Research on how far into the future adolescents' thinking and personal goals extend supports these results. Nurmi (1989b), for example, found that adolescents expected their education-related goals to be actualized at about the age of 19, occupation-related goals to be actualized at about the age of 23, and goals related to having a family at about the age of 26.

Research has also shown gender-differences in these conceptions of future lives. For example, girls tend to anticipate forming a partnership, establishing a family, and having children earlier than boys (Malmberg, 1996). Among girls their expectations about the timing of educational and occupational transition are closely connected to their expectations about family formation, but this is not the case for boys (Crockett & Bingham, 2000). This finding is thought to reflect the fact that girls take into account the role conflicts present in these two domains more than boys do (Hogan, 1985).

A variety of institutional transitions and tracks also provide a basis for adolescents' future-oriented goals. For instance, Klaczynski and Reese (1991) found that college-preparatory high school students held more career-oriented values and educational goals, and projected their future goals further into the future, than did vocational school students. By contrast, the vocational school students' goals focused more on preparation for adulthood and the attainment of adult status than the college-preparatory high-school students.

Interestingly, changes in cultural values, beliefs, and social opportunities have been found to be reflected in adolescents' future orientation. For example, Liberska (2002) compared the hopes of three cohorts of Polish adolescents. One was examined in 1987 before the collapse of the communist bloc, one in 1991 just after the changes, and one in 1999, by which time the social situation had stabilized. Her results showed substantial differences in hopes across the three cohorts. For example, with time, there was a steep increase in the number of goals relating to having a high income and material wealth. At the same time, hopes relating to permanent employment and the maintenance of good health increased. By contrast, the percentage of hopes concerning travel, being accorded social respect, and professional excellence, decreased across the three cohorts. Interestingly, hopes related to involvement in political activities first increased but then decreased to the same level they used to be before the political changes.

There were also substantial changes in the mean age the Polish adolescents expected to face major role transitions: whereas getting married was expected in 1987 to happen about the age of 23, in 1999 the expected age was 27. The same figures for having the first child were 24 in 1987 and 29 in 1999. By contrast, in 1987 the adolescents expected that they would be earning a high income at the age of 36, whereas in 1999 the expected age was 28. These are interesting results because they show that major changes in society, and in dominant cultural beliefs, values, and opportunities have an immediate impact on the ways in which young people perceive their future lives.

Adolescents' fears and worries about their future typically concern three major topics (Nurmi, 1991). Many young people report concerns related to dealing with normative developmental tasks, such as becoming unemployed, failing at school, and facing a divorce in the future (Gillies, 1989; Solantaus, 1987). Some adolescents are concerned about possible negative life events that may happen to their parents and family members, such as health problems and divorce (Gillies, 1989). Several adolescents are also concerned about events on the level of society, such as nuclear war (Solantaus, 1987), environmental problems (Poole & Cooney, 1987), or AIDS (Gillies, 1989). The latter kinds of concerns also change rapidly as the topics discussed in the media change (Wilkins & Lewis, 1993).

Research also suggests that a majority of adolescents are relatively optimistic about their future and believe that they have personal control over it (e.g., Brown & Larson, 2002; Nurmi, 1989b). They also deploy defensive illusions in support of such optimism. For example, they consider negative life events, such as divorce (Blinn & Pike, 1989), alcoholism, and unemployment (Malmberg & Norrgård, 1999) to be less likely in their own future life compared to that of their peers.

As mentioned earlier, children have generally acquired basic planning skills by the age of 10 to 11. However, such skills seem to continue to develop up to the early 20s (Dreher & Oerter, 1987; Pasupathi, Staudinger, & Baltes, 2001). In addition, future-related knowledge and strategic complexity have been shown to increase with age (Nurmi, 1989b).

Future orientation was also assumed to contribute to the ways in which individuals direct their future lives. Along with this assumption, Schoon and Parsons (2002) found that adolescents' aspirations at the age of 16 predicted their occupational aspirations during young adulthood. Similarly, concrete college goals have also been found to predict subsequent college attendance (Pimentel, 1996). Stattin and Kerr (2001) showed that adolescents who reported self-focused values (personal satisfaction and enjoyment) were more likely in later periods to become engaged in risky behaviors, such as norm breaking, risky sex, smoking, and drinking, and to associate with delinquent friends, compared with adolescents who have other-focused values (concern for others' well-being and the common good). Adolescents' aspirations, such as educational expectations, have also been found to predict teenage pregnancy (Hockaday, Jasper Crase, Shelley, & Stockdale, 2000).

Adulthood

Relatively little research has been carried out on the ways in which adults think about their future. As the existing studies give a different view depending on the stage of adulthood, the results are summarized separately for early, middle, and late adulthood.

Early Adulthood. The ways in which young adults see their future closely resembles that of adolescents. For example, when Nurmi (1992) investigated 19- to 64-year-old Finnish adults, he found that among the 19- to 24-year-olds the most often mentioned personal goals were those concerning education, family, and self, whereas among the 25- to 34-year-olds these were property, family, and self. Both age groups also frequently mentioned occupation-related goals. Similarly, Cross and Markus (1991) found that the hoped-for-selves of young adults up to the late 30s most often concerned occupation and family. As with adolescents, so too in the case of young adults future-oriented goals reflect the major developmental tasks of this life phase (Havighurst, 1948).

Salmela-Aro, Aunola, and Nurmi (2003) recently reported results obtained from a longitudinal study in which university students were examined every 2nd year across a 10-year period. Their results showed that, after the first 2 or 3 years at university, education-related goals assumed less importance. In contrast, personal goals concerning work, family, and children became more typical. Salmela-Aro et al. also found that the young adults' life situation was significantly associated with their goals. For example, those who had children reported fewer goals concerning leisure activities, but more occupational and family-related goals.

One feature typical of young adulthood is that during this life phase individuals are faced with several developmental transitions. Going through such transitions has also been found to have an impact on young adults' future orientation. For example, Salmela-Aro and Nurmi (1997) found in their study that young adults' life situation, such as being married and having children, predicted their subsequent family-related goals. By contrast, being single predicted turning to self-focused, existential goals. Moreover, young adults appear not only to construct goals that are in accordance with age-graded normative environments, but also to reconstruct their personal goals to fit in with the different stages of a particular transition they are going through. For example, Salmela-Aro, Nurmi, Saisto, and Halmesmäki (2000) showed that women who were facing the transition to parenthood reconstructed their goals to match the various stages of this transition: Women's personal goals changed from being achievement-related to pregnancy, then to the birth of their child, and finally to taking care of their child and motherhood.

When young adults are asked about their future-related concerns and fears, they typically report topics involving education, occupation, and self (Cross & Markus, 1991; Nurmi, 1992). Fears related to health and societal problems become more frequent only in middle adulthood.

Future orientation has been assumed to play an important role in the ways in which people select their future environments. In line with this assumption, Nurmi, Salmela-Aro, and Koivisto (2002) found that the more young adults emphasized the importance of work-related goals and the more progress they thought they made in the achievement of these goals, the more likely they were to find work that was commensurate with their education and the less likely they were to be unemployed after graduation. Similarly, young adults' family-related goals predict their subsequent transitions toward marriage or cohabitating (Salmela-Aro & Nurmi, 1997). By contrast, young adults' self-focused, existential types of goals have been found to predict subsequent negative life events, such as the breaking up of an intimate relationship.

The life-span theory of future orientation suggests that personal goals that match the developmental tasks of a particular age are adaptive, and that they subsequently contribute to individual well-being (Nurmi, 1993, 2001). Both cross-sectional research (Emmons, 1991) and longitudinal studies (Salmela-Aro & Nurmi, 1997) have shown that young adults who report interpersonal and family-related goals also show a higher level of well-being than other young people. Moreover, Salmela-Aro, Nurmi, Saisto, and Halmesmäki (2001) found that women who were facing a transition to parenthood, and who adjusted their personal goals to correspond to the particular stage-specific demands of this transition, involving domains related to family, spouse, and the birth of the child, showed a decrease in depressive symptoms; those who disengaged from such goals showed an increase in depressive symptoms.

Thinking about issues relating to self and identity has been assumed to be a natural part of adolescence and young adulthood (Bosma & Kunnen, 2001; Erikson, 1959; Marcia, 1980). However, there is considerable evidence that young adults have only a limited number of self-focused goals, and when they do, these kinds of goals are closely are associated with a low level of well-being (Salmela-Aro et al., 2001). Moreover, an increase in self-focused goals has been found to lead to an increase in depressive symptoms (Salmela-Aro & Nurmi, 1997; Salmela-Aro et al., 2001). However, most research on self-focused goals has been carried out among young adults. Consequently, it would be interesting to examine whether this pattern is also true for adolescence.

Middle Adulthood. When people move from early to middle adulthood, there are clear changes in their future orientation. For example, Nurmi (1992) found that, whereas 25- to 34-year-olds often mentioned future-oriented goals concerning family and self, 35- to 44-year-olds reported goals related to their children's lives and travel, 45- to 54-year-olds mentioned goals concerning health, their children's lives, and leisure activities, and 55- to 64-year-olds listed many goals concerning health, leisure activities, and world politics. All age groups had many goals related to occupation and property. Pulkkinen, Nurmi, and Kokko (2002) found that 36-year-old adults most often mentioned goals that concerned health followed by family, childrearing, livelihood, occupation, and lifestyle. Cross and Markus (1991) showed that the possible selves of the middle-aged were often related to family (including children's future) and health issues. Nurmi (1992) found a similar pattern for people's fears: Starting from the mid-30s, health-related fears became more frequent in adults' future concerns. Fears related to societal problems, such as war, were found to increase in the later years of middle adulthood. As was found among young adults, in middle adulthood self-related goals ("to be happy," "to grow old with dignity") have also been found to be associated with low life satisfaction (Cross & Markus, 1991), whereas high life satisfaction was associated with goals related to occupation and family.

Little research has been carried out on gender differences in adults' future orientation. In one study, Nurmi (1992) found that women more frequently reported goals in the domains of education, self, and travel, and concerns about the health of significant others. Men, however, expressed more interest in leisure activities and global societal issues (Nurmi, 1992).

Late Adulthood. Research suggests that people's future orientation changes again when they move from middle to late adulthood. Cross and Markus (1991), for example, found that elderly people particularly mentioned topics related to health and lifestyle. Smith and Freund (2002) showed that elderly people aged from 70 to 100 reported hoped-for-selves that were related to health, in particular, but also to personal characteristics, identity, and social relationships. In another study, Takkinen and Ruoppila (2001) found that the reasons given for meaning in life in elderly people included human relations, respect for life, religion, hobbies, and good health. Prager (1996) found that elderly people particularly valued preserving values, service to others, and religious activities (see also Orbach, Iluz, & Rosenheim, 1987). Many researchers have emphasized the role of religion in giving meaning to life in elderly people. Findings concerning its role as providing a basis for well-being are, however, contradictory (McFadden, 1995).

Only a few studies have examined the changes in thinking about the future during old age. Lawton, Moss, Winter, and Hoffman (2002) found that the importance of personal goals concerning spiritual and moral issues increases with age after the age of 70, whereas those related to home planning decrease. The only kind of elderly people's goals that were associated with a high level of well-being were interpersonal goals. Little work has been done comparing elderly women's and men's future orientation. When comparisons have been attempted, no gender differences have been found (Orbach et al., 1987).

A few studies have also examined the possible reasons why people's goals change in later life. Lang and Carstensen (2002) suggested recently that, as individuals grow older, they increasingly perceive their future as limited. This constricted time perspective leads older people to value personal goals concerning generativity (helping others, influencing the world) and emotion regulation. By contrast younger people, who see the future as open-ended, place more value on goals concerning autonomy (independence, work, education) and social acceptance (friends, lack of loneliness). Sheldon and Kasser (2001) found a somewhat similar pattern when they examined the strivings of adults aged between 17 and 82. Their results showed that the older participants more frequently emphasized generativity strivings (helping others, giving something to the younger generations, symbolic immortality), whereas younger people more often reported identity strivings (self-understanding, role conflicts, and autonomy).

Overall, the review of adults' future orientation showed that people in different age groups have reported a somewhat different pattern of future goals, hopes, and expectations. As in the case of adolescents, these age differences reflected the age-graded developmental tasks and life situations that are typical of a particular period of adulthood. These findings are also in accordance with the results of Settersten and Hägestadt (1996a, 1996b), who examined how adults of different ages perceive the normative deadlines for a variety of educational, occupational, and family transitions. They found that the majority of adults perceived cultural age deadlines for most of the educational, work (Settersten & Hägestadt, 1996b), and family transitions (Settersten & Hägestadt, 1996b), although cultural thinking about age timetables was relatively loose and flexible. The perceived deadlines also mirror well the transition patterns at the demographic level, although the deadlines cited by the respondents were slightly later than most people actually experience. Deadlines were generally mentioned more often in the family than work and education domains. Interestingly, there were no differences between the age groups/cohorts in this cultural thinking.

A few interesting *differences* were found between men and women. For example, deadlines for educational, occupational (Settersten & Hägestadt, 1996b), and family transitions (Settersten & Hägestadt, 1996a) were mentioned more often by men than women, but the range of deadlines for men was also larger. Moreover, men's lives were more rigidly structured by chronological age, whereas women's lives were more fluid, unpredictable, and discontinuous (Settersten & Hägestadt, 1996a). For example, in the educational and work sphere, a man's age was considered significantly more important in relationship to finishing schooling, settling on a career, and reaching the peak of the work trajectory compared to that of women (Settersten & Hägestadt, 1996b). Several differences were also found for a variety of background variables. For both men's and women's lives, non-Whites, nonprofessionals, and those with lower educational levels not only mentioned cultural age deadlines more often but also gave earlier deadlines.

Little research has been done on age differences in other aspects of future orientation than goal contents. In one study, Smith (1999) compared the kinds of strategies young adults, middle-aged individuals, and elderly people reported in the context of life planning. The results showed that elderly people considered time management (making lists and schedules, setting priorities) to be more important than did young adults, who particularly emphasized interpersonal management strategies (persuasion tactics, communication, reciprocal support).

Similarly, few studies have examined age differences in how far into the future adults' thoughts and interests extend. The research done suggests that there is no evidence of major changes in overall temporal extension across adulthood (Nurmi, 1992). However, if the changes in temporal extension are examined separately in different life domains, a decrease in how far people extend their goals is found in many domains of life. This decrease reflects the fact that, as people grow older, they draw close to the major age-graded life events in which their personal goals are embedded, which is then reflected among other things in decreasing temporal extension (Nurmi, 1991).

Summary

Although relatively young children are able to distinguish events in the near future from those in the immediate future, it is only during preadolescence that individuals develop the ability to construct conceptions of the more distant future. Accordingly, planning the distant future only becomes general when children reach the age of 11–12 years.

Research findings concerning the contents of future orientation suggest that individuals' goals, interests, and plans generally reflect the developmental tasks and life situation that typically belong to a particular age phase. Up to the age of 7 children's plans focus on everyday activities and daily chores. At the age of 9, children become interested in interpersonal relationships and achievement-related topics. At the age of 11, more future-oriented topics, such as studies, career, marriage, leaving the parental home, and the future of society, emerge in children's thinking. Adolescents and young adults typically report topics such as future education, occupation, family, and leisure activities. In middle adulthood, individuals begin to mention more goals related to their children's lives, property, and leisure activities. Later on, these topics are complemented by issues related to health and broader societal issues. Elderly people mention, in particular, health, lifestyle, service to others, and religious topics.

COUNSELING AND INTERVENTION

People are faced with a variety of challenges, demands, and problems at different stages of their lives. These include, for example, decisions concerning their future education and occupation, going to college at an adult age, finding a new career after losing a job, getting married, getting divorced, saving and buying a house, and decisions concerning their children's education and issues related to retirement. It might be assumed that future orientation plays an important role in the ways in which people deal with these kinds of challenges and demands. To be successful in handling them, an individual needs to be motivated by the future, to be aware of his or her own interests and values, to have information concerning future opportunities, to be able to construct effective plans, and to have strategies for dealing with the problems that may arise in the course of handling a particular challenge. It is not only that the kinds of challenges and demands people face in their current life situations that vary according to their age, but also the experience and knowledge they have concerning how to deal with them.

Because people do not always succeed in dealing with the major life decisions that face them, some of them may benefit from various kinds of counseling and interventions. The theory and research on future orientation might be assumed to be helpful in designing such efforts. First, the future orientation framework provides a tool that helps to conceptualize the processes taking place during counseling or intervention. Second, it helps to identify the kinds of problems that people have in dealing with a particular life situation. Third, the framework itself together with the research findings can help a counselor to devise tools to support people in making their life decisions. Finally, research on future orientation provides measures for the evaluation of people's progress in their life planning and decision making during a particular intervention program. A few empirical examples of the role of future orientation in counseling and intervention are reviewed next.

Future Orientation and Counseling

One context in which a future-orientation framework can be applied is vocational counseling, either for adolescents or for adults who wish to change their career or find a new one. The model provides a basis for identifying the different stages of vocational choice that a person has to go through:

1. Counseling starts by identifying a person's motivational basis for career decisions: What kind of activities is a person interested in? In what

kinds of situations does he or she feel comfortable? What are his or her future priorities regarding work (e.g., salary, working hours, helping people, prestige, etc.)?

2. The next stage of counseling includes a comparison of a person's motivation to the opportunities that are available to him or her given his or her academic achievement, previous education, work history, and in some cases his or her possibilities to finance further education.

3. After identifying a particular occupational goal, the counseling turns to providing information about the person's current prospects of attaining the goal. This typically includes providing information about different forms of schooling, educational institutions, or types of apprenticeship. As such a planning phase needs to be followed by people's own efforts, in many cases they will benefit from strengthening their efficacy beliefs in the educational and occupational domains (Vuori, Silvonen, Vinokur & Price, 2002).

4. Sometimes a person may benefit from thinking ahead about what may happen in the course of aiming at a particular education or occupation. Things do not always turn out as planned. It has been suggested, for example, that preparing for setbacks is an important stage in any kind of life planning (Vuori et al., 2002).

As there are substantial differences in the ways in which secondary and tertiary education, including vocational training, is organized in different countries (Hurrelmann, 1994; Jensen Arnett, 2002), educational and vocational counseling must always to be tailored to the local situation. Due to the differences in institutional structures and trajectories between different countries, the problems faced by adolescents will differ substantially. Unfortunately, only a few cross-national comparisons have been carried out with the aim of learning about these differences (Schnabel, Alfed, Eccles, Koeller, & Baumert, 2002).

Besides occupational choice that takes place often in adolescence and young adulthood, people are also faced with other kinds of life decisions in different stages of their lives about which they may benefit from counseling. These are typically situations in which an individual has to deal with a life transition that requires a substantial amount of reorientation in his or her thinking and behavior. Retirement is a good example of such a transition. In such a situation counseling may help individuals to identify major goals and meanings of their lives, and help them to think about the best ways to structure their lives in accordance with these goals. But there are other situations as well, such as facing the birth of the first child in the family during early adulthood, the empty nest in middle age after the children have left home, facing divorce, and so on.

Future Orientation and Interventions

Besides counseling, a future-orientation framework may provide a basis for building up different intervention programs that focus on helping people to deal successfully with a variety of life decisions. For example, Whan Marko and Savickas (1998) developed a short intervention procedure to increase high school and college students' interest in career planning. The intervention consisted of three phases. The orientation phase sought to increase students' future orientation and optimism by administering a set of future-related tasks (e.g., arranging three circles representing "past," "present," and "future"), and then discussing the tasks in groups. The differentiation phase consisted of attempts to make the future feel real, to reinforce positive attitudes toward planning, and to prompt goal setting. To this end the students were asked to think about and respond to different questions concerning their long-range life expectations and life planning. Finally, the integration phase attempted to link participants' present behavior to future outcomes, to provide practice with their planning skills, and to heighten career awareness. The structure of the intervention closely resembles the theoretical model presented in this chapter. The results obtained by Whan Marko and Savickas also showed that the intervention group, compared with the control group, exhibited an increase in future orientation, optimism, and positive attitudes toward career planning. No effects were found, however, on planning outcomes.

In another study, Vuori et al. (2002) used the Job Search Program (Caplan, Vinokur, Price, & van Ryn, 1989) among Finnish unemployed adults. The 5-day group intervention aimed at the enhancement of unemployed adults' job search skills. The training was designed to increase the participants' self-efficacy beliefs and motivation, enhance their recognition of their marketable skills, encourage the identification and use of social networks to find job openings, enhance their contacts with promising employers, and teach them to draw up job applications and prepare for job interviews. Interestingly, the activities on which this intervention program focused are in accordance with the major processes described in the future-orientation framework introduced in this chapter. As an additional aim, the participants discussed possible problems that may crop up in finding a job as a way of "vaccinating against setbacks." Vuori et al.'s results showed that, at the 6-month follow-up, the program had a beneficial impact on the quality of reemployment and also decreased psychological distress among the participants.

A future-orientation framework can also be used to understand what is happening in more clinical types of counseling. For example, Salmela-Aro, Näätänen, and Nurmi (in press) examined the effectiveness of an intervention program for employees suffering from severe burnout symptoms. In this intervention study, individuals who suffered from burnout were given the possibility to participate in two kinds of group therapy lasting for a period of 1 year, one using psychoanalytic techniques and another using an experiential method. The pre-, middle-, and postmeasurements not only included a burnout inventory but also a measure of personal goals. The results showed that both therapeutic interventions led to a decrease in participants' burnout symptoms. Moreover, the intervention groups, compared to control groups, showed a decrease in the number of work-related goals, reflecting perhaps a move from overexcessive work focus to a broader personal interest in different domains of life. Furthermore, intervention groups reported a decrease in negative emotional appraisals of work-related goals. Interestingly, it was particularly those participants who reported an increase in the progress of their work-related goals, and a decrease in their negative goal-related emotions, that benefited most from the intervention.

Although these examples show that a future-orientation framework is useful in devising interventions for problems encountered in making a vocational choice, reemployment efforts, and burnout, there are many other contexts in which such a framework can be used to help individuals with problems they face in their lives.

FUTURE DIRECTIONS

Future orientation was conceptualized in the present chapter as a sequential process that consists of constructing future-oriented goals, finding means for their attainment, and dealing with possible setbacks. Moreover, a variety of developmental tasks, role transitions, and institutional opportunities typical of a particular life phase were assumed to provide an "opportunity space" for people's thinking about and acting upon the future. Using this model as a theoretical guideline, previous research on future orientation was reviewed. Most of this research has focused on the kinds of future-oriented goals, possible selves and fears people typically report at different times in their lives. The review revealed that the majority of people's goals reflected the developmental tasks, institutional tracks, and life situations typical of a specific phase of life. Although at certain ages people often reported similar kinds of goals and interests, individual differences in this future orientation were also found to direct their future lives and to contribute to their well-being. For example, future-oriented goals that focused on dealing with major age-graded challenges and demands were found to be related to a high level of well-being, whereas self-focused, existential types of goals were associated with a low level of well-being.

The previous research on future orientation has, however, many limitations that should be taken into account when planning new research in the field. First, little is known, besides the content of goals, about other processes involved in future orientation, such as optimism, control beliefs, life planning, and the evaluation of one's own future, particularly among adults. Consequently, there is a need to complement previous research by including these aspects in future studies. As some studies have been done on these aspects of future orientation among adolescents, this methodology can be easily applied among other age groups.

Second, very few studies have examined future orientation using a cross-lagged longitudinal procedure. Hence little is known about the developmental dynamics of people's thinking and acting upon the future. For example, we don't know how stable the key characteristics of future orientation are across a longer time period. There is some research to show, for example, that the contents of individual goals in those life domains in which people are not going through major transitions are relatively stable, in contrast to those domains in which they are facing a transition (Nurmi & Salmela-Aro, 2002). It might be assumed, however, that overall interest in the future, optimism about the future and planning skills may show considerable stability over time. However, there is no empirical evidence for this hypothesis.

Another limitation due to the lack of longitudinal data is that relatively little is known about the major antecedents and consequences of future orientation. It has been assumed, for example, that positive self-concept and high self-esteem provide a basis for future optimism, and related beliefs in internal control, but as the previous research has been cross-sectional, it is possible that future orientation influences self-related beliefs. Similarly, parents and their adolescent children have been shown to share similar views about the future of the latter (Hallinan & Williams, 1990). However, without cross-lagged longitudinal data we cannot be sure whether this similarity is due to parents' tutoring, advice, or being role models for their children, or whether it is due to the fact that adolescents' conceptions of their future are also reflected in what parents think (Nurmi, 2001, 2004). Particularly little is know about the antecedents and consequences of how adults and elderly people think about and act upon the future.

As there is an increasing amount of variation across adulthood in the timing and sequencing of the role transitions and life events people face, one alternative for age group comparisons and traditional longitudinal studies would be to follow people across particular transitions, such as parenthood, facing the "empty nest" after children have left home, retirement, and divorce. In such research on "critical transitions," measurements are timed according to the changes in participants' developmental context rather than their age (Nurmi, 2004). Such a research design provides a tool for examining in detail the processes involved in the ways in which people deal with such transitions.

The third limitation of the previous research is that the majority of theories and studies have conceptualized future orientation as wholly individual thinking and behavior. One may argue, though, that thinking about the future is not solely an outcome of individual cognitive and motivational processing but is shared by other people, such as spouses, parents, friends, colleagues, and peers (Nurmi, 2001). For example, Meegan and Berg (2001) showed that many goals that people own are shared by other people. Moreover, other people often provide resources, help, and support, as well as advice and guidance, when people are in the process of constructing personal goals and trying to find ways of attaining them (Diener & Fujita, 1995). Malmberg (2001) found that, although the most typical situation in which adolescents reported thinking about and planning their future was when they were alone, a substantial amount of future planning was also reported during time spent with peers, time spent at home and at school, and when consuming mass media. One important line of research to be conducted in the future is examining the extent to which individuals, like children and their parents, spouses, and members of work teams, share their views of the future, to what extent people count on other people's support in their future orientation, and what role such a "shared" future orientation has in people's well-being and their interpersonal relations.

Research on thinking and acting upon the future has a long history originating in the early work on future-time perspective (Lewin, 1942) and time orientation (Hoornaert, 1973) through studies on future orientation (Nurmi, 1991; Trommsdorff et al., 1982) to more recent work on personal goals (Little, 1983; Nurmi & Salmela-Aro, 2002; Salmela-Aro et al., 2000) and possible selves (Cross & Markus, 1991). All this work suggests that future orientation (i.e., people's expectations, personal goals, life planning, optimism, and control beliefs) plays an important part in people's thinking, and also has consequences for their future lives and well-being. Despite previous efforts to understand thinking about and acting upon the future, there is a clear need for further research on the ways in which people direct their future lives, and the ways in which they adjust to changes in their developmental environments and the outcomes of their previous future-oriented efforts. As the major focus of interest should be the developmental dynamics and contextual links of future orientation, future research should include both cross-lagged longitudinal and cross-cultural studies.

REFERENCES

Arlin, M. (1990). What happens to time when you sleep? Children's development of objective time and it's relation to time perception. *Cognitive Development, 5,* 71–88.

Blinn, L. M., & Pike, G. (1989). Future time perspective: Adolescents' predictions of their interpersonal lives in the future. *Adolescence, 24,* 289–301.

Bosma, H. A. (1985). *Identity development in adolescence. Coping with commitments.* Groningen, the Netherlands: Rijksuniversiteit Te Groningen.

Bosma, H. A., & Kunnen, E. S. (2001). Determinants and mechanisms in ego identity development: A review and synthesis. *Developmental Review, 21,* 39–66.

Brown, B. B., & Larson, R. W. (2002). Kaleidoscope of adolescence: Experiences of the world's youth at the beginning of the 21st century. In B. B. Brown, R. W. Larson, & T. S. Saraswathi (Eds.), *The world's youth: Adolescence in eight regions of the globe* (pp. 1–20). New York: Cambridge University Press.

Caplan, R. D., Vinokur, A. D., Price, R. H., & van Ryn, M. (1989). Job seeking, reemployment, and mental health: A randomized field experiment in coping with job loss. *Journal of Applied Psychology, 74,* 759–769.

Crockett, L. J., & Bingham, C. R. (2000). Anticipating adulthood: Expected timing of work and family transitions among rural youth. *Journal of Research on Adolescence, 10,* 151–172.

Cross, S., & Markus, H. (1991). Possible selves across the life span. *Human Development, 34,* 230–255.

Diener, E., & Fujita, F. (1995). Resources, personal strivings, and subjective well-being: A nomothetic and idiographic approach. *Journal of Personality and Social Psychology, 38,* 668–678.

Dreher, M., & Oerter, R. (1987). Action planning competencies during adolescence and early adulthood. In S. H. Friedman, E. Kofsky Scholnick, & R. R. Cocking (Eds.), *Blueprints for thinking: The role of planning in cognitive development* (pp. 321–355). Cambridge, England: Cambridge University Press.

Elder, G. H., Jr. (1985). Perspectives on the life course. In G. H. Elder, Jr. (Ed.), *Life course dynamics* (pp. 23–49). Ithaca, NY: Cornell University Press.

Emmons, R. A. (1991). Personal strivings, daily life events and psychological and physical well-being. *Journal of Personality, 59,* 455–472.

Erikson, E. H. (1959). *Identity and the life cycle.* New York: International Universities Press.

Friedman, W. J. (2000). The development of children's knowledge of the times of future events. *Child Development, 71,* 913–932.

Gardner, W., & Rogoff, B. (1990). Children's deliberateness of planning according to task circumstances. *Developmental Psychology, 26,* 480–487.

Gillies, P. (1989). A longitudinal study of the hopes and worries of adolescents. *Journal of Adolescence, 12,* 69–81.

Hägestadt, G. O., & Neugarten, B. L. (1985). Age and the life course. In R. H. Binstock & E. Shanas (Eds.), *Handbook of aging and the social sciences* (pp. 35–61). New York: Van Nostrand Reinhold.

Hallinan, M. T., & Williams, R. A. (1990). Students' characteristics and the peer influence process. *Sociology of Education, 63,* 122–132.

Havighurst, R. J. (1948). *Developmental tasks and education* (3rd ed.). New York: McKay.

Hockaday, C., Jasper Crase, S., Shelley, M. C., & Stockdale, D. F. (2000). A prospective study of adolescent pregnancy. *Journal of Adolescence. Special Issue: Adolescents and Risk-Taking, 23,* 423–438.

Hogan, D. P. (1985). Parental influences on the timing of early life transitions. *Current Perspectives on Aging and the Life Cycle, 1,* 1–59.

Hoornaert, J. (1973). Time perspective: theoretical and methodological considerations. *Psychologica Belgica, 13,* 265–294.

Hudson, J. A., & Fivush, R. (1991). Planning in the preschool years: The emergence of plans from general event knowledge. *Cognitive Development,* 6, 393–415.

Hurrelmann, K. (Ed.). (1994). *International Handbook of Adolescence.* Westport, CT: Greenwood Press.

Jensen Arnett, J. (2002). Adolescents in Western countries in the 21st century: Vast opportunities—for all? In B. B. Brown, R. W. Larson, & T. S. Saraswathi (Eds.), *The world's youth: Adolescence in eight regions of the globe* (pp. 307–343). New York: Cambridge University Press.

Klaczynski, P. A., & Reese, H. W. (1991). Educational trajectory and "action orientation": Grade and track differences. *Journal of Youth and Adolescence, 20,* 441–462.

Kreitler, S., & Kreitler, H. (1987). Conceptions and processes of planning: The developmental perspective. In S. L. Friedman, E. K. Scholnick, & R. R. Cocking (Eds.), *Blueprints for thinking: The role of planning in cognitive development* (pp. 205–272). Cambridge, England: Cambridge University Press.

Lang, F. R., & Carstensen, L. L. (2002). Time counts: Future time perspective, goals, and social relationships. *Psychology and Aging, 17,* 125–139.

Lanz, M., & Rosnati, R. (2002). Adolescents' and young adults' construction of the future: Effects of family relations, self-esteem, and sense of coherence. In J. Trempala & L.-E. Malmberg (Eds.), *Adolescents' future-orientation: Theory and research* (pp. 17–34). Frankfurt am Main, Germany: Peter Lang.

Lawton, M. P., Moss, M. S., Winter, L., & Hoffman, C. (2002). Motivation in later life: Personal projects and well-being. *Psychology & Aging, 17,* 539–547.

Lessing, E. E. (1972). Extension of personal future time perspective, age, and life satisfaction of children and adolescents. *Developmental Psychology,* 6, 457–468.

Lewin, K. (1942). Time perspective and morale. In G. Watson (Ed.), *Civilian morale* (pp. 48–70). New York: Houghton Mifflin.

Liberska, H. (2002). Life perspectives of adolescents in the context of social and economic changes in Poland. In J. Trempala & L.-E. Malmberg (Eds.), *Adolescents' future-orientation: Theory and research* (pp. 51–63). Frankfurt am Main, Germany: Peter Lang.

Little, B. R. (1983). Personal projects: A rationale and method for investigation. *Environment and Behavior, 15,* 273–309.

Malmberg, L.-E. (1996). How do Finnish students prepare for their future in three school types? The relation between content of plans, information gathering and self-evaluations. *British Journal of Educational Psychology,* 66, 457–469.

Malmberg, L.-E. (2001). Future orientation in educational and interpersonal contexts. In J.-E. Nurmi (Ed.), *Navigating through adolescence: European perspectives* (pp. 119–140). New York: Routledge Falmer.

Malmberg, L.-E., & Norrgård, S. (1999). Adolescents' ideas of normative life-span development and personal future goals. *Journal of Adolescence, 22,* 33–47.

Marcia, J. E. (1980). Identity in adolescence. In J. Adelson (Ed.), *Handbook of adolescent psychology* (pp. 159–187). New York: Wiley.

Markus, H., & Wurf, E. (1987). The dynamic self-concept: A social psychological perspective. *Annual Review of Psychology, 38,* 299–337.

Mayer, K. U. (1986). Structural constraints on the life course. *Human Development, 29,* 163–170.

McCabe, K. M., & Barnett, D. (2000). The relation between familial factors and the future orientation of urban, African American sixth graders. *Journal of Child and Family Studies,* 9, 491–508.

McCallion, A., & Trew, K. (2000). A longitudinal study of children's hopes, aspirations and fears for the future. *Irish Journal of Psychology, 21,* 227–236.

McFadden, S. H. (1995). Religion and well-being in aging persons in an aging society. *Journal of Social Issues. Special Issue: Religious Influences on Personal and Societal Well-Being, 51,* 161–175.

Meegan, S. P., & Berg, C. A. (2001). Whose life task is it anyway? Social appraisal and life task pursuit. *Journal of Personality, 69,* 363–389.

Neugarten, B. L., Moore, J. W., & Lowe, J. C. (1965). Age norms, age constraints, and adult socialization. *American Journal of Sociology, 70,* 710–717.

Nurmi, J.-E. (1989a). *Adolescents' orientation to the future: Development of interests and plans, and related attributions and affects, in the life-span context.* (Commentationes Scientiarum Socialium, Vol. 39.) Helsinki: The Finnish Society for Sciences and Letters.

Nurmi, J.-E. (1989b). Development of orientation to the future during early adolescence: A four-year longitudinal study and two cross-sectional comparisons. *International Journal of Psychology, 24,* 195–214.

Nurmi, J.-E. (1991). How do adolescents see their future? A review of the development of future orientation and planning. *Developmental Review, 11,* 1–59.

Nurmi, J.-E. (1992). Age differences in adult life goals, concerns, and their temporal extension: A life course approach to future-oriented motivation. *International Journal of Behavioral Development, 15,* 487–508.

Nurmi, J.-E. (1993). Adolescent development in an age-graded context: The role of personal beliefs, goals, and strategies in the tackling of developmental tasks and standards. *International Journal of Behavioral Development, 16,* 169–189.

Nurmi, J.-E. (2001). Adolescents' self-direction and self-definition in age-graded sociocultural and interpersonal contexts. In J.-E. Nurmi (Ed.), *Navigating through adolescence: European perspectives* (pp. 229–250). New York & London: Routledge Falmer.

Nurmi, J.-E. (2004). Socialization and self-development: Channeling, selection, adjustment, and reflection. In R. Lerner & L. Steinberg (Eds.), *Handbook of Adolescent Psychology* (pp. 85–124). Hoboken, NJ: Wiley.

Nurmi, J.-E., Poole, M. E., & Kalakoski, V. (1994). Age differences in adolescent future-oriented goals, concerns, and related temporal extension in different sociocultural contexts. *Journal of Youth & Adolescence, 23,* 471–487.

Nurmi, J.-E., & Salmela-Aro, K. (2002). Goal construction, reconstruction and depressive symptomatology in life-span context: The transition from school to work *Journal of Personality, 70,* 385–420.

Nurmi, J.-E., Salmela-Aro, K., & Koivisto, P. (2002). Goal importance and related achievement beliefs and emotions during transition from vocational school to work: Antecedents and consequences. *Journal of Vocational Behavior, 60,* 241–261.

Oppenheimer, L. (1987). Cognitive and social variables in the plan of action. In S. H. Friedman, E. Kofsky Scholnick, & R. R. Cocking (Eds.), *Blueprints for thinking: The role of planning in cognitive development* (pp. 356–392). Cambridge, England: Cambridge University Press.

Oppenheimer, L., & van der Wilk, R. (1984). *Children's interests: Development and gender differences.* Unpublished manuscript, University of Amsterdam.

Orbach, I., Iluz, A., & Rosenheim, E. (1987). Value systems and commitment to goals as a function of age, integration of personality, and fear of death. *International Journal of Behavioral Development, 10,* 225–239.

Pasupathi, M., Staudinger, U. M., & Baltes, B. P. (2001). Seeds of wisdom: Adolescents' knowledge and judgment about difficult life problems. *Developmental Psychology, 37,* 351–361.

Pimentel, E. F. (1996). Effects of adolescent achievement and family goals on the early adult transition. In T. T. Mortimer & M. D. Finch (Eds.), *Adolescents, work, and family: An intergenerational developmental analysis.* Thousand Oaks, CA: Sage.

Poole, M. E., & Cooney, G. H. (1987). Orientations to the future: A comparison of adolescents in Australia and Singapore. *Journal of Youth and Adolescence, 16,* 129–151.

Prager, E. (1996). Exploring personal meaning in an age-differentiated Australian sample: Another look at the Sources of Meaning Profile (SOMP). *Journal of Aging Studies, 10,* 117–136.

Pulkkinen, L., Nurmi, J.-E., & Kokko, K. (2002). Individual differences in personal goals in mid-thirties. In L. Pulkkinen & A. Caspi (Eds.), *Paths to successful development: Personality in the life course* (pp. 331–352). New York: Cambridge University Press.

Russell, G., & Smith, J. (1979). "Girls can be doctors ... can't they?": Sex differences in career aspirations. *Journal of Social Issues, 14,* 91–102.

Salmela-Aro, K., Aunola, K., & Nurmi, J.-E. (2003, October). *Personal projects and depressive symptoms during emerging adulthood: A 10-year longitudinal study.* Poster session presented at the meeting of the Emerging Adulthood Special Interest Group of the Society for Research on Adolescence, Cambridge, MA.

Salmela-Aro, K., Näätänen, P., & Nurmi, J.-E. (in press). The role of personal projects during two burnout interventions: A longitudinal study. *Work & Stress.*

Salmela-Aro, K., & Nurmi, J.-E. (1997). Goal contents, well-being, and life context during transition to university: A longitudinal study. *International Journal of Behavioral Development, 20,* 471–491.

Salmela-Aro, K., Nurmi, J.-E., Saisto, T., & Halmesmäki, E. (2000). Women's and men's personal goals during the transition to parenthood. *Journal of Family Psychology, 14,* 171–186.

Salmela-Aro, K., Nurmi, J.-E., Saisto, T., & Halmesmäki, E. (2001). Goal reconstruction and depressive symptoms during the transition to motherhood: Evidence from two cross-lagged longitudinal studies. *Journal of Personality and Social Psychology, 81,* 1144–1159.

Scheier, M. F., Carver, C. S., & Bridges, M. W. (1994). Distinguishing optimism from neuroticism (and trait anxiety, self-mastery, and self-esteem): A reevaluation of the Life Orientation Test. *Journal of Personality & Social Psychology, 67,* 1063–1078.

Schnabel, K. U., Alfred, C., Eccles, J. S., Koeller, O., & Baumert, J. (2002). Parental influence on students' educational choices in the United States and Germany: Different ramifications—same effect? *Journal of Vocational Behavior. Special Issue on the Transition From School to Work: Societal Opportunities and Individual Agency, 60,* 178–198.

Schoon, I., & Parsons, S. (2002). Teenage aspirations for future careers and occupational outcomes. *Journal of Vocational Behavior, 60,* 262–288.

Seginer, R. (1988). Social milieu and future orientation: The case of kibbutz vs. urban adolescents. *International Journal of Behavioral Development, 11,* 247–273.

Settersten, R. A., Jr., & Hägestadt, G. O. (1996a). What's the latest? Cultural age deadlines for educational and work transitions. *The Gerontologist, 36,* 602–613.

Settersten, R. A., Jr., & Hägestadt, G. O. (1996b). What's the latest? Cultural age deadlines for family transitions. *The Gerontologist, 36,* 178–188.

Sheldon, K. M., & Kasser, T. (2001). Getting older, getting better? Personal strivings and psychological maturity across the life span. *Developmental Psychology, 37,* 491–501.

Shell, D. F., & Husman, J. (2001). The multivariate dimensionality of personal control and future time perspective beliefs in achievement and self-regulation. *Contemporary Educational Psychology, 26,* 481–506.

Slee, P. T., & Cross, D. G. (1989). Living in the nuclear age: An Australian study of children's and adolescent's fears. *Child Psychiatry and Human Development, 19,* 270–278.

Smith, J. (1999). Life planning: Anticipating future life goals and managing personal development. In J. Brandtstädter & R. M. Lerner (Eds.), *Action and self-development: Theory and research through the life span* (pp. 223–255). Thousand Oaks, CA: Sage.

Smith, J., & Freund, A. M. (2002). The dynamics of possible selves in old age. *Journals of Gerontology: Psychological Sciences, 57B,* 492–500.

Solantaus, T. (1987). Hopes and worries of young people in three European countries. *Health Promotion, 2,* 19–27.

Stattin, H., & Kerr, M. (2001). Adolescents' values matter. In J.-E. Nurmi (Ed.), *Navigating through adolescence: European perspectives* (pp. 21–58). New York: Routledge Falmer.

Strough, J., Berg, C. A., & Sansone, C. (1996). Goals for solving everyday problems across the life span: Age and gender differences in the salience of interpersonal concerns. *Developmental Psychology, 32,* 1106–1115.

Sundberg, N. D., Poole, M. E., & Tyler, L. E. (1983). Adolescents' expectations of future events: A cross-cultural study of Australians, Americans, and Indians. *International Journal of Psychology, 18,* 415–427.

Takkinen, S., & Ruoppila, I. (2001). Meaning in life in three samples of elderly persons with high cognitive functioning. *The International Journal of Aging and Human Development, 53,* 51–73.

Trommsdorff, G., Burger, C., & Fuchsle, T. (1982). Social and psychological aspects of future orientation. In M. Irle (Ed.), *Studies in decision making* (pp. 167–194). Berlin: de Gruyter.

Vuori, J., Silvonen, J., Vinokur, A. D., & Price, R. H. (2002). The Tyoehoen Job Search Program in Finland: Benefits for the unemployed with risk of depression or discouragement. *Journal of Occupational Health Psychology, 7,* 5–19.

Whan Marko, K., & Savickas, M. L. (1998). Effectiveness of a career time perspective intervention. *Journal of Vocational Behavior, 52,* 106–119.

Wilkins, R., & Lewis, C. (1993). Sex and drugs and nuclear war: Secular, developmental and Type A influences upon adolescents' fears of the nuclear threat, AIDS and drug addiction. *Journal of Adolescence, 16,* 23–41.

Wyman, P. A., Cowen, E. L., Work, W. C., & Kerley, J. H. (1993). The role of children's future expectations in self-esteem functioning and adjustment to life stress: A prospective study of urban at-risk children. *Development and Psychopathology. Special Issue: Milestones in the Development of Resilience, 5,* 649–661.

Yowell, C. M. (2000). Possible selves and future orientation: Exploring hopes and fears of Latino boys and girls. *Journal of Early Adolescence, 20,* 245–280.

Zimbardo, P. G., Keough, K. A., & Boyd, J. N. (1997). Present time perspective as a predictor of risky driving. *Personality & Individual Differences, 23,* 1007–1023.

4

Time and Terror: Managing Temporal Consciousness and the Awareness of Mortality

Clay Routledge
Jamie Arndt
University of Missouri–Columbia

> If there is one constant in this universe, it is death. . . . We're all going to die sometime, its just a matter of how, of when. Aren't you beginning to feel time gaining on you? Its like a predator—its stalking you. Oh you can try and outrun it with doctors, medicines, and new technologies, but in the end time is going to hunt you down and make the kill.
>
> —Rick Berman & David Carson, *Star Trek: Generations* (1997)

For many people, each morning begins with the jolt of an alarm clock, the temporal starting pistol that initializes a day of routine activities and scheduled events. *Homo sapiens* are unique temporally conscious creatures who live in a world of reflections on the past and dreams about the future. The highly evolved ability to think in terms of time has allowed humans to harness their physical environment and has thus helped people gain an upper hand in the struggle to thrive as organisms. However, the benefits of temporal consciousness come at a psychological price. Whereas we can pause and think about the day when we will win an Oscar or a Nobel Prize, we can also think about the day when we will die and begin our decomposition into fertilizer. That is, being able to think about the future means being able to think about a future without us in it. People can envision an untimely splat of their bodies against an oncoming truck, or a visit to the doctor's office where they learn that a bump on the body is in fact cancerous. Of course, if they are fortunate, they can look forward to the slow decay and degeneration of the body as they age. This then is a fundamental problem confronting the human condition: How does one live in the face of the awareness of inevitable mortality and the potential anxiety such awareness engenders? As many scholars have noted, the answer to this question may reveal important insights about the needs and nature of human social behavior (e.g., Becker, 1971; Berger & Luckmann, 1967; Brown, 1959; Goldschmidt, 1990; Lifton, 1983; Rank, 1932/1989; Yalom, 1980).

In this chapter we use terror management theory (TMT; e.g., Solomon, Greenberg, & Pyszczynski, 1991), a social psychological theory that builds from the work of Ernest Becker (1971, 1973) and other existential traditions, to explicate how the awareness of time is related to concerns about mortality. We then review a program of research that suggests people manage concerns about death by defending a socially constructed world of meaning and value in which the physical reality of mortality can be transcended either symbolically or literally via adherence to culturally delineated beliefs and values. We specifically consider the form and function of these different defenses, the temporal progression through which they manifest themselves, and the implications they have for people's ability to peacefully coexist with those who are different. Furthermore, we present new programs of research that more specifically focus on the ways in which people manage the awareness of time. Consistent with other findings derived from TMT, these new lines of research suggest that people defend themselves from an awareness of time and its logical connection to death by investing in beliefs that make time itself meaningful, turning to the past or future to garner a sense of personal value, and distancing from those people who are often most associated with death, the elderly.

TERROR MANAGEMENT THEORY: TIME, DEATH, AND PSYCHOLOGICAL DEFENSE

> Whatever became of the moment when one first knew about death? There must have been one, a moment, in childhood when it first occurred to you that you don't go on for ever. It must have been shattering—stamped into one's memory. And yet I can't re-

member it. It never occurred to me at all. What does one make of that? We must be born with an intuition of mortality. Before we know the words for it, before we know that there are words, out we come, bloodied and squalling with the knowledge that for all the compasses in the world, there's only one direction, and time is its only measure. (Tom Stoppard, *Rosencrantz and Guildenstern Are Dead,* 1966)

Temporal Consciousness as Existential Crisis

Most people would agree that there is something unique about the way people navigate through life. As noted by comparative and cultural anthropologists (e.g., Becker, 1971; Erchak, 1992; Goldschmidt, 1990), for most if not all animals, the world exists as a stream of sensory experience that enables little more than short-term acquisition of life-sustaining resources and basic social interaction (Goldschmidt, 1990). However, humans are quite different in this regard. In the context of a multitude of means by which organisms adapt to environmental circumstances (e.g., fish gills, skunk stench, cactus thorns) to facilitate survival and reproduction, the uniquely human adaptation is our complex cognitive capabilities, which provide the profound capacity for self-awareness. In 1972, Duval and Wicklund introduced mainstream psychology to the notion that people's attention can be directed either externally on the world or internally on the self. From this perspective, the capacity for self-directed attention leads to self-regulatory processes by which people can move their present selves toward their future goals.

In addition to the capacity for self-reflective thought, Becker (1971) highlighted the interconnection of two other critically important human abilities: the capacity for abstract and symbolic thought and the capacity to think in terms of time. People can reflect on what they did last Thursday, what they are doing this Thursday, and furthermore, what they may be doing on the third Thursday of next month. In Becker's terms, the *ego* allows humans to be a self-aware point of temporal reference, fully capable of indexing and reflecting on past events and contemplating future events with planned calculation, thus endowing people with an unparalleled freedom of reactivity from instinctually elicited responses. Moreover, humans can think at levels of abstraction that render them capable of contemplating distant future outcomes, and furthermore, transforming those visions into experienced reality. Whereas our single-celled brethren, the amoeba, can only approach and consume a glucoselike substance, people, if exercising restraint, can eye some frosting formed from sugar, some angel food-cake, and fashion themselves a Twinkie. People attend college, go on diets, learn new skills, and engage in many other planned behaviors all because of the ability to place oneself in a hypothetical future.

The time-conscious human thus uses the self as a temporal reference point that allows for reflection upon the past and planning of the future (Becker,

1971). Becker drew heavily from the psychoanalytic historian Norman O. Brown (1959) in making this point. For Brown, following Freud, an important function of the ego (or self) is to repress that which is threatening to the conscious mind. Of course, Brown, like Becker, differed from Freud in what he viewed as the threatening content that people are motivated to repress. However, the implications of the mechanism of repression are important. Because people are able to operate without conscious intrusion of threatening thoughts, the conscious mind is then afforded the luxury of projecting itself into the past and imagining the future (Brown, 1959). Yet, as we explain later, the future is inevitably not a pleasant one and thus this luxury may perhaps ironically increase the burdens to be repressed.

Although these sophisticated cognitive capabilities engender humans with unparalleled self-regulation and behavioral plasticity, there are some unsettling consequences produced by awareness of one's self and one's time and place in the world. Humans are aware that regardless of all efforts to thrive as life forms, time cannot be stopped and the passing of time means the passing of life. People need only look around themselves to see time having its effect on life. The grass grows but then the leaves fall. Temporal consciousness cannot be divorced from the awareness of death. Therefore, regardless of all the advantages thinking temporally provides, this enlightenment is potentially problematic as the juxtaposition of self-preservation motives and the awareness of mortality creates the potential for paralyzing death anxiety (e.g. Brown, 1959; Goldschmidt, 1990; Solomon et al., 1991). And what is more, humans have developed methods of measuring and labeling time, rendering the passing of time, and ultimately the mortal condition of life, an integral and quantifiable component of daily existence. The calendar marks the amount of life that has passed and the amount that likely remains. Life expectancy can be calculated, thus making time an always-draining hourglass.

Managing the Problem of Time and Death

The assertion that being aware of time is psychologically problematic makes time awareness appear to be a paradoxical evolutionary development. Time can be used to make logical predictions, plan for future events, create a social order, and even increase longevity. However, in addition to all the adaptive benefits, understanding time helps people understand the certainty of death. Moreover, no matter how long or pleasant the plot, humans know that the movie can end at any point. Death can come for reasons that people can neither anticipate nor control. If this rather gloomy contention is accurate, why do people seem relatively unaffected psychologically by the passing of time? Very few people are cowering in the corner as twitching blobs of biological protoplasm, completely petrified by the realization that they will inevitably die at some point in the future. Indeed, most individuals function with relative equanimity, consumed and engaged by their daily activities and challenges. How do humans manage the existential dilemma of knowing about a future of certain death?

The advanced human cognitive abilities of abstract, self-reflective, and temporal thought may have rendered people aware of mortality; however, perhaps ironically, following Becker (1971, 1973), TMT (e.g., Solomon et al., 1991) asserts that these same mental resources that create the problem also facilitate the solution. That is, the human capacity for symbolic thought allows for the social construction of a meaningful and controllable world. It is to this end that TMT posits that *cultural worldviews* are humanly created beliefs about the nature of reality shared by groups of people that developed as a means by which individuals manage the potential for terror (hence the term *terror management*) engendered by the human awareness of mortality. In short, the reduction of death anxiety lies within the successful construction of, and adherence to, a cultural belief system that provides the sense that one's existence transcends the physical confines of biological life.

Cultural worldviews provide this protection from existential fear by offering answers to basic and universal cosmological questions ultimately connected to time (e.g., When and how did my time begin?; What do I do with the time I have?; What happens after my time is up?) and structuring the human perception of reality in a manner that imbues the universe with meaning, order, and permanence. Cultures also provide prescriptions for value—for what constitutes good and bad behavior—that when fulfilled, promise protection and ultimately death transcendence, either literally through such beliefs as heaven or reincarnation, or symbolically by identification with entities beyond oneself. As we later explore in more detail, although religion is often a major component of a cultural worldview, this analysis views a variety of identifications, such as investment in one's romantic relationship, corporation, political party, or even basketball team, as ultimately (fictional) socially constructed conceptions of reality that make it possible for people to be valued and significant members of a meaningful universe; in other words (following Becker, 1971, 1973) for them to have *self-esteem*. Indeed, one might also note the tenacity with which scientists often cling to their theories as reflections of the trenchant need to sustain faith in a system of beliefs that provides the basis for individuals' self-worth.

However, each person enters the world unadorned by these cultural dressings, and it is only as the capacities for abstract and temporal thought develop that the protection from mortality concerns becomes a fundamental motivating force in human behavior. Following a range of theorists (e.g., Becker, 1971,

1973; Bowlby, 1969; Horney, 1937; Mead, 1934/1968; Sullivan, 1953; Yalom, 1980), TMT proposes that the association between a sense of symbolic value and anxiety protection begins early in life, given the child's profound immaturity, and in the context of the socialization process. Upon birth, the human is virtually helpless, unable to roll over by itself let alone able to navigate him or herself through shopping malls with charge card in hand. The child thus quickly develops dependence on the caregiver in order to survive as an organism. And it is here that the critical association between appropriate conduct and security is formed. The child learns that as he or she meets the symbolic standards of value espoused by mom (e.g., "don't play in the flowerbed"), the child enjoys the security and comfort of the parents' love and protection. However, when the child veers from these parental standards of behavior (e.g., and does in fact play in the flowerbed), the child experiences, at the least, the absence of the overt affection that renders secure functioning possible. Thus to be good is to be safe and secure, and not to be good is to be exposed to anxiety. This social training quickly weaves the child into the cultural fabric and is reinforced by a range of cultural teachings that associate adherence to values of goodness with security (cf. Lerner, 1980).

As the child continues to mature, progressing toward adolescence and adulthood, he or she begins to depend less on caregiver approval and more on societal approval (Becker, 1971, 1973; Rosenberg, 1981; Solomon et al., 1991). Concurrently, as the social network expands, so too do the child's cognitive abilities and the realization that the parents cannot provide the omnipotent protection from death and vulnerability. The security blanket thus broadens (a process that Becker describes as the ultimate transference) to include a range of cultural and social affiliations. Whereas the child first develops a sense of self-esteem by pleasing the parents, he or she must eventually please society (or at least the subculture of society with which he or she identifies) in order to maintain the feelings of self-esteem and meaning that provide existential security. The child eventually advances to adulthood and continues to navigate through life within a cultural matrix that provides protection from the always-looming reality of death.

An important implication of this analysis is that cultural worldviews and the sense of self-esteem we derive from them are in fact culturally relative. That is, what constitutes value and meaning to one culture or group of people does not necessarily do so for another. For example, Americans put a shortstop on the cover of magazines and pay him over $20 million a year because his athletic prowess is so highly valued by people who share the dominant worldview in our culture. Conversely, there is no national praise or lucrative endorsement opportunities for the scholar who outpublishes all of his or her peers in a single academic year. Yet within an academic subculture such accomplishments are highly regarded. In other cultures, far different behaviors may bring honor and status to individuals (see, e.g., Erchak, 1992; Goldschmidt, 1990). For the Sambia of New Guinea, the prestige of male adolescents of the tribe is tied to their acquisition of *jergunda*, or strength, which may be obtained either by drinking a particular type of tree sap or by performing oral sex on the male elders of the village (Herdt, 1987).

Thus, cultural worldviews and the self-esteem that people derive from following their prescriptions are socially constructed and effective only to the extent that an individual can maintain faith in their validity (see also Berger & Luckmann, 1967). As Festinger (1954) and many others observed, such confidence in a particular view of the world is fortified when others believe similarly and thus corroborate our conceptions of reality. When others believe differently, even if only implicitly, they threaten the integrity of these structures and undermine their ability to provide psychological security. Thus, TMT posits that these two structures, faith in a cultural worldview, and self-esteem, which results from living up to the standards of value associated with that worldview, are essential to sustain psychological equanimity. To the extent that others may threaten these structures, people are strongly motivated to maintain them, and defend them. From this perspective, then, human beings' lurid and long-standing traditions of hostility and disdain toward those who are different can be understood in part as egregious manifestations of the defenses marshaled against the threats different others pose to one's vital beliefs about the nature of reality.

TERROR MANAGEMENT RESEARCH: DEFENSES AGAINST THE IMPLICATIONS OF FUTURE DEATH

> Civilization originates in delayed infancy and its function is security. It is a huge network of more or less successful attempts to protect mankind against the danger of object-loss, the colossal efforts made by a baby who is afraid of being left alone in the dark. (Geza Roheim, *The Origin and Function of Culture*, 1943, p. 131)

TMT was first proposed by Greenberg, Pyszczynski, and Solomon in 1986 and since that time the theory has inspired a considerable amount of empirical research, most of it well beyond the scope of the current chapter. The first hypothesis generated by the theory was based on the premise that cultural worldviews provide psychological defense from the awareness of death, and that reminding people of their mortality should therefore engender the need to adamantly protect the absolute integrity of their worldviews (for a review of this research, see Greenberg, Solomon, & Pyszczynski, 1997). This contention, known as the *mortality salience hypothesis*, has been examined and supported in more than 150 studies to date.

In the most typical paradigm, individuals participate in a study purportedly on the relationship between personality attributes and interpersonal judgments or decisions. During the study, among other filler questionnaires, participants are asked to answer two open-ended questions about their death or a control topic, such as thoughts about experiencing intense physical pain. Participants later evaluate targets who in some way impinge on the viability of an aspect of their belief system. For example, the first set of mortality salience studies found that whereas municipal court judges in a nondeath control condition recommended a $50 bond for an alleged prostitute, those judges who were reminded of their mortality set bond at an average exceeding $400 (Rosenblatt, Greenberg, Solomon, Pyszczynski, & Lyon, 1989). From the perspective of TMT, this effect occurred because an awareness of death leads people to reaffirm or bolster their conceptions of meaning, which for judges includes a belief in the sanctity of the law. This study was the first demonstration of what is now known as *worldview defense*, a reaction to mortality salience in which participants become more negative toward those people or things that threaten their beliefs and, as shown in subsequent work, more positive toward those people or things that support their beliefs (see Greenberg et al., 1997). Later we review worldview defense studies that directly bear on both the literal and symbolic denial of time's progression to death, but first it is important to note the reliability, generalizability, and specificity of these effects.

Mortality salience effects have been obtained in at least nine other countries (e.g., Israel—Florian & Mikulincer, 1997; the Netherlands—Dechesne, Greenberg, Arndt, & Schimel, 2000). A wide variety of geographically and demographically diverse samples have also evidenced these effects, ranging from fifth graders (Florian & Mikulincer, 1998b), to college students (e.g., Greenberg et al., 1990), to elderly Americans (McCoy, Pyszczynski, Solomon, & Greenberg, 2000), to Israeli soldiers (e.g.,Taubman Ben-Ari, Florian, & Mikulincer, 1999), to Australian Aborigines (Halloran & Kashima, 2004), to municipal court judges (Rosenblatt et al., 1989), to pedestrians in Colorado and Germany (Pyszczynski et al.,1996).

Convergent support for the unique role of death in these effects comes from multiple operationalizations of mortality salience in addition to the traditional open-ended questions about one's death (e.g., using standardized fear of death scales—Greenberg, Simon, Harmon-Jones et al., 1995; showing participants fatal-accident footage—Nelson, Moore, Olivetti, & Scott, 1997). For example, in one series of studies, Pyszczynski et al. (1996) found that pedestrians interviewed in front of the funeral parlor were more apt to think that other people shared their beliefs, relative to pedestrians interviewed 100 yards before walking by the funeral parlor. Thus, this study makes the important point that relatively subtle everyday reminders of death can exacerbate worldview defense reactions. Indeed, Arndt, Greenberg, Pyszczynski, and Solomon (1997) exposed participants to subliminal presentations of the word *death* (i.e., flashed the word on a computer screen for 28 ms) and found that such presentations increased bias for a pro-American target over an anti-American target.

These and other studies (see, e.g., Harmon-Jones et al., 1997; Simon et al., 1997) also demonstrate that the same conditions that elevate the accessibility of death-related cognitions increase worldview defense. In contrast, these effects have not emerged in response to other aversive thoughts and nor are they explainable by the notion that reminders of death simply increase negative affect or arousal (see Greenberg et al., 1997, for a review of this research). For example, thoughts of intense physical pain, meaninglessness, social exclusion, uncertainty, future worries, failing an important exam, dental pain, or an actual failure experience, have not been found to reproduce these effects. In addition, a typical mortality salience manipulation does not produce physiological arousal (as measured by pulse rate, peripheral blood volume, and skin conductance) or self-reports of anxiety or negative affect, and there is no indication that experienced affect is involved with these responses (Arndt, Allen, & Greenberg, 2001).

Literal Immortality Worldviews

From the perspective of TMT, because of the fundamental problems posed by the awareness of time and mortality, belief in a literal transcendence of mortal existence should be particularly compelling to people. For example, religious worldviews have long been an integral part of civilization for many reasons. One important aspect of such beliefs is that they typically posit physical life as merely one stage of existence. Heaven, Nirvana, and Moksha, in theologically distinct ways, offer some form of metaphysical death defiance that refutes the view that physical termination represents the end of existence. This being the case, it is not surprising that immortality worldviews, even in largely secular modern societies, play a significant role in the defense mounted against the awareness of death.

Indeed, some of the early worldview defense studies examined the extent to which people will defend their religious ideas against even very subtle threats after mortality has been rendered salient. For example, Greenberg et al. (1990, Study 1) exposed Christian participants to a mortality salience or a control topic induction, and then had them evaluate two other students who were quite similar except for the religious affiliation of the students' parents (and by inference, the students themselves): Christian or Jewish. In the control condition, the results indicated no difference between ratings of the Christian and Jewish targets by the Christian participants. However, following mortality salience, participants increased their affection for the Christian target and had more negative reactions to the now significantly less desirable Jewish target.

Further highlighting the relationship between death and religion, Greenberg, Simon, Porteus, Pysczynski, and Solomon (1995) conducted a study in which participants were instructed to write about either death or watching television and were then given a creative problem-solving task that involved hanging a hard plastic crucifix on the wall. Half of the participants were given a wooden block that could be used to hammer a nail into the wall in order to hang the crucifix. The other half of the participants were not given the wooden block and thus had to use the crucifix itself to hammer the nail into the wall. For those participants who did not write about death, having to use the crucifix in an inappropriate manner (i.e., as a hammer) was not problematic. However, participants who wrote about their own death and had to use the crucifix as a hammer took significantly longer to complete the task and reported significantly more discomfort than all other participants. This study further demonstrates the importance of religion in the face of concerns about mortality. Using the crucifix in an inappropriate manner was of little consequence to all except those who had been asked to think about their own mortality.

Just as beliefs that feature a literal sense of immortality will be defended in the face of reminders of mortality, so too can they insulate people from the need for further defense by mitigating the impact of such reminders. Specifically, in Dechesne et al. (2003), the veracity of perceptions of immortality was manipulated to determine if such beliefs could mitigate the defensive responses that are typically evoked by reminders of death. Before a standard mortality salience or control manipulation (in some studies the topic was watching television and in others dental pain) manipulation, participants read a compelling article stating that "near death experiences" could be completely explained with neurobiological science or that such experiences could not be explained scientifically and thus provide evidence for some form of continued consciousness after death. Across a series of studies, after thinking about death, participants who read the article bolstering beliefs in literal immortality showed lower levels of defensiveness than participants who read the article that did not promote immortality beliefs (e.g., punitive judgments toward worldview violators, self-esteem biases), presumably because the indoctrination of an immortality worldview buffered the effects of thoughts about death and thus alleviated the need to engage in symbolic forms of existential defense.

Symbolic Immortality Worldviews

In the preceding pages we have alluded to some worldview defense reactions that do not involve faith in literal immortality beliefs, but instead increased investment in what may be seen as symbolic immortality beliefs. Thus, for example, we earlier considered findings that, following mortality salience, municipal-court judges became more punitive against a legal transgressor (Rosenblatt et al., 1989). As further examples, mortality salience effects have been documented with regard to identifications involving participants' university (Dechesne et al., 2000), nationality (e.g., Arndt, Greenberg, Solomon, Pyszczynski, & Simon, 1997), ethnic and gender group (Arndt, Greenberg, Schimel, Pyszczynksi, & Solomon, 2002), academic major (Arndt & Greenberg, 1999), political party (McGregor et al., 1998), and even sports team (Dechesne et al., 2000). In all of these cases, and many others, reminders of death are seen as increasing people's investment in beliefs that imbue the world with a sense of meaning and importance. To the extent that such identifications provide a socially validated consensus of meaning, they offer the possibility of affiliating with entities that extend beyond the self and, to the extent that such beliefs persist over time, offer a way for the self to divorce itself from its temporal limitations.

There are, of course, a number of other avenues through which people can try to maintain their belief that they in fact have some sense of individual symbolic value in the broader context of this meaningful view of the world. Recent terror management research suggests this may occur as people strive to achieve a sense of symbolic distinction from other animals, invest in other pursuits that imply symbolic continuance beyond death, as well as through the multifarious ways in which people obtain self-esteem. We consider briefly later each of these different domains and how they are impacted by reminders of one's finitude.

One need only spend an afternoon watching the lion pursue his prey on The Discovery Channel to come to the conclusion that being an animal means being physically vulnerable to an uncertain demise. Though it is unlikely that death will come via a hungry predator, awareness of one's physical nature exposes just how fragile and uncertain life is for all animals, including humans. Thus, as Becker (1973) asserts, humans create a cultural world in which they are not merely members of the organismic rat race for survival, but are cosmically significant beings worth more than the sum of their biological parts. In line with this idea, Goldenberg et al. (2001) found that in response to thoughts about death, people engaged in efforts to distance themselves from the physicality of the body. For example, in Study 2 of this package, participants either wrote about death or dental pain and were then asked to read and evaluate an essay arguing that humans either are similar to animals or are quite distinct from animals. In the dental-pain condition, evaluations of the humans-as-similar-to-animals essay and the humans-as-distinct-from-animals essay did not differ. In contrast, in the mortality salience condition, participants were significantly more favorable toward the humans-as-distinct essay relative to the humans-as-similar-to-animals essay. This study supports the assertion that the physical nature of the human body is ultimately problematic when concerns about mortality are made salient.

Beyond attempts to symbolically disengage from the mortal confines of one's animal nature, Lifton (1983) argued that the awareness of death leads some people to search for a more explicit sense of symbolic immortality. Lifton thus describes how events like passing on one's genes, making societal contributions, and viewing one's own existence as merely a part of a larger cosmic whole provide humans with an explicit way to fight back against concerns about mortality. In one series of studies, Florian and Mikulincer (1998a) examined the idea that humans protect themselves from death awareness by explicitly investing is various modes of death transcendence. For example, in one study, participants completed a scale designed to specifically assess one's level of investment in the distinct modes of death transcendence described by Lifton (e.g., "It is important for me to write, create, or build something that will exist after my death" and "I would do almost anything to ensure the future of my children"; Mathews & Kling, 1988). In addition, participants completed a scale that assessed levels of fear regarding death (Florian & Kravetz, 1983). Supporting Lifton's assertions, scores on the symbolic immortality scale and the fear of death scale were significantly negatively correlated. That is, a high sense of symbolic immortality was related to lower levels of death fear. Another study found that people high in a sense of symbolic immortality, compared to people scoring low on this scale, displayed decreased levels of worldview defense after mortality was made salient. Previously, we described the studies in which people who were provided with evidence for literal immortality displayed lower levels of symbolic defensiveness, presumably because such defenses were not needed when death could be defied via an afterlife. In a conceptually similar fashion, this study shows how one well-established method for defense (i.e., a strong sense of symbolic death transcendence) can reduce the need to turn to other symbolic defenses (i.e., cultural worldviews) after thinking about death.

Though TMT maintains that belief in a meaningful conception of the world is an important ingredient toward protecting oneself from the mortal implications of time, it is not sufficient. A person must also sustain the perception that he or she a significant player in the cultural drama (Becker, 1971). This notion, that self-esteem serves an anxiety-buffering function, and moreover, that people will strive to maintain self-esteem when reminded of the source of existential anxiety, has also garnered a significant amount of empirical attention. The first incantation of this proposition assessed (and found) that high levels of self-esteem reduce self-reported and physiological anxiety in response to threat as well as cognitive distortions that aim to deny one's vulnerability (Greenberg et al., 1992, 1993).

Terror management studies further showcased the defensive role that self-esteem plays by finding that high self-esteem reduces worldview defense after mortality salience. For example, in Harmon-Jones et al. (1997), participants with either dispositionally high levels of self-esteem (Study 1) or experimentally induced high self-esteem (Study 2) did not display the typically found elevated levels of worldview defense after mortality was made salient. Thus, a person's sense of self-worth appears to buffer the impact of the awareness of death and reduces the need to respond with defense of one's meaning providing beliefs.

Further supporting the assertion that self-esteem protects humans from the awareness of death, studies have also shown that participants react to mortality salience with increased efforts to boost their self-esteem. For example, mortality salience has been shown to increase risky driving behavior among individuals whose driving ability plays a major role in their sense of self-esteem (Taubman et al., 1999) and intentions to exercise among those whose physical fitness is an important component of self-esteem (Arndt, Schimel, & Goldenberg, 2003). People may be routinely confronted with reminders that time and life are passing; however, self-esteem allows people to feel like more than an insignificant organism whose life is barely a blink in the cosmic scheme.

The Architecture of Terror Management: A Dual-Defense Model

The defenses that we have been describing may in many ways seem quite removed from the reactions one might expect to thoughts of death, as they often bear no logical connection to such cognitions. That is, for example, identifying with social institutions will not reduce one's vulnerability to physical demise. However, as we have outlined, the connection lies in the capacity of such beliefs to provide ways in which humans can identify with symbolic structures that transcend the confines of mortal existence. Of course, people also experience very tangible fears concerning mortality-related thoughts and often engage in direct efforts to combat specific threats to their existence. For example, telling people that smoking causes cancer may motivate a direct effort to cease smoking. Thus, there appear to be distinct ways in which people protect themselves from explicit death-related concerns and over the last few years an increasing amount of research has been devoted to understanding the cognitive and temporal processes that underlie the different ways in which people manage conscious and unconscious concerns about death (see, e.g., Arndt, Cook, & Routledge, 2004; Pyszczynski, Greenberg, & Solomon, 1999, for reviews).

Experiments that have borrowed methods from cognitive psychology indicate that explicit death thoughts first provoke *proximal defenses* designed to remove death-related cognitions from conscious awareness. Thus, when initially consciously presented with a reminder of death, people may seek to avoid self-reflective thought (Arndt, Greenberg, Simon, Pyszczynski, & Solomon, 1998), suppress thoughts of death (Arndt, Greenberg, Solomon, et al., 1997), deny vulnerability to relevant risk factors (Greenberg, Arndt, Simon, Pyszczynski, & Solomon, 2000), or increase intentions to improve one's health (when such intentions are assessed immediately after mortality reminders; Arndt, Schimel, et al., 2003). As the earlier examples imply, in certain cases (e.g., improving fitness) these reactions can have proactive beneficial effects on improving one's life span, but in others (e.g., vulnerability denial), these reactions can engender maladaptive reactions that could decrease life-span expectancy.

Different concerns emerge, however, when the individual confronts thoughts of death outside of conscious awareness. In situations involving explicit reminders of one's mortality, with the passage of time comes the emergence of the worldview and self-esteem defenses that we have been describing. When the proximal defenses just described are relaxed, there is a delayed increase in death thought accessibility (the extent to which death-related thoughts are cognitively available but outside of focal attention; Greenberg, Pyszczynski, Solomon, Simon, & Breus, 1994) that spreads to increase the accessibility of constructs that serve the self-protective goal of buffering the individual from existential concerns (Arndt, Greenberg, & Cook, 2002). It is then, when thoughts of mortality are outside of consciousness, that increases in worldview defense are observed. For example, across a series of studies, Arndt, Greenberg, Pyszczynski, and Solomon (1997) found that subliminal presentations of death reminders increased defense of one's patriotic identifications. Successful defense of these worldview beliefs (i.e., *distal defenses*) then serves to reduce the elevated death thought accessibility that lead to their occurrence (Arndt, Greenberg, Solomon, et al., 1997; Greenberg, Arndt, Schimel, Pyszczynski, & Solomon, 2001; Mikulincer & Florian, 2000). Thus, the symbolic investment in death denying worldview beliefs occurs when such thoughts are outside of conscious attention. Emerging research is starting to lead to understanding not only of the way in which beliefs about a cultural reality can offer transcendence from the temporal limitations of human existence, but in the more precise temporal dynamics through which such defenses emerge.

APPLICATIONS OF TERROR MANAGEMENT THEORY: FINDING MEANING IN TIME

> And you run and you run to catch up with the sun, but its sinking
> And racing around to come up behind you again
> The sun is the same in a relative way but you're older
> Shorter of breath and one day closer to death.
>
> (Pink Floyd, *Time*, 1973)

We have so far described how people identify with cultural worldviews that give life meaning, order, and permanence to symbolically or even literally transcend the mortal and temporally limited nature of existence. Furthermore, people strive to be meaningful members of the cultural world in order to gain a sense of personal significance. Whereas these strategies to manage the awareness of death may vary from person to person, all involve the social creation of symbolic or literal beliefs about what constitutes a meaningful and valuable life.

In addition to making life meaningful, people can make time itself meaningful. Humans may realize that the passing of time equals the passing of life; however, if the passing of time can also be perceived as an important and valuable part of existence, then the experience of passing through time can be viewed as a positive and significant part of life. In the pages that follow, we explore some ways in which time itself can be shaped in existentially meaningful ways by discussing the social construction of time, the defense of time-related worldviews, and the extraction of meaning from the past and future. Unfortunately, other people's progression down the temporal highway of the life cycle can remind one of the inevitable exit of mortality and thus pose a threat against which people will often defend themselves. In this light, we also consider how people react to the aging process and its logical connection to death.

The Construction of Meaning in Time

The awareness of time may provide humans with the knowledge of inevitable death but the social construction of time provides a method for dealing with the fact that the future will bring this unfortunate reality. In short, time has been integrated into a meaningful cultural belief system. From the perspective of astrophysics, a person can view one rotation of the Earth as merely a routine physical event, a mechanical phenomenon indicative of all celestial bodies within a solar system. However, to humans, this event represents a meaningful unit of time. Seven days make 1 week, several weeks make a month, and each time the Earth revolves around the sun, starting with a culturally meaningful but physically arbitrary date, people explode the corks from their champagne and celebrate with

Dick Clark the dropping of the ball and the dawning of another year. The social world has been structured around a calendar that serves to give life order and predictability (Solomon et al., 1991). People can take comfort in knowing that their world is stable across time. Monday, as much as we often dread it, comes every 7 days. The bank opens at the same time each morning. Christmas takes place on the same date each year. People have transformed purely physical time into culturally meaningful time and in this way defend against the terror of temporal consciousness.

Because beliefs about time are part of the fabric of people's perceptions of meaning, people should be motivated to defend these conceptions, and particularly so after being reminded of death. In line with this assertion, a recent study found that thoughts about death can lead to unfavorable evaluations of people who violate time-related worldviews (Routledge & Arndt, 2003). In this study, participants either wrote about death or feelings of uncertainty and then read a vignette about a person who refuses to wear a watch or use a calendar. The purpose of this vignette was to present an individual who does not subscribe to the socially constructed rules about time that provide order and predictability. After reading the vignette, participants answered questions concerning how much they liked the individual described in the vignette. It was found that mortality salience participants liked the time transgressor described in the vignette less than participants who wrote about uncertainty. Again, the assertion is that socially constructed time provides existential protection and thus, after reminders or mortality, people are motivated to defend their socially constructed temporal world.

Looking to the Past and Future to Cope With Death

It is important to keep in mind that, according to TMT, people invest in cultural worldviews at least in part because of their capacity to imbue life with meaning and thereby offer protection from existential fears. And indeed, research indicates that reminders of mortality can stimulate mildly depressed individuals to identify more strongly with cultural beliefs and as a result increase their perceptions of meaning in life (Simon, Arndt, Greenberg, Pyszczynski, & Solomon, 1998). Considering this press for meaning in the face of the awareness of death in turn suggests some interesting connections with how people can perceive time in terms of a past, a present, and a future to distill meaning from life.

It is a common societal joke that senior citizens are always talking about the good old days. Young children may dread the trip to grandpa's house because they fear the day will be consumed with story after story about the war, the greatest World Series ever, or the wonderful days before computers took over the world. Though such accounts of interactions with the elderly may be stereotypical exaggerations, from an existential perspective, it makes sense that as one approaches her or his final years, she or he can take comfort in reflecting on the past as a source of meaning (Yalom, 1980). Although we suspect that proclivity for nostalgic reflection is a more common meaning making endeavor for the elderly, research also indicates there are individual differences in the extent to which people are prone to reflect positively on the past (Zimbardo & Boyd, 1999). To the extent this is the case, from a terror management perspective, one might expect that when reminded of mortality individuals with such proneness would be particularly likely to use the opportunity to reflect on the past as a way of seeing positive meaning.

A recent study by Arndt, Hunt, Sheridyn, King, and Routledge (2003) supports this reasoning. Undergraduate students completed the Time Perspective Inventory (TPI; Zimabrdo & Boyd, 1999), were reminded of mortality or dental pain, and then were asked to write briefly about their last couple of years of high school. The TPI includes, among other subscales, a measure of how positively people tend to view the past. The results indicated a significant interaction between participants' disposition for positive past reflection and reminders of death on what participants wrote. Specifically, participants more positively oriented to the past tended to respond to mortality salience, relative to dental pain, by writing more vividly and focusing more on interpersonal and personal accomplishment themes, whereas less positively oriented participants tended to decline in these domains of expression after mortality salience. Thus, although further research is clearly needed, preliminary evidence speaks to the possibility that at least some people can use reflection on the past as a way to imbue life with meaning so as to protect themselves from the implications of their finitude.

In addition to looking to the past for meaning, people can also look to the future. Time is not something that humans have control over and no matter how many times one injects her or himself with botox in order to look younger, the human cellular system continues to decay. The future cannot be stopped, but the future can be meaningful and filled with hope. Both Brown (1959) and Yalom (1980), in different ways, discuss how people utilize future-oriented behavior as a way to cope with the mortal implications of passing time. For example, Brown describes how societies have developed monetary systems predicated on the hopeful assumption of a meaningful future. Financial investments and savings that yield profit returns and compounded interest provide people with a way to derive meaning and value from the passing of time. Yalom echoes this theme in his discussion about how people gain much comfort from the sense that they are not squandering the time that they have, but are planning for the future through education, career planning, and other forms of personal development:

> Not too long ago I was taking a brief vacation at a Caribbean beach resort. One evening I was reading, and from time to time I glanced up to watch the bar boy who was doing nothing save languidly staring out to sea—much like a lizard sunning itself on a warm rock, I thought. The comparison I made between him and me made me feel very smug, very cozy. He was simply doing nothing—wasting time; I was, on the other hand, doing something useful, reading, learning. I was, in short, getting ahead. All was well, until some internal imp asked the terrible question: Getting ahead of what? How? And (even worse) why? Those questions were, and are still, deeply disquieting. What was brought home to me with unusual force was how I lull myself into a death defeating delusion by continually projecting myself forward in time. I do not exist as a lizard exists; I prepare, I become, I am in transit. (p. 124)

Whereas we suspect that reflection on the past may be more likely among older persons, we similarly suspect that bolstering hope about one's future might be an important meaning-making strategy of younger people. Kasser and Sheldon (2000) found that college students who were reminded of mortality (as compared to a control topic of listening to music) evidenced higher fiscal expectations for their life 15 years in the future, both in terms of their overall worth and especially in the amount they expected to spend on various luxury items. This suggests that concerns about mortality can indeed lead people to look to the future for what they may perceive as providing meaning and happiness.

Life Expectancy and the Denial of Aging

In addition to using time to create an illusion of order and predictability, and utilizing reflections on the past and goals for the future to find meaning, people also construct worldviews about the relationship between humans and time. People, at least in the Western industrialized world, believe that humans have the inherent right to a life that extends beyond 70 years. Even though such a worldview might have more empirical legitimacy than other cultural beliefs (i.e., most people live at least 70 years), the subjective implications of this worldview are like the implications of many other worldviews. That is, a life expectancy worldview provides a sense of order and predictability. Spock's favorite salutation, "Live long and prosper," nicely reflects how humans have transformed their observations regarding average life expectancy into a belief system about human entitlement to longevity.

Indeed, most people would likely agree that the death of child is worse than the death of an elderly person. We often feel that the elderly person has lived out his or her entitled time whereas the child has been temporally short-changed. Of course, this attitude is not perceived as socially problematic or undesirable.

On the other hand, an understanding of life expectancy can create attitudes that are more socially problematic when they involve individuals that are most closely associated with death, the elderly. To the extent that the elderly are often associated with death, the mere presence of elderly persons may activate thoughts of death and therefore motivate people to distance themselves from the elderly.

Recent studies have begun to elucidate associations between thoughts about the elderly and concerns about mortality (Martens, Greenberg, & Schimel, in press). Specifically, in one study, participants were given pictures of different people and asked to make some evaluations (e.g., How attractive is the person? How tall is the person?). Half of the participants were given pictures of young people and half were given pictures of both young and elderly people. After participants evaluated the pictures, they completed a word fragment measure of how accessible death-related themes are outside of focal attention (e.g., completing COFF_ _ as either *coffin* or *coffee*; Greenberg et al., 1994). Supporting the idea that the elderly remind people of mortality, participants who were exposed to pictures of both young people and elderly completed the word fragments with significantly more death-related words than participants who were exposed to pictures of only young people.

A second study built upon this finding by testing the assertion that people asked to think about their own mortality will attempt to differentiate themselves from elderly persons. Participants were told that the purpose of the research was to explore similarities and differences between people of different generations. First, participants rated their agreement with various statements (e.g., I like poetry). Next, participants wrote about either death or dental pain. Finally, participants were given a picture of an elderly woman in a nursing home and a teenager at summer camp and were asked to make ratings on the same statements they completed at the start of the experiment. However, this time, they were asked to make the ratings thinking about how the average person of the generation represented in each picture (i.e., elderly and teenagers) would respond to the statements. Also, these ratings were made in separate columns on the same page that participants made the self-ratings. This allowed participants to consider their personal ratings when making ratings that represent the elderly and teenagers. It is worth noting that the participants in this study were drawn from freshman psychology classes and were thus for the most part teenagers themselves (mean age = 18). To construct a measure of how similar to teenagers and elderly participants they perceived themselves, their self-ratings on the various statements (e.g., I like poetry) were correlated with their ratings for teenagers and elderly. Thus, high positive correlations between self and other ratings represent increased feelings of similarity, whereas low negative correlations between self and other ratings represent decreased feelings of similarities (see, e.g., Arndt, Greenberg, Cook, 2002). Concerning feelings of similarity to teenagers, it was found that both mortality salient and dental-pain participants showed similar high levels of feelings of similarity, presumably because the participants were teenagers themselves and being a teenager is not temporally threatening when mortality is salient. However, a different pattern altogether emerged concerning feelings of similarity to the elderly. Both mortality salient and dental-pain participants perceived themselves as being dissimilar from the elderly (i.e., both groups had low, negative correlations between self and elderly ratings). Nonetheless, it was the mortality salient participants, relative to the dental-pain control participants, who differentiated themselves most from the elderly. Thus, after thinking about their own mortality, these participants were most motivated to deny any similarity to the group of people most vulnerable to death (i.e., the elderly). Again, this study highlights the notion that people have beliefs about the human relationship with time and death and strive to distance themselves from those who serve as reminders of the temporal limitations of life.

In sum, several new lines of research are beginning to directly investigate connections between the awareness of time and concerns about mortality. These studies suggest that there are a variety of ways in which people can defend themselves from the realization that the passing of time equals the passing of life. Specifically, people invest in and defend cultural worldviews pertaining to time, look to the past and future in the search for meaning, and engage in efforts to distance themselves from those who serve as a reminder that humans are in fact mortal.

CONCLUSION AND FUTURE DIRECTIONS

> To-morrow, and to-morrow, and to-morrow, Creeps in this petty pace from day to day To the last syllable of recorded time, And all our yesterdays have lighted fools The way to dusty death. Out, out, brief candle!
>
> Life's but a walking shadow, a poor player That struts and frets his hour upon the stage And then is heard no more: it is a tale Told by an idiot, full of sound and fury, Signifying nothing.
>
> (William Shakespeare, *Macbeth*)

What are the implications of being a temporally conscious animal? In this chapter, we have argued that an understanding of time, though clearly instrumental in humans' rise to the top of the food chain, helps allow for an understanding of death, and an understanding of death is psychologically problematic for humans who, like all living organisms, are hard-wired to strive for self-preservation in the service of reproduction. As the quotes throughout this chapter illustrate, the connection between time and death is not only of interest to social scientists, but has long fascinated creative writers, from Shakespeare to Pink Floyd. We suggest that the ubiquity of this recognition reveals a fundamental element of the human condition.

People may have the mental capacity required to realize that their time on life's stage is temporally limited; however, they also have the ability to manage this realization using a variety of methods. In this chapter we explored how humans harness their sophisticated, highly evolved cognitive abilities to literally or symbolically transcend the mortal limits of time. In sum, we believe that whereas the awareness of time and death is problematic to the human animal striving to maintain life, people possess and utilize the mental resources required to buffer the negative psychological consequences of this awareness. Unfortunately, many of the defensive strategies utilized by humans are often problematic to others who are perceived as threatening to one's existential world.

There may be, however, methods to manage the problem of time that are less problematic for those who are different. For example, numerous worldviews adopted by people provide meaning without necessitating a need to derogate those who are different. For example, Greenberg et al. (1992) found that priming the value of tolerance reduced the typically found worldview defense after mortality salience. This finding suggests that people may be able to coexist more peacefully with different others if an open-mindedness to diversity became a more dominant worldview. Part of the effectiveness of incorporating tolerance beliefs into one's worldview may stem from recognizing that tolerance can facilitate improved intergroup social relations in the future. Indeed, considering the future consequences of one's actions predicts aggressive behavior (Joireman, Anderson, & Strathman, 2003). In this way, incorporating measures of the extent to which people consider future consequences (Strathman, Gleicher, Boninger, & Edwards, 1994) in terror management research might reveal ways in which people can derive existential meaning and purpose that fosters positive relations with others as well as with the environment (Joireman, Lasane, Bennett, Richards, & Solaimani, 2001).

It may also be possible to manage the problem of time and avoid creating social conflict by having the sense that one is outside of time. Engaging in a mentally enthralling activity may allow one to temporarily step outside of the experience of passing time. Brown (1959) alludes to such a possibility when he describes intrinsically pleasurable activities as being "outside of time." Specifically, Brown suggests that intrinsically pleasurable activity allows one to escape the subjective experience of passing time. Consistent with this idea, Lifton (1983), in what he calls the experiential mode of symbolic death transcendence, suggests that the individual can step outside of the experience of passing time by becoming immersed in the flow of some vitalizing activity. In a way, Lifton's experiential mode is a method to suspend time.

Of course, physical time cannot be altered; however, one's experience of time can vary and being psychologically consumed by an intrinsically engaging activity may provide a reprieve from the experience of passing time, and the subsequent psychological consequences temporal thought engenders. For example, in a recent study, Routledge, Arndt, and Sheldon (2004) found that after being reminded of death, participants who were given a task in which they were instructed to be as creative as possible showed lower levels of worldview defense than participants who were given other tasks that did not allow for creative engagement. Thus, the Routledge et al. (2004) study begins to explore the idea that certain enthralling or consuming behaviors (e.g., creativity) can temporarily reduce the burden of concerns about death. In line with this idea, Csikszentmihalyi (1990) discusses the experience of flow as being so intensely engaged in a personally meaningful activity that events occurring peripherally are drowned out. In a state of flow, people forget about other activities and are consumed by the experience of their current engagement. Writing a chapter, painting a picture, and participating in a musical jam session are just a few examples of activities that an intrinsically interested person could engage in as a way to temporarily leave behind all of the baggage of being a cognitive being aware of time and death. Such activities in a way serve as a time-out from time consciousness and could thus reduce the need to adamantly cling to the social identifications and institutions that symbolically defy death, often at the expense of others.

REFERENCES

Arndt, J., Allen, J. J. B., & Greenberg, J. (2001). Traces of terror: Subliminal death primes and facial electromyographic indices of affect. *Motivation and Emotion, 25,* 253–277.

Arndt, J., Cook, A., & Routledge, C. (2004). The blueprint of terror management: Understanding the cognitive architecture of psychological defense against the awareness of death. In J. Greenberg, S. L. Koole, & T. Pyszczynski (Eds.), *Handbook of experimental existential psychology* (pp. 35–53). New York: Guilford Press.

Arndt, J., & Greenberg, J. (1999). The effects of a self-esteem boost and mortality salience on responses to boost relevant and irrelevant worldview threats. *Personality and Social Psychological Bulletin, 25,* 1331–1341.

Arndt, J., Greenberg, J., & Cook, A. (2002). Mortality salience and the spreading activation of worldview-relevant constructs: Exploring the cognitive architecture of terror management. *Journal of Experimental Psychology: General, 131,* 307–324.

Arndt, J., Greenberg, J., Pyszczynski, T., & Solomon, S. (1997). Subliminal exposure to death-related stimuli increases defense of the cultural worldview. *Psychological Science, 8,* 379–385.

Arndt, J., Greenberg, J., Schimel, J., Pyszczynski, T., & Solomon, S. (2002). To belong or not to belong, that is the question: Terror management and identification with gender and ethnicity. *Journal of Personality and Social Psychology, 83,* 26–43.

Arndt, J., Greenberg, J., Simon, L., Pyszczynski, T., & Solomon, S. (1998). Terror management and self-awareness: Evidence that mortality salience provokes avoidance of the self-focused state. *Personality and Social Psychological Bulletin, 24*, 1216–1227.

Arndt, J., Greenberg, J., Solomon, S., Pyszczynski, T., & Simon, L. (1997). Suppression, accessibility of death-related thoughts, and cultural worldview defense: Exploring the psychodynamics of terror management. *Journal of Personality and Social Psychology, 73*, 5–18.

Arndt, J., Hunt, D., Sheridyn, O., King, L., & Routledge, C. (2003). *Nostalgia and terror management: The effect of mortality salience on reflections of the past.* Manuscript in preparation.

Arndt, J., Schimel, J., & Goldenberg, J. L. (2003). Death can be good for your health: Fitness intentions as a proximal and distal defense against mortality salience. *Journal of Applied Social Psychology, 33*, 1726–1746.

Becker, E. (1971). *The birth and death of meaning* (2nd ed.). New York: The Free Press.

Becker, E. (1973). The denial of death. New York: The Free Press.

Berger, P. L., & Luckmann, T. (1967). *The social construction of reality.* Garden City, NY: Anchor.

Bowlby, J. (1969). *Attachment and loss: Vol. 1. Attachment.* New York: Basic Books.

Brown, N. O. (1959). *Life against death: The psychoanalytical meaning of history.* Middletown, CT: Wesleyan Press.

Csikszentmihalyi, M. (1990). *Flow: The psychology of optimal experience.* New York: Harper & Row.

Dechesne, M., Greenberg, J., Arndt, J., & Schimel, J. (2000). Terror management and sports fan affiliation: The effects of mortality salience on fan identification and optimism. *European Journal of Social Psychology, 30*, 813–835.

Dechesne, M., Pyszczynski, T., Arndt, J., Ransom, S., Sheldon, K. M., van Knippenberg, A., & Janssen, J. (2003). Literal and symbolic immortality: The effect of evidence of literal immortality on self-esteem striving in response to mortality salience. *Journal of Personality and Social Psychology, 84*, 722–737.

Duval, S., & Wicklund, R. A. (1972). *A theory of objective-self-awareness.* New York: Academic Press.

Erchak, G. M. (1992). *The anthropology of self and behavior.* New Brunswick, NJ: Rutgers University Press.

Festinger, L. (1954). A theory of social comparison processes. *Human Relationships, 1*, 117–140.

Florian, V., & Kravetz, S. (1983). Fear of personal death: Attribution, structure, and relation to religious belief. *Journal of Personality and Social Psychology, 44*, 600–607.

Florian, V., & Mikulincer, M. (1997). Fear of death and the judgment of social transgressions: A multidimensional test of terror management theory. *Journal of Personality and Social Psychology, 73*, 369–380.

Florian, V., & Mikulincer, M. (1998a). Symbolic immortality and the management of the terror of death. *Journal of Personality and Social Psychology, 74*, 725–734.

Florian, V., & Mikulincer, M. (1998b). Terror management in childhood: Does death conceptualization moderate the effects of mortality salience on acceptance of similar and different others. *Personality and Social Psychology Bulletin, 24*, 1104–1112.

Goldenberg, J. L., Pyszczynski, T., Greenberg, J., Solomon, S., Kluck, B., & Cornwell, R. (2001). I am not an animal: Mortality salience, disgust, and the denial of human creatureliness. *Journal of Experimental Psychology: General, 130*, 427–435.

Goldschmidt, W. (1990). *The human career: The self in the symbolic world.* Cambridge, MA: Blackwell.

Greenberg, J., Arndt, J., Schimel, J., Pyszczynski, T., & Solomon, S. (2001). Clarifying the function of mortality-salience induced worldview defense: Renewed suppression or reduced accessibility of death-related thoughts? *Journal of Experimental Social Psychology, 37*, 70–76.

Greenberg, J., Arndt, J., Simon, L., Pyszczynski, T., & Solomon, S. (2000). Proximal and distal defenses in response to reminders of one's mortality: Evidence of a temporal sequence. *Personality and Social Psychology Bulletin, 26*, 91–99.

Greenberg, J., Pyszczynski, T., & Solomon, S. (1986). The causes and consequences of a need for self-esteem: a terror management theory. In R. F. Baumeister (Ed.), *Public self and private self* (pp. 189–212). New York: Springer-Verlag.

Greenberg, J., Pyszczynski, T., Solomon, S., Pinel, E., Simon, L., & Jordan, K. (1993). Effects of self-esteem on vulnerability-denying defensive distortions: Further evidence of an anxiety-buffering function of self-esteem. *Journal of Experimental Social Psychology, 29,* 229–251.

Greenberg, J., Pyszczynski, T., Solomon, S., Rosenblatt, A., Veeder, M., Kirkland, S., & Lyon, D. (1990). Evidence for terror management II: The effects of mortality salience on reactions to those who threaten or bolster the cultural worldview. *Journal of Personality and Social Psychology, 58,* 308–318.

Greenberg, J., Pyszczynski, T., Solomon, S., Simon, L., & Breus, M. (1994). Role of consciousness and accessibility of death-related thoughts in mortality salience effects. *Journal of Personality and Social Psychology, 67,* 627–637.

Greenberg, J., Simon, L., Harmon-Jones, E., Solomon, S., Pyszczynski, T., & Chatel, D. (1995). Testing alternative explanations for mortality effects: Terror management, value accessibility, or worrisome thoughts? *European Journal of Social Psychology, 12,* 417–433.

Greenberg, J., Simon, L., Porteus, J., Pyszczynski, T., & Solomon, S. (1995). Evidence of a terror management function of cultural icons: The effects of mortality salience on the inappropriate use of cherished cultural symbols. *Personality and Social Psychology Bulletin, 21,* 1221–1228.

Greenberg, J., Solomon, S., & Pyszczynski, T. (1997). Terror management theory of self-esteem and social behavior: Empirical assessments and conceptual refinements. In M. P. Zanna (Ed.), *Advances in experimental social psychology* (Vol. 29, pp. 61–139). New York: Academic Press.

Greenberg, J., Solomon, S., Pyszczynski, T., Rosenblatt, A., Burling, J., Lyon, D., Pinel, E., & Simon, L. (1992). Assessing the terror management analysis of self-esteem: Converging evidence of an anxiety-buffering function. *Journal of Personality and Social Psychology, 63,* 913–922.

Halloran, M. J., & Kashima, E. S. (2004). Social identity & worldview validation: The effects of ingroup identity primes and mortality salience on value endorsement. *Personality and Social Psychology Bulletin, 30,* 915–925.

Harmon-Jones, E., Simon, L., Greenberg, J., Pyszczynski, T., Solomon, S., & McGregor, H. (1997). Terror management theory and self-esteem: Evidence that increased self-esteem reduces mortality salience effects. *Journal of Personality and Social Psychology, 72,* 24–36.

Herdt, G. (1987). *The Sambia: Ritual and culture in New Guinea.* New York: Holt, Rinehart & Winston.

Horney, K. (1937). *The neurotic personality of our time.* New York: Norton.

Kasser, T., & Sheldon, K. M. (2000). Of wealth and death: Materialism, mortality salience, and consumption behavior. *Psychological Science, 11,* 348–351.

Joireman, J. A., Anderson, J., & Strathman, A. (2003). The aggression paradox: Understanding links among aggression, sensation seeking, and the consideration of future consequences. *Journal of Personality and Social Psychology, 84,* 1287–1302.

Joireman, J. A., Lasane, T. P., Bennett, J., Richards, D., & Solaimani, S. (2001). Integrating social value orientation and the consideration of future consequences within the extended norm activation model of proenvironmental behaviour. *British Journal of Social Psychology, 40,* 133–155.

Lerner, M. J. (1980). *The belief in a just world: A fundamental delusion.* New York: Plenum Press.

Lifton, R. J. (1983). *The broken connection: On death and the continuity of life.* New York: Basic Books.

Martens, A., Greenberg, J., & Schimel, J. (in press). Ageism and death: Evidence that mortality salience increases distancing from the elderly. *Personality and Social Psychology Bulletin.*

Mathews, R., & Kling, K. (1988). Self-transcendence, time perspective, and prosocial behavior. *Journal of Voluntary Action Research, 17,* 4–24.

McCoy, S. K., Pyszczynski, T., Solomon, S., & Greenberg, J. (2000). Transcending the self: A terror management perspective on successful aging. In A. Tomer (Ed.), *Death attitudes and the older adult: Theories, concepts, and applications* (pp. 37–63). Philadelphia: Brunner-Routledge.

McGregor, H., Lieberman, J. D., Solomon, S., Greenberg, J., Arndt, J., Simon, L., & Pyszczynski, T. (1998). Terror management and aggression: Evidence that mortality salience motivates aggression against worldview threatening others. *Journal of Personality and Social Psychology, 74,* 590–605.

Mead, G. H. (1968). *Mind, self, and society.* Chicago: University of Chicago Press. (Original work published 1934)

Mikulincer, M., & Florian, V. (2000). Exploring individual differences in reactions to mortality salience: Does attachment style regulate terror management mechanisms? *Journal of Personality and Social Psychology, 79,* 260–273.

Nelson, L. J., Moore, D. L., Olivetti, J., & Scott, T. (1997). General and personal mortality salience and nationalistic bias. *Personality and Social Psychology Bulletin, 23,* 884–892.

Pyszczynski, T., Greenberg, J., & Solomon, S. (1999). A dual-process model of defense against conscious and unconscious death-related thoughts: An extension of terror management theory. *Psychological Review, 106,* 835–845.

Pyszczynski, T., Wicklund, R. A., Floresky, S., Gauch, G., Koch, S., Solomon, S., & Greenberg, J. (1996). Whistling in the dark: Exaggerated estimates of social consensus in response to incidental reminders of mortality. *Psychological Science, 7,* 332–336.

Rank, O. (1989). *Art and artist: Creative urge and personality development.* New York: Knopf. (Original work published 1932)

Rosenberg, M. (1981). The self-concept: Social product and social force. In M. Rosenberg & R. H. Turner (Eds.), *Social psychology: Sociological perspectives* (pp. 591–624). New York: Basic Books.

Rosenblatt, A., Greenberg, J., Solomon, S., Pyszczynski, T., & Lyon, D. (1989). Evidence for terror management theory I: The effects of mortality salience on reactions to those who violate or uphold cultural values. *Journal of Personality and Social Psychology, 57,* 681–690.

Routledge, C., & Arndt, J. (2003). *The terror of time: Mortality salience and time-related worldviews.* Manuscript in preparation.

Routledge, C., Arndt, J., & Sheldon, K. M. (2004). Task engagement after mortality salience: The effects of creativity, conformity, and connectedness on worldview defense. *European Journal of Social Psychology, 34,* 477–487.

Simon, L., Arndt, J., Greenberg, J., Pyszczynski, T., & Solomon, S. (1998). Terror management and meaning: Evidence that the opportunity to defend the worldview in response to mortality salience increases the meaningfulness of life in the mildly depressed. *Journal of Personality, 66,* 359–382.

Simon, L., Greenberg, J., Harmon-Jones, E., Solomon, S., Pyszczynski, T., Arndt, J., & Abend, T. (1997). Cognitive-experiential self-theory and terror management theory: Evidence that terror management occurs in the experiential system. *Journal of Personality and Social Psychology, 72,* 1132–1146.

Solomon, S., Greenberg, J., & Pyszczynski, T. (1991). A terror management theory of social behavior: The psychological functions of self-esteem and cultural worldviews. In M. P. Zanna (Ed.), *Advances in experimental social psychology* (Vol. 24, pp. 93–159). New York: Academic Press.

Strathman, A., Gleicher, F., Boninger, D. S., & Edwards, C. S. (1994). Consideration of future consequences: Weighing immediate and distant outcomes of behavior. *Journal of Personality and Social Psychology, 66,* 742–752.

Sullivan, H. S. (1953). *The interpersonal theory of psychiatry.* New York: Norton.

Taubman Ben-Ari, O., Florian, V., & Mikulincer, M. (1999). The impact of mortality salience on reckless driving—A test of terror management mechanisms. *Journal of Personality and Social Psychology, 76,* 35–45.

Yalom, I. D. (1980). *Existential psychotherapy.* New York: Basic Books.

Zimbardo, P., & Boyd, J. (1999). Putting time in perspective: A valid, reliable individual-differences metric. *Journal of Personality and Social Psychology, 77,* 1271–1288.

5

Time Perspective, Health, and Risk Taking

John N. Boyd
Philip G. Zimbardo
Stanford University

Over 15 million minutes ago, in the slower days of the early 1970s (Glieck, 1999), a remarkable experiment on time press and helping behavior was conducted by social psychologists John Darley and Dan Batson (1973). They observed how individual Princeton seminary students behaved in preparation for giving a speech on the parable of the Good Samaritan. The speech was to be given in a building across campus and would be evaluated by the seminarians' supervisors. As each student completed his preparation, he was told that: (a) He was late for his presentation, that he had been expected a few minutes ago; (b) he had plenty of time before his presentation, but that he might as well head on over; or (c) an in-between time message, that he was on time and should head over now. The researchers were curious about the effect, if any, of this manipulation of time press on helping behavior among those whose religious preparation and assumed altruistic inclination may have predisposed them to helping others.

As each student walked alone from the preparation classroom to the presentation studio, he encountered a lone person (a confederate of the experimenters), slumped and coughing in an alleyway, obviously in need of assistance. Here was the opportunity to put into practice what they were about to preach. Would it make any difference what they had been told about the various time constraints given the obvious need to help this stranger in distress?

The majority of those students who perceived that they had sufficient time to get to their destination stopped to help, consistent with these students' choice of vocation. However, remarkably, in the "late" condition, 90% of these religious students failed to stop and help the victim in distress!

Given the sample of seminarians, it is hard to argue that this failure of bystander intervention is the product of cynicism, selfishness, or indifference to the needs of others—the standard dispositional excuses. Rather, it is more probable that the culprit is time. Most of those who believed they had enough time to both stop to help and to fulfill their obligation to the researchers did stop. In dramatic contrast, the vast majority of those who believed they were late did not stop to help. Their sense of time press resulted in very surprising behaviors from individuals who were devoting their lives to helping others. Time press caused good, well-intentioned people to behave in ways that put their immediate concerns ahead of the welfare of another who was obviously in need of assistance, to behave in ways that the seminarians themselves would probably find contemptible.

But the lesson is actually even more personal: As each of us makes our way through life, not only do we not stop to help others, but also we often do not stop to help *ourselves*. We avoid diagnostic exams, such as annual physical and dental checkups, that we know are critical to our health; we postpone physical exercise to complete other work; and we fail to engage in a variety of health maintenance behaviors essential for our well-being. In addition, whereas these forms of personal neglect *may* have negative future consequences for us, many people ensure negative outcomes by engaging in behaviors that are actually harmful, such as smoking, overeating, using drugs, and taking unnecessary physical risks.

As our world continues to accelerate, as our pace of life continues to move faster and faster (Levine, 1997), our relationship with time, to the currency of our lives, will undoubtedly become more and more central to who we are and how we behave both toward ourselves and toward others. The social relationship that we develop with time will partially determine whether or not we stop to help those in need around us and whether those around us stop to help us when we are in need. The personal relationship that we develop with time will also determine whether we stop to help ourselves by living more fully in the present, without putting ourselves at risk for future negative consequences.

What does this accelerating future promise to bring? Only time can tell for certain, but it appears likely that the often underappreciated influence of time on our lives will become more critical as we and the world in which we live continue to pass through a dynamically changing temporal dimension. A recurring and defining characteristic of the human condition is the balance that each of us strikes between the certainty of the present, the promise of the future, and the memory of the past in this temporal trinity.

TIME PERSPECTIVE: THEORETICAL HISTORY

The phenomenon of time has a long and controversial history among psychologists, philosophers, and scientists. William James, the father of American psychology, thought the concept of time important enough to include an entire chapter on "time perception" in *The Principles of Psychology* (1950/1890). Behaviorists such as John Watson (1913) and B. F. Skinner (1987) disagreed and questioned the importance of consciousness and the conception of time to psychology. Although the behaviorists tried to stop time, at least the psychological study of it, they were unsuccessful due in part to the influence of physics and philosophy. Albert Einstein (1931) demonstrated that time is not a constant, but rather subject to context and frame of reference effects. Even through the eyes of a physicist, time is a social phenomenon. Although Einstein believed time to be constrained by such things as relative position, velocity, and the speed of light, social psychologists may recognize it as a social construction (Moscovici, 1984). Similarly, Kant (1781/1965) believed the conception of time to be an "innate ability," and felt that it colored the way that we experience the world. Existential philosophers and psychologists expounded upon Kant's notion of time (Heidegger, 1927; Husserl, 1964), and this thinking influenced the Gestalt psychologists, most notably Kurt Lewin, the pioneer of modern social psychology (1942).

Unlike the behaviorists, Lewin (1951) allowed for the influence of both the past and the future on behavior and defined "time perspective" as, "the totality of the individual's views of his psychological future and psychological past existing at a given time" (p. 75). Joseph Nuttin (1985) continued this Lewinian conception and asserted that, "future and past events have an impact on present behavior to the extent that they are actually present on the cognitive level of behavioral functioning" (p. 54). Nuttin argued that the future is our primary motivational space and de-emphasized the importance of the past. Of behaviorism, Nuttin (1964) observed that, "One cannot help expressing his astonishment at the fact that man's behavior, which is so strongly characterized by a restless striving towards something new in the future, is now to be explained entirely or mainly as a function of what he has previously done" (p. 61).

Time is a difficult phenomenon to study, even for the physicists psychologists so envy. Though time itself can be measured relatively accurately with a stopwatch, capturing the subjective experience of time is a more difficult proposition closely associated with the assessment of values and attitudes. Previous attempts to quantify time perspective include use of the Thematic Apperception Test (TAT), circles tests (Cottle, 1976), the Motivational Induction Method (Nuttin, 1985), and time lines (Rappaport, 1990). None of these methods has been widely accepted, and the study of time has remained hindered by disparate definitions and varying assessment methods of questionable utility.

Our research, also in the Lewinian tradition, has been influenced by the thinkers that have come before us, and has been shaped by each of our unique temporal experiences and personal histories. Recently our work has attempted to bring empirical tools and new assessment methods to bear on this complex and controversial subject. It began in the early 1970s (just before the time of Darley and Batson's 1973, study) with an experiment on the hypnotic induction of present time perspective in college students who were generally future oriented. We documented profound changes in our subjects' thoughts, feelings, and behavior simply by having them psychologically enter an expanded present time zone, where past and future were distant (Zimbardo, Marshall, & Maslach, 1971). That initial research culminated nearly 30 years later in the publication of the Zimbardo Time Perspective Inventory (ZTPI), which we developed to address one of the long-standing challenges in the field of time perspective research, that of the need for reliable and valid measurement of the elusive construct of time perspective (Zimbardo & Boyd, 1999).

DEFINITION OF TIME PERSPECTIVE

Our research has focused on what we term "time perspective," the often nonconscious subjective manner in which each of us relates to time and the process whereby the continual flow of personal and social experiences is assigned to, parceled into, temporal categories, or time frames, that help to give order, coherence, and meaning to those events. These cognitive frames may reflect cyclical, repetitive temporal patterns or unique, nonrecurring linear events in our lives. They are used in encoding, storing, and recalling experienced events, as well as in forming expectations, goals, contingencies, and imaginative scenarios. Between the abstract, psychological constructions of prior past and anticipated future events lies the concrete, empirically centered representation of the present action moment.

When a tendency develops to habitually overemphasize past, future, or present temporal frames when making decisions, it serves as a cognitive temporal "bias" toward being primarily past, future, or present oriented. When chronically elicited, this bias becomes a dispositional style, or individual difference variable, that is characteristic and predictive of how the individual will respond across a host of daily life choice situations. Temporal bias may include either habitual over- or underuse of one or more of these temporal frames. Such limiting biases are in contrast with a "balanced time perspective," an idealized mental framework that allows individuals to flexibly switch temporal frames between past, future, and present depending on situational demands, resource assessments, or one's personal and social appraisals.

The Zimbardo Time Perspective Inventory (ZTPI)

The ZTPI was developed in an effort to provide researchers with a valid and reliable metric for use in documenting the critical role that attitudes toward time and cognitive styles organized around time play in our lives (Zimbardo & Boyd, 1999). The ZTPI has served an important role in integrating research on time perspective, and much of the research presented in this chapter is based on this metric. The ZTPI comprises five factorially distinct subscales, each of which captures a coherent time perspective dimension.

The Past-Negative Scale. This factor of the ZTPI reflects a generally negative, pessimistic view of the past, and is labeled the "past-negative." Items that comprise this factor include, "I think about the bad things that have happened to me in the past," "I think about the good things that I have missed out on in my life," and "I often think of what I should have done differently in my life." Because of the reconstructive nature of the past, we cannot be certain to what extent this negative attitude is due to the actual earlier experience of negative events or to the current negative reconstruction of benign events. From a psychological perspective, what individuals believe happened in the past is as important to present thoughts, feelings, and behavior as what actually did happen.

The Past-Positive Scale. This factor reflects an attitude toward the past that is very different than that captured by the first factor. Whereas the first factor suggests trauma, pain, and regret, the past positive reflects a warm, sentimental, nostalgic attitude toward the past. Items that load on the past-positive factor include, "It gives me pleasure to think about my past," "I get nostalgic about my childhood," "I enjoy stories about how things used to be in the 'good old times,' " and, "I like family rituals and traditions that are regularly repeated." As with all factors, the past positive captures a respondent's belief about the past without speculating on the veracity and accuracy of these beliefs.

The Present-Hedonistic Scale. This factor reflects a hedonistic, risk-taking, "If it feels good do it" attitude toward time. It includes such diverse items as, "Taking risks keeps my life from becoming boring," "I do things impulsively," "I often follow my heart more than my head," and, "When listening to my favorite music, I often lose all track of time." These items suggest an orientation toward present pleasure or immediate benefit with little concern for future consequences.

The Present-Fatalistic Scale. This factor reveals a fatalistic, helpless, and hopeless attitude toward the future and life. Items that comprise the present-fatalistic factor include, "My life path is controlled by forces I cannot influence," "You can't really plan for the future because things change so much," and, "Often luck pays off better than hard work." Items on this factor suggest little relation between present behavior and future costs or benefits.

The Future Scale. This factor reflects a general future orientation. Items that load on the future factor include, "I complete projects on time by making steady progress," "I am able to resist temptations when I know that there is work to be done," "It upsets me to be late for appointments," and, negatively, "I take each day as it is rather than try to plan it out." The future scale suggests that present behavior is dominated by a striving for future goals and rewards. In contrast to the present-hedonistic scale, these items suggest an orientation away from focusing on immediate benefits and toward calculating future gains and costs.

The Transcendental-Future Time Perspective. Although the ZTPI consists of five subscales, there appears to be at least one additional time perspective. The seminary students we met earlier may have been high on this sixth time perspective factor, the transcendental-future time perspective (Boyd & Zimbardo, 1996). The transcendental-future is composed of beliefs about the period after the death of the physical body. Items that comprise the transcendental-future time perspective include, "Only my physical body will ever die," "I will be held accountable for my actions on earth when I die," and "Death is just a new beginning." Although psychologists have traditionally shied away from investigating issues surrounding life after death, beliefs held in this time perspective may influence present behavior, just as beliefs about the traditional future—expectations—feed back to influence current behavior. The transcendental-future time perspective may prove useful in explaining some extreme behaviors, such as suicide bombings.

ONE-DIMENSIONAL APPROACHES TO TIME

The factor analysis statistical procedure through which the ZTPI was created produces factors that are theoretically orthogonal. It is therefore possible for an individual to score high on all five factors, or to score low on all of them. Although this theoretical independence of the subscales is usually not found in practice, it is an important theoretical point. Time perspective is a multifaceted construct, and it is important that assessment metrics reflect this complexity.

Despite the acknowledged complexity of time perspective and the flexibility offered by assessment instruments such as the ZTPI, most previous research, including our own, has focused on one or at most two time perspectives. For example, Zuckerman (Zuckerman, 1994; Zuckerman, Eysenck, & Eysenck, 1978) developed a sensation-seeking scale that is closely related to present orientation; Zaleski (1996) has created a future anxiety scale, and Strathman, Gleicher, Boninger, and Edwards (1994) developed a Consideration of Future Consequences scale that measures another singular aspect of the future. Although these scales may be improvements over previous attempts to measure time perspective, they are literally one-dimensional, focusing on only the present or future.

RESEARCH

General Research on Time Perspective

Time perspective, or specific facets of it, has been related to numerous psychological constructs and behavior. Different aspects and operational definitions of time perspective have been shown to be related to gender (Gjesme, 1979; Lamm, Schmidt, & Trommsdorff, 1976; Tismer, 1985; Von Wright & Von Wright, 1977), social class (Lamm et al., 1976), academic achievement (DeVolder & Lens, 1982), attributions (Miller & Porter, 1980), locus of control (Koenig, 1979), motivation (Atkinson, 1964; McClelland, 1985; Raynor, 1969; Raynor & Entin, 1983), mental health (Cohen, 1967; Melges, 1982), and even the time during the term at which students sign up to participate in experiments (Harber, Zimbardo, & Boyd, 2003). Future orientation has consistently been related to higher social class, academic achievement, higher motivation, less risk taking, and less substance abuse, whereas present orientation appears associated with lower social class, reduced motivation, lower academic achievement, greater risk taking, and increased substance abuse.

The aspect of time perspective that has fostered the most research is its relationship to juvenile delinquency. According to Oyserman and Markus' (1990a,

1990b) theory of "possible selves," juveniles without well-developed possible selves are more likely to become delinquent. Several other researchers have found similar cognitive biases or aberrations in the psychological conception of the future by juvenile delinquents (Barndt & Johnson, 1955; Brock & Del Giudice, 1963; Landau, 1975, 1976; Stein, Sarbin, & Kulik, 1968). Because of the strong relationship between time perspective and juvenile delinquency, one possible application of time perspective research may be in rehabilitation programs designed to program more future-oriented ways of thinking into present-oriented juveniles who are delinquents.

Time Perspective, Health, and Risk-Taking Research

Past-Negative Time Perspective. Less research has been done on the past time perspectives than on the present and future time perspectives, and much of the relevant research has not been framed specifically in terms of time perspective. For example, Lyubomirsky and Nolen-Hoeksema (1993, 1995) found that negative rumination about past events was associated with depression, which is closely associated with our finding that past-negative time perspective is strongly associated with depression, anxiety, unhappiness, and lack of emotional stability. See Table 5.1. These findings are also consistent with findings from Holman and Zimbardo (2003) that suggest that individuals high in past-negative time perspective cope with life stress differently than individuals low in past-negative time perspective. For example, those in the former group talk about their stress less frequently, experience more social conflict while coping, and experience more negative emotions while coping than do individuals low in past-negative time perspective. These less than ideal coping strategies may be the result of actually experiencing more severe negative events in the past or may be the result of negative constructions and beliefs about their past experiences.

Past-Positive Time Perspective. The past-positive time perspective is the least researched time perspective, but it may also be the one that each of us would most like to have. Our research has shown it to be related to happiness, self-esteem, lower anxiety, and friendliness (Zimbardo & Boyd, 1999). See Tables 5.1 and 5.2.

In addition to a positive construction of the past, individuals high in past-positive time perspective tend to cope with stressful situations more effectively than individuals lower in past-positive time perspective. Holman and Zimbardo (2003) found that individuals high in past-positive time perspective enjoyed more social support, demonstrated more emotional growth coping, and

TABLE 5.1

Convergent Validity for the Five Scales on the Zimbardo Time Perspective Inventory

	Past-Negative	*Past-Positive*	*Present-Hedonistic*	*Present-Fatalistic*	*Future*
Aggression	.49***	–.16***	.29***	.39***	–.31***
Beck Depression	.59***	–.17*	.20**	.37***	–.19**
Conscientiousness	–.18**	.05	–.22**	–.34***	.63***
CFC	–.19**	.02	–.31***	–.55**	.52***
Ego Control	.26***	–.04	.60***	.29***	–.39***
Impulse Control	–.34***	–.01	–.25***	–.23**	.29***
Need for Consistency	–.10	.09	–.41***	–.16*	.47***
Novelty Seeking	.29***	–.03	.57***	.28***	–.41***
Reward Dependence	.01	.18*	–.01	–.13	.37***
Self-Esteem	–.48***	.28***	.11	–.28***	.13*
Sensation Seeking	.05	–.05	.57***	.17*	–.31***
Trait Anxiety	.62***	–.25***	.07	.38***	–.14*

Note. CFC = consideration of future consequences. $N = 205$ (from College of San Mateo).
* $p < .05$. ** $p < .01$. *** $p < .001$. "Ego Control" is actually scored as ego under control.

experienced less social conflict while coping than did individuals low in past-positive time perspective.

As with the past-negative time perspective, the positive construction of the past may reflect positive events that were actually experienced, or it may reflect an attitude that allows individuals to make the best of even very difficult situations. Nietzsche's famous saying, "that which does not kill me makes me stronger," captures this attitude well. It would seem that those who actually experience aversive events but recode them in positive terms become resilient children and perhaps optimistic adults.

Present-Hedonistic Time Perspective. Although it is tempting to conclude that present time perspective is the absence of future time perspective, this is not necessarily the case. Future time perspective and present time perspective are distinct theoretical constructs, which in practice means that an

TABLE 5.2

ZTPI and Single Self-Report Item Correlations

	Past- Negative	*Past-Positive*	*Present-Hedonistic*	*Present-Fatalistic*	*Future*
Age	–.08	.01	–.10*	–.08*	.23***
GPA	–.05	.07	–.07	–.08*	.21***
Hours Study/Week	.06	.01	–.15**	.02	.28***
How Creative	–.06	.13***	.28***	–.11*	.09*
How Happy	–.41***	.36***	.16***	–.23***	.01
How Often Steal	.12*	.04	.16**	.13*	–.02
How Often Tell Lies	.18***	.03	.16***	.17***	–.20***
How Shy	.20***	–.13**	–.16**	.13**	.00
Temper	.18***	–.06	.05	.18***	–.08

Note. $N = 566$ (from College of San Mateo and San Francisco State University).
$^{*} p < .05$. $^{**} p < .01$. $^{***} p < .001$.

individual's score on one construct is unrelated to his or her score on the other construct.

Despite this theoretical distinction, research has shown that individuals high in present hedonism consistently behave in ways similar to that of individuals low in future time perspective. For example, women high in present hedonism are less likely to have regular breast exams (Guarino, DePascalis, DiChiacchio, & Zimbardo, 1999); individuals high in present hedonism are less likely to practice safe sex (Hutton et al., 1999; Rothspan & Read, 1996), are more likely to use alcohol and other substances (Alvos, Gregson, & Ross, 1993; Keough, Zimbardo, & Boyd, 1999), are more likely to take risks while driving (Zimbardo, Keough, & Boyd, 1997), and are more likely to have heated arguments than individuals lower in present hedonism (Keough et al., 1999).

In addition to an indifference to future consequences, there is a definite pleasure-seeking aspect as well. Individuals high in present time perspective appear to be governed by the situational pressure of their immediate environment, or by what Freud referred to as the "pleasure principle" (Freud, 1950). Hutton et al. (1999) found that women prisoners that were high in present hedonism were more likely to have engaged in prostitution. Our research suggests that individuals high in present hedonism are more likely to ride roller coasters, skateboard, and binge eat than individuals lower on present-hedonism time perspective. See Tables 5.1 and 5.2 for related examples.

Individuals high in present hedonism also seem to cope with negative consequences differently than individuals lower in present hedonism. Holman and Zimbardo (2003) found that individuals high in present hedonism tended to talk about the stress of traumatic events but tended to use an avoidant coping style more than individuals lower in present hedonism. This finding is consistent with research that showed that homeless individuals who are high in present hedonism spend more time watching TV and eating, and less time working (Epel, Bandura, & Zimbardo, 1999). This research suggests that individuals who score high on the present-hedonism scale talk about their problems, but then avoid doing anything constructive to cope with them.

Present-Fatalistic Time Perspective. Individuals high in present-fatalism do not believe in contingent relationships between their present actions and future consequences. They believe that what they do today has little influence on what happens tomorrow, in terms of both positive and negative future consequences. This extreme lack of personal agency may be partially responsible for the anxiety and depression that is strongly associated with this time perspective. See Table 5.1.

Although individuals high in present fatalism do share a lack of concern for future consequences with individuals that score high in present hedonism, they lack the pleasure-seeking aspect that defines present hedonism. For example, like those high in present hedonism, those high in present fatalism are more likely to engage in unsafe sex (Hutton et al., 1999; Rothspan & Read, 1996) and use drugs than are individuals low in present hedonism, but they may do so for different reasons than individuals high in present hedonism. Individuals high in present hedonism may have unsafe sex and use drugs because they engage in such pleasurable behaviors with reckless abandon and "mindless" spontaneity. In contrast, individuals high in present fatalism may have unsafe sex and use drugs, because they do not believe that practicing safe sex or refraining from drug use will make any difference in the future consequences that they experience. Researchers have found that drug users who are high in present fatalism are more likely to share needles than are individuals low in present fatalism (Hutton et al., 1999).

Finally, individuals high in present fatalism cope with stressful events differently than individuals lower in present fatalism. Across event types, individuals high in present fatalism tend to talk less about their stress, cope with their stress less actively, and to avoid coping more frequently than individuals lower in present fatalism (Holman & Zimbardo, 2003). Unlike individuals high in present hedonism who talk about their stress but do nothing about it, individuals high in the present fatalism do not even talk about their life stressors. It is as if they put such aversive events out of mind because they are believed to be unmodifiable.

Future Time Perspective. Future time perspective has been consistently related to obtaining positive future consequences and to the avoidance of negative future consequences. This finding is not surprising, because individuals high in future time perspective are by definition those who are most concerned about the future. More surprisingly, individuals high in future time perspective are willing to work harder in the present than individuals lower in future time perspective, and they are better able to cope with unavoidable negative future consequences than individuals lower in future time perspective. It appears that individuals that are high in future time perspective plan for, practice for, and cope with the consequences of the future better than individuals lower in future time perspective. They are extremely sensitive to the fact that their present behavior will have future consequences, and may be governed by what Freud referred to as the "reality principle."

For example, research has demonstrated that women high in future time perspective are more likely to have regular breast examinations than women lower in future time perspective (Guarino et al., 1999), and individuals high in future time perspective are more likely to practice safe sex than individuals lower in future time perspective (Hutton et al., 1999; Rothspan & Read, 1996). We have found that they also eat healthy foods more often, and study more hours per week than individuals lower in future time perspective. Each of these behaviors represents a proactive response that is likely to have positive future consequences. See Tables 5.1 and 5.2 for selected results from previous research (see Zimbardo & Boyd, 1999, for a fuller discussion of the outcomes associated with the ZTPI scales).

In addition to performing behaviors that can be predicted to be associated with positive future consequences, individuals high in future time perspective also avoid behaviors that are likely to be associated with negative future consequences. For example, high future time perspective has been associated with lower risk taking (Zimbardo et al., 1997), lower alcohol and substance use (Alvos et al., 1993; Keough et al., 1999), and even such concrete actions such as using slower speeds while driving.

A third finding regarding future time perspective is perhaps the most surprising of all. It is reasonable to expect that individuals high in future time perspective would proactively seek positive future consequences and proactively avoid negative future consequences. What is not obvious is that individuals high in future time perspective would be better able to cope with crises and negative events when they are unavoidable and occur naturally. For example, Holman and Zimbardo (2003) found that individuals high in future time perspective coped with traumatic events in a more active and analytic manner and showed less social conflict in their coping behavior than individuals lower in future time perspective across a wide range of traumatic events, some of which may have been at least partially under the individual's control and some of which were natural disasters completely outside of the individual's control, such as firestorms.

These findings are corroborated by other research as well. Research has shown that women inmates who are higher in future time perspective complete a vocational training program more often than inmates lower in future time perspective (Chubick, Rider, Owen, & Witherspoon, 1999). Such programs are likely to lead to more satisfactory adjustments to life after release from prison. Moreover, individuals high on future time perspective experienced shorter periods of homelessness and enrolled in training programs more frequently than individuals lower in future time perspective (Epel et al., 1999). Similar results are found for situations in which individuals are assumed to have no control in the negative outcome that they are experiencing. Mann, Kato, Figdor, and Zimbardo (1999) studied children with cancer and found that children higher in future time perspective were more optimistic than children lower in future time perspective. Such positive attitudes may contribute to improved health outcomes.

Finally, although future time perspective has been associated in general with healthy functioning, there is at least one downside to the work-oriented, time press focus of some future-oriented individuals. Recent research suggests that pace of life, a rough proxy for future time perspective, is associated with increased risk for coronary heart disease among men (Levine, 1997).

THEORY AND FUTURE DIRECTIONS

Stopping to Help, and Stopping to Smell the Roses

One potential conclusion that could be reached after reviewing previous research on time perspective, health, and risk taking is that future time perspective is the "preferred" time perspective. Individuals high in future time perspective engage in risky behaviors less often than individuals high in present time perspective, cope with negative life situations more effectively, and tend to have greater success in academic and vocational settings. The preponderance of evidence suggests that efforts to enhance future time perspective and decrease present time perspective should be encouraged and perhaps even mandated. This obvious conclusion, however, ignores one important fact: Life is lived in the present. The present is where we live, love, and play. A life that focused exclusively on the future to the exclusion of the present would miss out on much of what it means to be human. For some "missing out" may be better termed escape, and for such individuals an extreme future time perspective may offer some relief from an unpleasant past or present, but, for most of us, the present is what makes us who we are, and makes our lives real. As social psychologists, we also are aware that human relationships exist in a present life space, and that currently an excessive focus on future orientation among many educated people comes at the expense of making and sustaining friendships. We see later that although it is better to be highly future oriented than highly present oriented in many situations, both are part of an idealized balanced time perspective that also includes the past-positive.

As one example, many of the behaviors that could accurately be labeled "risky" or "unhealthy" from a future time perspective could also be labeled "adventurous" and "fun" from other perspectives. Despite differences in the time course of future consequences, activities such as skydiving, drinking alcohol, and eating unhealthy food share a common trait: They are enjoyable in the present, at least to some people. Suggesting that individuals delay gratification indefinitely runs the risk that gratification will never occur at all. A more pragmatic approach may be to temper high present time perspective with high future time perspective. Regarding behaviors that are pleasurable in the present but have potential negative future consequences, high future time perspective may serve both a preparatory and an amelioratory function. In the example of skydiving, future-oriented participants may spend more time thinking about and preparing for the risks inherent in skydiving. In effect, they may be better prepared to cope with the risk when it arrives in the present. Individuals that enjoy drinking alcohol and that are also high in future time perspective may only drink in situations in which they are unlikely to drive under the influence, may be more likely to designate a nondrinking driver, and may be more likely to care for their physical bodies so that they are better able to cope with the potential long-term health consequences. Finally, individuals who enjoy eating high-fat/high-sugar snack and fast (unhealthy) foods, and who are also high in future time perspective may be influenced to eat such unhealthy foods less frequently, may prepare for their hedonistic indulgence by eating well or exercising before or afterward to "make up for it."

Time Perspective "Profiles"

The aforementioned examples suggest that focusing on a single time perspective may provide an incomplete and potentially distorted picture of individuals who are high and low on each dimension. Examining "profiles" of how individuals score across more than one time perspective may add depth to our understanding of human behavior and may suggest profiles that foster both enjoyment of the present and realistic preparation for the future. This approach is similar to what is done for personality profiles on instruments such as the Meyers–Briggs and the Big Five Questionnaire (Caprara, 1999; Caprara, Barbaranelli, Borgogni, & Perugini, 1993), and initial research on time perspective profiles has been promising (F. Sutherland, personal communication, December 5, 2003).

High Present-Hedonism—Low Future. Individuals high in present-hedonism and low in future time perspective enjoy life, but do so at the expense of the future. They are likely to be outgoing, creative free spirits who have never committed to academic or career paths that may lead to traditional definitions of success. They are the life of the party, but may be less comfortable in academic or work environments. Such individuals are also more likely to suffer from the consequences of excessive enjoyment of activities, such as alcoholism, substance abuse, and lower academic achievement.

Low Present-Hedonism—High Future. Individuals low in present-hedonism and high in future time perspective are all work, but no play. They may be successful in many traditional senses, but may not enjoy their success themselves. Their vocational successes may also not transfer to social relationships, and thus they may be more prone to the phenomenon of "midlife crisis." Others may never experience a "crisis" at all and may travel hurriedly through life only to realize once they have reached the end that they never stopped to smell the roses. Some surveys reveal that a majority of successful middle managers report that their lives seem empty and without significant meaning, and that if they could replay their lives they would spend less time working and invest more in developing their social lives.

High Present-Hedonism—High Future. A balanced time perspective that includes high present-hedonism and high future time perspective may be part of an ideal time perspective. The present-hedonism allows individuals to enjoy life, whereas the future component helps individuals prepare for the future by avoiding risks and delaying gratification. The critical issue is whether individuals are able to balance these complementary time perspectives so that they use the appropriate time perspective for their immediate situation. Using a future time perspective in a present-hedonistic situation, such as while at a party, may lead to lack of enjoyment, anxiety, and guilt about playing when they could be working. Using a present-hedonistic time perspective in a future situation may undermine planning and perseverance and lead to negative future consequences. It is important to note that it is possible for highly future-oriented individuals to get into the "flow" of their creative/productive work and get a measure of hedonistic pleasure from it (Csikszentmihalyi, 1990).

High Present-Hedonism—High Present-Fatalism. Individuals who are high in both present-hedonism and present-fatalism may embody the "rebel without a cause" mentality. They may enjoy the present, but see little sense in preparing for the future and avoiding risks. They may be most vulnerable to the perils of risk-taking behavior, because they may enjoy the excitement of the risk but have little incentive to avoid risks through preparation. They may also see little benefit in engaging in any maintenance or preparatory behavior that has a present cost, such as hard work, because they see little or no contingency between the cost and any possible future consequences, such as rewards.

High Present-Fatalism—Low Present-Hedonism—Low Future. Individuals high in present-fatalism and low in present-hedonism and future time perspective are in a painful bind. Unlike the present-hedonists, they do not enjoy the present, and, unlike individuals high in future time perspective, they do not see a reason to prepare for the future. They are therefore likely to be apathetic and to be swayed by other situational factors such as desire for respect rather than desire for pleasure.

Time Perspective Profiles

The time perspective profiles postulated earlier are one potential avenue of fruitful future research on time perspective. The relationship between the present and the future in particular suggests that examining one perspective in isolation could be misleading and incomplete. Recognizing that the subscales of the ZTPI are theoretically independent, that an individual can be high or low on any or all of the subscales is also important. It is also important to remember that we exist in the present, but that both the past, as emphasized by behaviorists, and the future, as emphasized by humanists, have a profound influence on our thoughts, feelings, and actions. Research on time perspective and health can benefit from incorporating multiple time perspective dimensions, as captured by the ZTPI.

How the Situation Affects Time Perspective

The Good Samaritan study discussed in the introduction is an excellent example of how situational factors such as time press can affect behavior. As the pace of life continues to accelerate, it will be increasingly important to recognize which situational factors influence behavior most dramatically and to understand how these factors can be controlled to promote prosocial and prohealth behaviors. As technology and society continue to operate more quickly, research may reveal that there is an optimal pace of live at which planning, productivity, and pleasure are optimally balanced. Faster may not always prove to be better.

Increasing Future Time Perspective

Although we have argued that future time perspective alone is not the answer to risk taking and unhealthy behaviors, it may be an important part. Educational programs that include mental simulation of the future (Zimbardo & Boyd, 1999) have shown promise in instantiating future time perspective in individuals who have heretofore not possessed it. Examination of how and when some individuals learn future time perspective may also provide clues about how individuals that are low in future time perspective can be taught to turn on that future light in their lives. Although there is no research directly on the issue, our hypothesis is that family life, educational systems, and religious groups may all be adapted to play important roles in promoting the internalization of useful time perspectives. These social institutions teach us how to relate to each other, and it seems reasonable that they teach us how to relate to time as well.

The Idealized Balanced Time Perspective

The notion of an "ideal" time perspective suffers from obvious personal biases, along with learned values and morals concerning work, success, indulgence, relationships, and more. Such an "ideal" is likely to be shaped by culture, economic, religious, and political conditions. Despite these constraints, all the data we have gathered in the past decade or more prompt us to offer a combination of three time frames, which in flexible balance should provide individuals and society with the best outcomes. For Western society, or at least for the United States, we believe that this optimal profile of time perspectives includes: moderate levels of future and present-hedonism, blended with high levels of past-positive time perspective. Any one person who had internalized this blend of time perspectives would be able to cognitively switch seamlessly between them depending on situational demands and available resources (see Boniwell & Zimbardo, 2003). It should be clear that in such a total time package, one would want only the lowest levels of past-negative and present-fatalism. Let's review what each of the components of this trilogy offers to such an idealized time perspective.

Future Time Perspective. The high future component of this hypothetical ideal time perspective ensures that individuals will avoid risks, when possible, will prepare for the negative consequences of unavoidable risks, and will temper their enjoyment of the present with an eye toward possible future consequences. Future time perspective also appears to provide a rationale and strength in coping with the inevitable challenges of life that even a strong future time perspective cannot help us avoid completely. Future focus also means taking care of business and the business of life, and increases the probability of attaining one's goals. However, overreliance on or exclusive use of future time perspective may block the full experience of life's emotions and the joys of playfulness and intimate human connections.

Present-Hedonism. Present-hedonism allows individuals to enter a greater range of daily "flow" experiences (Csikszentmihaly, 1990), to feel and experience life to the fullest. It enables adults to sustain some of the childlike wonder at novelty, and seek ever-new stimulation. This time frame enables us to take pleasure in nature, in our solitude, as well as in intense friendships and romance. But excessive reliance on present-hedonism not tempered by future time perspective can lead to many of the destructive behaviors we outlined earlier. Absence of present-hedonism, however, may rob an individual of the richness and depth of the human experience, depriving people from valuing every moment of life's journey before we all get to our ultimate destination.

Past-Positive. A strong past-positive time perspective is vital to the ideal time perspective for many reasons: It brings past happiness and positive sense of self into the present; it is a reminder that painful periods are transitory, and it provides positive, optimistic expectations for the future. To the extent that people have developed a past-positive time perspective they become their own positive role model, brimming with self-esteem and a high sense of self-efficacy. An additional interesting aspect of past-positive time perspective is its developmental history, of either being blessed with a supportive family and social environment in which more good than bad events were experienced, or cognitively rising above more negative or traumatic early life experiences by recasting them as learning experiences to be avoided or modified in the future.

In sum, we like to think of this idealized trilogy as giving each of us the three essentials of a good, full life: The future gives us the promise of an ideal world, guided by imagination and not restricted by present limitations or the status quo; the past gives us the shoulders of giants to stand upon and roots to connect us to people, places, and personal history; and the present gives us the energy to link the past and the future and to fully embrace the sights, sounds, smells, family, and friends that are life.

CONCLUSION

The study of time has a long and distinguished history among philosophers, poets, and psychologists. Although most have recognized the substantial influence that time exerts on our lives, its systematic empirical investigation is relatively new. During the last hundred years, time study has transitioned from largely theoretical treatises to laboratory studies. Some of the recent advancements in the research on the psychology of time are attributable partially to the increasing recognition that time and time press play vital roles within Western society. These advances can also be traced to the development of standardized temporal assessment instruments, such as the ZTPI. We have summarized some of the research from our laboratory and that of other psychological investigators that has related specific ZTPI factors to other established psychological variables and behaviors. We have focused heavily on the ways in which future time perspective is associated with positive future consequences, notably in the realm of health and risk. We have also pointed out the trade-offs common among those high on present-hedonism, who may live to enjoy the expanded present moment fully but then are more likely to suffer the consequences of failing to invest in health maintenance. We have also uncoupled dual aspects of present time orientation and past orientation. The present-fatalism and past-negative time zones predict a host of pathologies when present in otherwise healthy, young people functioning in affluent settings, like American high schools and colleges. Similarly, despite some negative outcomes associated with excessive hedonism, at least there is a measure of personal gratification involved—which is absent in present-fatalism. Decomposing the past orientation into the opposite valences of positive and negative allows us to see that looking back with rose-colored glasses may be preferable to looking at the rearview mirror with darkened glasses.

Although we hope that this research makes a significant contribution to the literature of risk taking and health behavior, many questions remain unanswered. We need to know more about the developmental history of each time perspective. How are they learned, maintained, strengthened, and weakened over time? Can our understanding of temporal dynamics governing behavior be enhanced by analysis based on time perspective profiles rather than the simplistic categorical analysis others and we have used? How do individuals who have developed a balanced time perspective differ from those with biased time perspectives emphasizing only one dimension of the trilogy we have postulated? As the pace of life in our society continues to accelerate, we need to understand how situational factors influence important behaviors, health outcomes, risk taking, and altruistic helping behavior—for starters.

One of the most important questions related to time perspective may be one that traditional research alone cannot answer. It can be framed empirically in terms of striking the optimal balance between present-hedonism and future time perspectives, but it can also be framed more personally in terms of the balance that each of us strikes between enjoying our present life space and preparing for future options. For some, the solution to this quandary is to reframe the question in terms of a pursuit of happiness, which is defined as a more chronic but perhaps less intensive form of pleasure (The Dalai Lama & Cutler, 1998). For others, the appropriate balance can be summed up in an old saying, "Live each day as if it is your last; plan as if you will live forever."

For each of us the ultimate choice is inevitably a very personal decision that strikes at the heart of what it means to be human. Perhaps the ultimate answer is not to be found in any optimal balance point between the abstract realms of present-hedonism and future time perspective, but rather in the daily conflicts we encounter as we observe ourselves in the mirror on our way to yet another busy day—could that person in the mirror use our help, and will we keep the time press of our postmodern world at bay long enough to give it?

REFERENCES

Alvos, L., Gregson, R. A., & Ross, M. W. (1993). Future time perspective in current and previous injecting drug users. *Drug and Alcohol Dependence, 31,* 193–197.

Atkinson, J. (1964). *An introduction to motivation.* Princeton, NJ: Van Nostrand.

Barndt, R., & Johnson, D. (1955). Time orientation in delinquents. *Journal of Abnormal and Social Psychology, 51,* 343–345.

Boniwell, I., & Zimbardo, P. G. (2003). Balanced time perspective and optimal time use. *The Psychologist. Special Issue: Putting Positive Psychology Into Practice, 16,* 129–131.

Boyd, J. N., & Zimbardo, P. G. (1996). Constructing time after death: The transcendental-future time perspective. *Time and Society, 6,* 35–54.

Brock, T., & Del Giudice, C. (1963). Stealing and temporal orientation. *Journal of Abnormal and Social Psychology, 66,* 91–94.

Caprara, G. V. (1999). The notion of personality: Historical and recent perspectives. *European Review, 7,* 127–137.

Caprara, G. V., Barbaranelli, C., Borgogni, L., & Perugini, M. (1993). The "Big Five Questionnaire": A new questionnaire to assess the Five Factor Model. *Personality and Individual Differences, 15,* 281–288.

Chubick, J. D., Rider, C. Y., Owen, S. M., & Witherspoon, A. D. (1999). Time perspective of female prisoners related to success in a training program. *Perceptual and Motor Skills, 88,* 648–650.

Cohen, J. (1967). *Psychological time and health and disease.* Springfield, IL: Thomas.

Cottle, T. J. (1976). *Perceiving time: A psychological investigation with men and women.* New York: Wiley.

Csikszentmihalyi, M. (1990). *Flow: The psychology of optimal experience*. New York: Harper & Row.

The Dalai Lama, & Cutler, H. C. (1998). *The art of happiness: A handbook for living*. New York: Riverhead Books.

Darley, J. M., & Batson, C. D. (1973). From Jerusalem to Jericho: A study of situational and dispositional variables in helping behavior. *Journal of Personality and Social Psychology, 27*, 29–40.

DeVolder, M., & Lens, W. (1982). Academic achievement and future time perspective as a cognitive-motivational concept. *Journal of Personality and Social Psychology, 42*, 566–571.

Einstein, A. (1931). *Relativity: The special and general theory* (R. W. Lawson, Trans.). New York: Crown.

Epel, E. S., Bandura, A., & Zimbardo, P. G. (1999). Escaping homelessness: The influences of self-efficacy and time perspective on coping with homelessness. *Journal of Applied Social Psychology, 29*, 575–596.

Freud, S. (1950). *Beyond the pleasure principle* (J. Strachey, Trans.). New York: Liveright. (Original work published 1920)

Gjesme, T. (1979). Future time orientation as a function of achievement motives, ability, delay of gratification, and sex. *Journal of Psychology, 101*, 173–188.

Gleick, J. (1999). *Faster: The acceleration of just about everything*. New York: Random House.

Guarino, A., DePascalis, V., DiChiacchio, C., & Zimbardo, P. G. (2003). *Breast cancer prevention, time perspective, and trait anxiety*. Unpublished manuscript, University of Rome.

Harber, K. D., Zimbardo, P. G., & Boyd, J. N. (2003). Participant self-selection biases as a function of individual differences in time perspective. *Basic and Applied Social Psychology, 25*(3), 255–264.

Heidegger, M. (1927). *Being and time*. Halle, Germany: Niemeyer.

Holman, E. A., & Zimbardo, P. G. (2003). *The social language of time: Time perspective and social relationships*. Manuscript submitted for publication.

Husserl, E. (1964). *Phenomenology of internal time consciousness*. (M. Heidegger, Ed.; J. Churchill, Trans.). Bloomington: Indiana University Press.

Hutton, H. H., Lyketsos, C. G., Hunt, W. R., Bendit, G., Harrison, R. B., Swetz, A., & Treisman, G. J. (1999). *Personality characteristics and their relationship to HIV risk behaviors among women prisoners*. Manuscript submitted for publication.

James, W. (1950). *The principles of psychology*. New York: Dover. (Original work published 1890)

Kant, I. (1965). *Critique of pure reason* (N. Smith, Trans.). New York: St. Martin's Press. (Original work published 1781)

Keough, K. A., Zimbardo, P. G., & Boyd, J. N. (1999). Who's smoking, drinking, and using drugs? Time perspective as a predictor of substance use. *Basic and Applied Social Psychology, 21*, 149–164.

Koenig, F. (1979). Future-orientation and external locus of control. *Psychological Reports, 44*, 957–958.

Lamm, H., Schmidt, R., & Trommsdorff, G. (1976). Sex and social class as determinants of future-orientation (time perspective) in adolescents. *Journal of Personality and Social Psychology, 34*, 317–326.

Landau, S. (1975). Future time perspective of delinquents and non-delinquents. *Criminal Justice and Behavior, 2*, 22–36.

Landau, S. (1976). Delinquency, institutionalization, and time orientation. *Journal of Consulting and Clinical Psychology, 44*, 745–759.

Levine, R., (1997). *A geography of time: The temporal misadventures of a social psychologist, or how every culture keeps time just a little bit differently*. New York: Basic Books.

Lewin, K. (1942). Time perspective and morale. In G. Lewin (Ed.), *Resolving social conflicts* (pp. 103–124). New York: Harper.

Lewin, K. (1951). *Field theory in the social science: Selected theoretical papers*. New York: Harper.

Lyubomirsky, S., & Nolen-Hoeksema, S. (1993). Self-perpetuating properties of dysphoric rumination. *Journal of Personality and Social Psychology, 65*, 339–349.

Lyubomirsky, S., & Nolen-Hoeksema, S. (1995). Effects of self-focused rumination on negative thinking and interpersonal problem solving. *Journal of Personality and Social Psychology, 69*, 176–190.

Mann, T., Kato, P. M., Figdor, E. J., & Zimbardo, P. G. (1999). *How to make a pessimist behave like and optimist: Future writing, optimism, and health behaviors in cancer survivors and HIV patients*. Unpublished manuscript, University of California, Los Angeles.

McClelland, D. (1985). How motives, skills, and values determine what people do. *American Psychologist, 40*, 812–825.

Melges, F. T. (1982). *Time and the inner future*. New York: Wiley.

Miller, D. T., & Porter, C. A. (1980). Effects of temporal perspectives on the attribution process. *Journal of Personality and Social Psychology, 39*, 532–541.

Moscovici, S. (1984). The phenomenon of social representations (S. Rabinovitch, Trans.). In R. Farr & S. Moscovici (Eds.), *Social representations* (pp. 3–69). Cambridge, England: Cambridge University Press.

Nuttin, J. R. (1964). The future time perspective in human motivation and learning. *Acta Psychologica, 23*, 60–83.

Nuttin, J. R. (1985). *Future time perspective and motivation: Theory and research method*. Hillsdale, NJ: Lawrence Erlbaum Associates.

Oyserman, D., & Markus, H. (1990a). Possible selves in balance: Implications for delinquency. *Journal of Social Issues, 46*, 141–157.

Oyserman, D., & Markus, H. (1990b). Possible selves and delinquency. *Journal of Personality and Social Psychology, 59*, 112–125.

Rappaport, H. (1990). *Marking time*. New York: Simon & Schuster.

Raynor, J. (1969). Future-orientation and motivation of immediate activity: An elaboration of the theory of achievement motivation. *Psychological Review, 76*, 606–610.

Raynor, J., & Entin, E. (1983). The function of future-orientation as a determinant of human behavior in step-path theory of action. *International Journal of Psychology, 18*, 463–487.

Rothspan, S., & Read, S. J. (1996). Present versus future time perspective and HIV risk among heterosexual college students. *Health Psychology, 15*, 131–134.

Skinner, B. F. (1987). *Upon further reflection*. Englewood Cliffs, NJ. Prentice-Hall.

Stein, K., Sarbin, T., & Kulik, J. (1968). Future time perspective: Its relation to the socialization process and the delinquent role. *Journal of Consulting and Clinical Psychology, 32*, 257–264.

Strathman, A., Gleicher, F., Boninger, D., & Edwards, C. (1994). The consideration of future consequences: Weighing immediate and distant outcomes of behavior. *Journal of Personality and Social Psychology, 66*, 742–752.

Tismer, K. (1985). Sex and age differences in personal and global future time perspective: A replication. *Perceptual and Motor Skills, 61*, 1007–1010.

Von Wright, J., & Von Wright, R. (1977). Sex differences in personal and global future time perspectives. *Perceptual and Motor Skills, 44*, 30.

Watson, J. (1913). Psychology as the behaviorist views it. *Psychological Review, 20*, 158–177.

Zaleski, Z. (1996). Future anxiety: Concept measurement and preliminary research. *Personality and Individual Differences, 21*, 165–174.

Zimbardo, P. G., & Boyd, J. N. (1999). Putting time in perspective: A valid, reliable individual-differences metric. *Journal of Personality and Social Psychology, 77*, 1271–1288.

Zimbardo, P. G., Keough, K. A., & Boyd, J. N. (1997). Present time perspective as a predictor of risky driving. *Personality and Individual Differences, 23*, 1007–1023.

Zimbardo, P. G., Marshall, G., & Maslach, C. (1971). Liberating behavior from time-bound controls: Expanding the present through hypnosis. *Journal of Applied Social Psychology, 1*, 305–323.

Zuckerman, M. (1994). *Behavioral expressions and biosocial bases of sensation seeking.* New York: Cambridge University Press.

Zuckerman, M., Eysenck, S. B. G., & Eysenck, H. J. (1978). Sensation seeking in England and America: Cross-cultural, age and sex comparisons. *Journal of Consulting and Clinical Psychology, 39*, 308–321.

6

Time Orientation and Economic Decision Making

Michael S. Finke
University of Missouri–Columbia

Imagine standing in line at a fast-food restaurant trying to decide whether to buy a thick hamburger and french fries or a grilled-chicken salad. The hamburger may taste better and might even keep you more satisfied through the afternoon, but still you consider the healthier alternative. What drives us to place value on the healthier alternative is recognition of the impact eating unhealthy foods will have upon the quality (and perhaps duration) of our lives in the future. The ability to defer gratification reveals a willingness to consider the consequences of placing too great a value on the present. Any decision that affects welfare in the present time period as well as welfare in any future time period requires the ability to compare outcomes over time. Some will give up potential future happiness to live well in the present, whereas others will consciously forego present happiness with an eye toward the future. The preference for outcomes across time can explain why some make perfectly rational decisions that seem to others impulsive and unwise.

To begin to grasp how economic theory and its basic framework of rationality and choice can prove useful to the analysis of time orientation, it is important to understand how economists define rationality and use it as a foundation for predicting and interpreting human behavior. Economics is often confused with the study of money, and criticized for failing to consider strong motivations outside the sphere of commerce. However, much of the new research and theory in behavioral economics is based not necessarily on money, but on the idea that people make decisions they believe will make them better off—happier, more satisfied, more stimulated, and so on. Any choice necessarily involves choices foregone, and it is within these foregone choices that economists measure and compare preferences (the values that define what will make us better off).

To an economist, all behaviors, even those that may appear on the surface to be irrational such as drug use or suicide, must be considered welfare maximizing and therefore reflect the preferences of the individual. Individuals who have developed an addiction to heroin will have a very different set of basic preferences than those who are not addicted (and may have a very different set of preferences than prior to addiction). Depressed individuals may likewise value their potential future well-being much differently than those free of depression. Because all individuals are assumed to make choices that they believe will make them better off, all actions are rational in the eyes of the person making the choice. Economics places no value judgment on what motivates a person—these preferences are considered only to the extent that they affect choice related to economic theory.

Because money is quantifiable and represents a handy means through which the value of a good or service may be measured and compared, it is used as a tool in economics. However, temporal economics is often discussed in terms of utility, or the relative satisfaction derived from an activity. Economists assume that individuals will always make the choice that provides more utility, and likewise all other possible choices would have provided less utility. Greater access to scarce resources (money, love, time) will provide greater utility.

The notion that we can estimate utility of a choice by looking at the value of what was given up (and thus provided less utility) is central to economic research. Economics may be better conceptualized as the science of trade-offs. If one is willing to give up more resources to obtain an ice-cream cone than a popsicle, then economists assume the ice-cream cone provides more utility. Many trade-offs involve giving up a variety of resources—for example, learning to play the piano involves time spent learning the skill, money to buy the piano, space to store the piano, and time spent playing in order to begin to derive utility. Because time and money are traded to produce pleasure through piano playing, it is assumed that no other use of time and money would have provided greater satisfaction if the piano player is acting rationally. Time adds a layer of complexity to the economic analysis of trade-offs. Because there is not an observable simultaneous exchange of resources, economists must try to compare the value of what is given up in the present and what is gained in the future.

Renewed interest in time orientation among economists has led to a resurrection of theory involving choice over time. Though identification and measurement of individual time orientation, or time preference, is in its infancy, advances in economic theory over the last 30 years have spurred a number of empirical analyses involving behaviors from smoking to saving for retirement (Frederick, Loewenstein, & O'Donaghue, 2002). The theory of time preference is quite simple, yet powerful in its ability to explain and predict a number of behaviors. It suggests that myopic decision making is rational, consistent, measurable, and applicable across decision-making domains.

This chapter provides a brief explanation of the economic theory of time orientation, or more accurately the preference for utility across periods of time. Economic terminology and more complex theoretical functions are intentionally kept to a minimum. Examples of recent extensions to time preference theory and applications to a variety of behaviors are presented. Explanation of methods by which individual time preference has been estimated are discussed, and suggested means through which time orientation may be appropriately modeled are compared using different empirical estimation techniques.

TIME PREFERENCE

If all individuals are rational, any trade-off of utility from one time period to another must be utility maximizing. The trade-off must make the individual, given their idiosyncratic set of preferences, better off. Many of the individual characteristics related to time orientation discussed elsewhere in this book are viewed by economists as factors impacting preferences—they affect how an individual estimates the utility of possible choices and are reflected in behavior involving temporal-resource trade-offs. For example, someone who is very impulsive may estimate an inheritance in the distant future to be of little value in terms of utility.

An individual's willingness to trade off future and present utility may be expressed not just as a general tendency, but as a single factor that is quantifiable and systematically applied to all changes in utility over time. This time preference factor is known to economists as the personal future discount rate, which is an important element of the fundamental theory guiding rational decision making across time periods—the discounted-utility model (Samuelson, 1937).

The discounted-utility model posits that when faced with decisions involving intertemporal (affecting multiple time periods) consequences, we will consider the impact of each choice upon total lifetime utility. Economists assume that utility is a function of consumption, and that greater consumption leads to greater utility. It is best to keep in mind that consumption is not necessarily limited to traditional goods and services, but includes any scarce resource used to increase welfare. We can designate consumption in the present period as c_t. Because we all have a finite number of years in which to maximize utility through consumption, we can look to consumption levels in the future up to time period T. The utility function over time can be represented as $U_t (c_t, \ldots, c_T)$. If we are rational, we will make choices that maximize our lifetime utility by considering the impact each outcome will have upon consumption in the future.

Weighing options that have future consequences involves comparison of changes in proximate and more distant time periods upon expected lifetime utility, however the process is complicated by the tendency of individuals to place a lower value on utility in the future than utility in the present. The natural and rational tendency to value future changes in satisfaction less than current changes in satisfaction may be explained as a response to future uncertainty, a finite life span, impatience, the existence of a positive real interest rate, or the expectation of rising wages. Instead of simply adding the positive or negative impact of each choice upon consumption in all future time periods, changes in future utility are discounted using the rate of time preference.

The discounted-utility model assumes that individuals consistently apply a single rate of future discounting to all intertemporal decisions. A higher rate of discounting leads an individual to place a lower value on any positive or negative change in future utility. This personal future discount rate (ρ) is then used to discount any expected change in future utility. Just as a \$50 savings bond is equivalent in the eyes of an investor to \$100 in 10 years through the use of market interest rates, a single unit of utility now is made equivalent to a larger amount of utility in the future through the personal future discount rate. To maximize lifetime utility, the individual looks from the present time period (t) to the final time period (T) and makes the choices that provide the highest discounted value of utility from consumption.

In order to make rational decisions consistent with maximum lifetime utility, an individual must weight the discounted present value of utility from one choice against the discounted present utility from an alternate choice. For example, a consumer considering a reliable car that will provide greater utility in future time periods and a more stylish car that will provider greater utility in nearer time periods will weigh the marginal impact of each car upon lifetime satisfaction in the discounted-utility model by considering the total discounted utility of each choice. Utility in more distant time periods is discounted relatively more than utility in the near future. Those with a higher rate of time preference (ρ) will discount future utility more than those with a lower rate of time preference (who will be more likely to buy a car that promises to provide future satisfaction; see Dreyfus & Viscusi, 1995).

Every day we are presented with choices that involve future consequences. Rational response to these consequences requires that future discounting be applied to all decision making domains consistently. An individual's rate of time preference should then be revealed through behaviors that require a comparison of utility over time. By discounting future changes in utility less, those with a smaller ρ will be more likely to prefer choices that reflect a less myopic (or impulsive) time orientation—whether they are choosing to save for retirement or to avoid cigarettes. To do otherwise would be economically irrational, because it would involve making great sacrifices in the present in order to increase utility in the future in one domain while simultaneously increasing current utility at great expense to future well-being in another.

Behaviors that may seem to violate this consistency of future discounting, for example, when a smoker goes to the gym, may be at least partially explained by the multidimensional nature of many activities. The smoker may go to a gym to improve physical attractiveness or to socialize—indicating a preference for present utility. Someone with a lower future discount rate, on the other hand, may go to the same gym in order to reduce the risk of future cardiovascular disease—a much more future-oriented activity. For this reason, variation in behaviors that involve a time-oriented utility trade-off may be explained partially by characteristics unrelated to the intertemporal nature of that activity. This is addressed later in the chapter when empirical estimation of time preference is discussed, but it is important to bear in mind that time preference may influence, but not fully explain, variation in choices involving intertemporal trade-offs.

If rates of individual time preference vary among a population, discounted-utility theory provides both an explanation for myopic behaviors and insight into possible correlations among them. This promise of consistency among decision-making domains may enable more precise empirical estimation of an individual's true rate of time preference.

EMPIRICAL ESTIMATION OF TIME PREFERENCE

Despite the theoretically predicted frequent use of temporal discounting in nearly all domains of decision making, the empirical estimation of an individual's ρ is not a simple task. Though we often know intuitively our willingness to be prudent or impulsive, few are able to provide their own numerical rate of time preference. Precisely estimating ρ as it is conceived in the discounted-utility model involves the comparison of potential consumption over multiple time periods—a process that can be measured either hypothetically or as the result of observed decision making. Empirical support of the discounted-utility model and the predicted influence of time preference upon decision making requires an estimation of ρ and comparison of its ability to influence decision making in multiple domains.

The most common method through which individual rates of time preference are estimated empirically involves comparing monetary tradeoffs between time periods. By asking individuals whether they prefer a dollar amount today to a dollar amount in a week, month, year, or decade in the future, researchers are able to determine ρ at the point where $U^0 = (1/(1 + \rho))^n * U^n$ where n is a designated future time period. For example, would you prefer \$100 right now to \$110 in 1 year? A preference for \$100 right now implies a ρ of greater than 10%, because \$100 now is preferred to $(1/(1 + 0.1))^1 * 110$, which equals \$100 today. If the respondent's personal future discount rate were 15%, then they would be indifferent between \$100 today and \$115 in a year. Note that this assumes utility is a function of consumption, which is a function of money available to use for consumption. Like increases in money imply like increases in utility across time periods.

This approach has yielded disappointingly inconsistent results (Frederick et al., 2002). Although the assumption that an increase in money necessarily results in an increase in utility is a basic foundation of economic thought, the relationship between money and utility over time is oversimplified by the comparison of specific dollar values. Estimates of ρ vary greatly among studies, and the relationship between rate of time preference and duration of reward has been found to be strongly hyperbolic (the rate is much higher in the near future than in more distant time periods).

Three basic problems may underlie the persistent inconsistency in direct measurement of absolute rates of time preference in a laboratory environment. The tendency toward hyperbolic measured rates of time preference may be due to the perceived probability of receiving future amounts and costs of accounting for money owed. Participants would much rather receive \$20 right now than a promise of \$25 in 1 month because they perceive a nonzero probability that the transfer won't occur. How often have we mailed in a rebate and not received a check as promised? Most of us, and particularly university students in a university lab environment, have lost full confidence in the ability of the institution to pay out when and in the amount promised. However, if they have to choose between money promised in 5 years versus money in 10 years, they are more likely to employ a rate of discounting independent of the probability effect.

Subjects may also anticipate a loss in utility from money received at a later date from psychic costs associated with keeping track of money owed. It appears that individuals perceive a cost of having to remember the promised future dollar transfer, and of having to remind the researcher to make good on the promise to make payment in the future. With a smaller amount of money, people would rather have it right now to avoid having to worry about a future payment. Most people when asked whether they'd prefer receiving $10 immediately or $11 next month would probably take the money now because the added dollar would mean such a small utility gain compared to the utility expended keeping track of the future promised payoff. However, taking the $10 right now implies a personal future discount rate of 214%. Asking the same question with much larger amounts, for example, $1,000 right now versus $1,100 in 1 year, may lead to more realistic results. Use of larger dollar amounts may avoid biased estimates of time preference influenced by psychic costs. Consider the following survey of 259 undergraduate students from an introductory psychology course at the University of Missouri conducted in the fall of 2002:

> Your uncle gives you a $1,000 savings bond for graduation. If you cash it in immediately, you can get $400. If you wait a year, you can cash it in for $500. If you wait 10 years, you can cash it in for the full $1,000. Would you rather wait the full 10 years, cash it in 1 year, or cash it now?

Option	*Frequency*	*Percent*	*Cumulative Percent*
wait 10 years for $1,000	187	72.2	72.2
cash it in now for $400	26	10.0	82.2
wait a year for $500	46	17.8	100.0
Total	259	100.0	

> Using the same preference choices as above, if the amounts were $50, $40, and $100, which option would you prefer?

Option	*Frequency*	*Percent*	*Cumulative Percent*
wait 10 years for $100	76	29.3	29.6
cash it in now for $40	108	41.7	71.6
wait a year for $50	73	28.2	100.0
Total	257	99.2	

When given the choice between receiving a lump sum right now or waiting to receive a larger amount in the future, 42% of respondents would prefer taking the money immediately if the amount was small ($40) whereas only 10% of respondents would cash it in if the amount was larger ($400). This wide variation in implied rates of future discounting among the same group of college students illustrates the empirical importance of considering the utility of each option rather than simply the dollar amount. The data clearly imply a loss in utility from either mental accounting or a probability effect that must be considered when computing a respondent's actual ρ.

A third problem with estimating time preference through comparison of money values is that the discount rate for nominal dollar amount over time periods is not the same as the discount rate for utility (Roelofsma, 2000). Given an expected increase in earnings and a utility function that reflects a decreasing marginal utility of income, individuals will naturally discount future dollar amounts at a higher rate than their true discount rate for utility. Marginal utility is the amount of satisfaction provided from the additional goods and services that can be bought by an added dollar of income. A classic example is of the college student and the CEO who each find \$100 lying on the ground. The goods and services the college student will buy with the \$100 provide her much greater satisfaction than the CEO because her income is not nearly as high, thus the marginal utility the student will receive from the \$100 will be much higher. If the college student graduates and increases her income considerably, her marginal utility of added income will fall. If you ask students to compare a dollar amount now (when marginal utility is very high) with a dollar amount in the future (when their marginal utility is much lower), their response may reveal a personal future discount rate for money that is much higher than their personal future discount rate for utility. Absolute dollar amounts, even if measured in a laboratory environment with precision and accounting for potential bias, may be a poor proxy for rate of time preference in a discounted-utility model.

Another method by which time preference may be measured empirically is through observed behavior. Economic theory suggests that rational decision making involving a utility trade-off over time periods is guided by the personal future discount rate. If this is true, then choices made across a wide realm of domains should yield consistent discounting of future utility. As mentioned previously, it is not rational to make great sacrifices to present utility in one domain in order to increase future welfare while simultaneously sacrificing future utility in another domain to increase present welfare. It follows that comparison of intertemporal decision making may yield more accurate estimates of ρ as it relates to discounted utility.

Though it may be impossible to place an absolute value on ρ using observed behavior, it is possible to obtain relative estimates by comparing decisions that imply a tendency to prefer future positive outcomes. Previous examples include choice of consumer durables (Gately, 1980), educational attainment (Lang & Ruud, 1986), and occupation choice (Moore & Viscusi, 1990). Results have shown a strong relationship between revealed rates of future discounting and consumption and risk-related behavior. Fuchs (1982) found a significant relationship between time preference and smoking. Subsequent research has established relationships between time preference and drug use (Bretteville-Jensen, 1999; Kirby, Petry, & Bickel, 1999), alcohol abuse (Vuchinich & Simpson, 1998), preventive medical procedures and willingness to accept influenza vaccinations (Chapman & Coups, 1999), and cancer screening.

Whereas many of the previous studies directly measure rates of time preference through hypothetical willingness to trade dollar amounts between time periods, it is possible to proxy relative time preference from revealed behavior. Evans and Montgomery (1994) and Munasinghe and Sicherman (2000) note the significance of smoking as a proxy for a high discount rate of future utility. Those who smoke have revealed that they are willing to give up health and longevity in the future in order to engage in an activity that provides gains in the present. Chapman (1996) found no difference between discount rates for health outcomes and discount rates for money received in different time periods.

The extent to which higher education proxies a lower future discount rate (due to the willingness to trade lost income and possible happiness in the present in order to earn a degree that might increase future earnings), versus the possibility that education merely proxies exposure to health-related information or improved skills through which better health can be produced, is addressed in Fuchs (1982) and Evans and Montgomery (1994), who find a strong relationship between education and personal future discount rate. Whether education itself leads to choices revealing a low rate of future, or whether those who have a low rate of time preference are far more likely to obtain higher education and are further more likely to make the sacrifices needed to graduate is not entirely clear. However, there is some evidence that time preference is not changed through the process of education (Bishai, 2001) and much evidence confirming correlations between education and other behavioral proxies for time preference such as smoking, obesity, and failure to use seat belts (Taylor, 1999), healthy food consumption behavior and exercise (Huston & Finke, 2003), and even savings behavior (Finke, Huston, & Weaver, 2003).

The decision to divert earnings into savings is arguably among the observed behaviors most strongly guided by an individual's personal future discount rate. Saving involves the purposeful reduction in present consumption in order to increase expected future consumption. Those (with a utility function indicating a decrease in marginal utility as total income increases) who wish to maximize utility over time are motivated to allocate financial resources from periods of higher income (lower marginal utility) to periods of lower income (higher marginal utility). Time preference will impact the weight an investor places on potential changes in future utility from financial resource transfers. An investor with a high personal future discount rate will discount potential gains in future consumption arising from increased savings today. This direct link between intertemporal wealth allocation and time preference is explained in Bernheim, Skinner, and Weinberg (1997) as a function of the expected consumption profile. An investor seeks to maximize his or her lifetime utility through present and future consumption. Those with a lower rate of time preference will borrow less while younger, save more during their peak earning years, and ultimately will accumulate a much higher level of peak wealth prior to retirement. Finke et al. (2003) find that time preference, measured indirectly through a factor of intertemporal behaviors, is a strong and consistent predictor of accumulated wealth among older preretirement households. The relationship between time preference and wealth should also be considered when ascribing health outcomes to increased access to financial resources. It is possible that positive health outcomes and financial resources may both be partially the result of individual time preference.

Revealed behaviors such as smoking, exercise, education, investment behavior, and food consumption all contain a strong intertemporal utility component. However, researchers should not expect to find perfect correlation between domains such as investment and food consumption behavior due to the nature of utility related to these activities. For example, food provides utility through a number of different characteristics—flavor, convenience, status, and so on—that may be unrelated to time preference but strongly related to other individual characteristics. Someone with a low rate of time preference and a busy schedule may place a much greater value on convenience and consume a lower-quality diet than someone with the same rate of time preference who has more free time.

The relationship between health and time preference is introduced theoretically in Grossman (1972) through a model that equates investment in health to an investment in human capital. Human capital can be viewed simplistically as one's ability to earn money now and in the future, or more accurately as one's capacity to derive utility from scarce resources. In Grossman, the endogenous nature of health is acknowledged, rather than taking one's longevity and quality of life as given. We can make choices to improve our health by getting regular checkups, taking vaccinations, exercising, eating healthy foods, and avoiding pollution. Whenever we are forced to make a choice between an alternative that may compromise our health and one that may improve our health, we measure the expected impact on future utility. Those who have a lower rate of time preference will place a greater value on reduced probability of premature death or increased quality of life in more distant time periods. Those who have a higher rate of time preference are less strongly motivated to invest in health, and may be expected to be less likely to engage in healthier behavior.

Measurement of time preference through cross-domain behaviors may be accomplished through the isolation of a common factor. Huston and Finke (2003) found a common factor from among intertemporal variables within the Health and Retirement Survey including prior flu vaccination, formal education, serum cholesterol testing, absence of smoking, and having a will or trust—all behaviors involving an intertemporal utility component. This finding suggests that there is one underlying construct that represents time preference over two decision-making domains (health and money). This result also lends support to the notion that personal future discounting is a singular construct that can be applied over all resource allocation decisions. The instrument created through factor analysis isolates the variation among these variables that is presumably the result of personal future discounting and not the result of other possibly confounding characteristics related to the observed behavior. The paucity of available data sets containing cross-domain intertemporal behaviors has limited research in this area.

Further evidence of time preference across domains among college students at the University of Missouri is provided by correlations between domains related only in their intertemporal component. Students who preferred to cash in a savings bond immediately rather than waiting to cash it in at a higher value in the future were more likely to reveal a higher future discount rate through other preferences (see Table 6.1).

Subjecting time preference–related variables to a principal-components analysis grouped the behaviors listed in Table 6.2 into a single component (Eigenvalue > 0.35).

Though many of the behaviors loading to the single factor were related to health, preference for style in an automobile was a strongly weighted component of the factor, as was willingness to cash in a savings bond immediately rather than in 1 or 10 years. The correlations among domains within this sample of students provides some insight into otherwise unrelated behaviors (such as

TABLE 6.1

Correlation With Willingness to Cash in a Savings Bond Immediately

	F Statistic	*Significance*
Preference for style versus reliability in an automobile	8.262	0.004
Would not be likely to accept a free flu vaccine	6.919	0.009
Does not plan to save for retirement after graduation	3.270	0.072
Consumed more than 30 alcoholic drinks in last month	8.678	0.004
Used marijuana in the last year	4.957	0.027

TABLE 6.2
Results of Subjecting Time Preference–Related Values to a Principal-Components Analysis

Variables	*Eigenvalue*
Preference for style versus reliability in an automobile	0.585
Would not be likely to accept a free flu vaccine	0.353
Does not always wear a seat belt when driving	0.563
Consumed more than 30 alcoholic drinks in last month	0.620
Used marijuana in the last year	0.597
Would cash in a savings bond immediately	0.401
Engaged in unprotected sex during the last year	0.547
Does not engage in regular exercise to improve health	0.385

style and risk taking). It also shows how a time preference factor may be useful as a measure of relative personal future discounting.

Use of a relative time preference factor rather than a measured discount rate offers some advantages. Prior numerical estimates of personal future discount rates have varied wildly (Frederick et al., 2002) due primarily to differences in measurement techniques. A single number representing the rate of time preference employed across domains through comparison of dollar amounts at different times will inevitably yield responses that reflect the financial sophistication of the respondents and their ability to match hypothetical money with hypothetical utility. Proxies of time preference, such as smoking, may be strongly related to the discounting of future utility, but may also be strongly related to the experiences and unrelated preferences of the individual.

Discounted-utility theory and the individual personal future discount rate provide a framework based on rational trade-offs that may be used to explain and predict a variety of intertemporal behaviors. Relationships between seemingly unrelated behaviors—the frequency of smoking among dancers, buying a reliable car and eating healthy food, increased wealth and having one's cholesterol checked—may be better understood as an inclination to place a greater value on future utility. Research exploring the relationships between psychological measures related to time orientation and economic measures of time preference have yielded inconsistent results (Frederick et al., 2002), however further research is needed to better understand an individual's willingness to make intertemporal trade-offs. The intractability of an individual's time preference over time and in response to experience or possible intervention is also not clear. The surprisingly strong relationship between a child's ability to defer consumption illustrated in the classic marshmallow test (being offered a single marshmallow in order to receive two marshmallows after 15–20 minutes; Goldman, 1996) with, among other characteristics, increased educational performance later in life, is evidence that individual time preference may not change over time.

Researchers should consider that the intertemporal utility component of many behaviors, for example, drinking and unprotected sex, may be at least partially correlated due to the underlying influence of time preference. Properly accounting for the mediating effect of time preference when studying relationships among risky behaviors will provide more accurate estimates of causation. If time preference proves stable among decision-making domains, and proves to be stable and resistant to change over time, then many behaviors that may appear inappropriate and irrational to those with lower rates of time preference may be perfectly rational and utility maximizing to those with higher rates of time preference. Because time preference is an innate component of our utility structure, forcing people to weigh future consequences to a greater extent than is dictated by their preferences will in fact make them worse off. For example, forcing those with a high rate of time preference to quit smoking or start wearing a helmet may be justified in terms of social welfare, but it will reduce the well-being of the individual.

Economic theory can also be used to help explain how consideration of utility over time can affect time orientation. If people indeed compare the potential lifetime utility of choices with outcomes in the present and future, the ability to imagine future utility may prove an important and malleable component of intertemporal behavior. It is hypothesized in Becker and Mulligan (1997) that intertemporal decision making is enhanced by the ability to imagine outcomes in the future. For example, spending time with an elderly relative living on a fixed income will provide insight into the consequences of failing to save money for retirement on future utility. This ability to imagine future welfare is almost certainly endogenous, and can be enhanced through education and experience. It may be that individuals making very nearsighted choices are revealing an inability to imagine consequences. Although they may be making choices that maximize their welfare according to what they currently know, had they been better educated about the consequences they may have made different choices.

The economic contribution to the study of time orientation may be to suggest that impulsive or otherwise nearsighted behavior is a result of rational intertemporal preferences, and may influence behaviors over a number of decision-making domains. It may also illustrate the limitations of efforts to modify behavior and may improve understanding of the root causes of risky or nearsighted choices. The recognition that rational behavior must involve consideration of future utility is central to intertemporal economics, and opportunities to increase welfare through intervention may be limited to improved understanding of the link between present choices and future consequences.

REFERENCES

Becker, G., & Mulligan, C. (1997). The endogenous determination of time preference. *Quarterly Journal of Economics, 112,* 729–758.

Bernheim, B. D., Skinner, J., & Weinberg, S. (1997). *What accounts for the variation in retirement wealth among U.S. households?* Unpublished manuscript, Stanford University, Stanford, CA.

Bishai, D. (2001). Lifecycle changes in the rate of time preference: Testing the theory of endogenous preferences and its relevance to adolescent substance use. In M. Grossman & C. R. Hsieh (Eds.), *The economics of substance abuse* (pp. 155–177). Northampton, MA: Edward Elgar Press.

Bretteville-Jensen, A. L. (1999). Addiction and discounting. *Journal of Heath Economics, 18,* 393–407.

Chapman, G. B. (1996). Temporal discounting and utility for health and money. *Journal of Experimental Psychology: Learning, Memory, and Cognition, 22,* 771–791.

Chapman, G. B., & Coups, E. J. (1999). Time preferences and preventive health behavior: Acceptance of the influenza vaccine. *Medical Decision Making, 19,* 307–314.

Dreyfus, M. K., & Viscusi, W. K. (1995). Rates of time preference and consumer valuations of automobile safety and fuel efficiency. *Journal of Law and Economics, 38,* 79–105.

Evans, W. N., & Montgomery, E. (1994). *Education and health: Where there's smoke there's an instrument* (Working Paper No. 4949). Washington, DC: National Bureau of Economic Research.

Finke, M., Huston, S. J., & Weaver, D. (2003). Time preference and intertemporal decision making: Evidence from the health and retirement study. In J. Lee (Ed.), *Consumer interests annual* (p. 49). New York: American Council on Consumer Interests.

Frederick, S., Loewenstein, G., & O'Donoghue, T. (2002). Time discounting and time preference: A critical review. *Journal of Economic Literature, 40,* 351–401.

Fuchs, V. (1982). Time preference and health: An exploratory study. In V. Fuchs (Ed.), *Economic aspects of health* (pp. 93–120). Chicago: University of Chicago Press.

Gately, D. (1980). Individual discount rates and the purchase and utilisation of energy-using durables: Comment. *Bell Journal of Economics, 11,* 373–374.

Goldman, D. (1996). *Emotional intelligence.* London: Bloomsbury.

Grossman, M. (1972). On the concept of health capital and the demand for health. *Journal of Political Economy, 82,* 223–255.

Huston, S. J., & Finke, M. S. (2003). Diet choice and the role of time preference. *Journal of Consumer Affairs, 37*(1), 143–160.

Kirby, K., Petry, N., & Bickel, W. (1999). Heroin addicts have higher discount rates for delayed rewards than non-drug-using controls. *Journal of Experimental Psychology, 128,* 78–87.

Lang, K., & Ruud, P. (1986). Returns to schooling, implicit discount rates and Black–White wage differentials. *Review of Economics and Statistics, 68,* 41–47.

Moore, M. J., & Viscusi, W. K. (1990). Models for estimating discount rates for long-term health risks using labor market data. *Journal of Risk and Uncertainty, 3,* 381–401.

Munasinghe, L., & Sicherman, N. (2000). *Why do dancers smoke? Time preference, occupational choice, and wage growth.* National Bureau of Economic Research Working Paper No. 7542. Retrieved August 26, 2003, from http://www.nber.org/papers/w7542

Roelofsma, P. (2001). Methodological and theoretical contributions to the study of time preference. Retrieved August 26, 2003, from http://www.mariecurie.org/annals/volume1/roelofsma.pdf

Samuelson, P. (1937). A note on measurement of utility. *Review of Economic Studies, 4*, 155–161.

Taylor, H. (1999). The obesity epidemic is getting even worse. The Harris Poll #11, March 6, 2002. Retrieved August 28, 2003, from http://www.harrisinteractive.com/harris_poll/index.asp?PID=288

Vuchinich, R. E., & Simpson, C. A. (1998). Hyperbolic temporal discounting in social drinkers and problem drinkers. *Experimental and Clinical Psychopharmacology, 6*, 290–305.

7

Future Orientation and Anxiety

Zbigniew Zaleski
Catholic University of Lublin

Human future orientation is a strong component of psychological activity. The prestigious institution of future tellers, the Delphi of ancient Greece, was created in the interest of revealing future events. This interest has not diminished in present times. Rather, it continues to increase. Commonly shared reflections on the big wars of the 20th century, the terrorist attack of September 11, 2001, and current tensions between the Muslim and Western worlds make people think and speak of the near as well as more distant future with deep concern and worry. What will it be like on a personal level? Will people's private plans and goals be realized or will some disasters hinder their achievement? Thus, on a mental level, the future is a large part of our deliberations. The beginning of the third millennium only intensifies this process, which may be accompanied by hope and/or anxiety.

People are interested in predicting the future for many reasons. They want to know what and where things will happen. We could refer to Heider's (1958) conception of attributional explanation of the surrounding world of events and to Kruglanski's (Kruglanski & Webster, 1996) idea of the need for closure, which in this context would be treated as making things known, bound together, limiting entropy. Second, a known or mentally conceived future allows an individual to direct his or her actions in ways that conserve energy. Third, the unknown is a source of tension, whereas knowledge reduces tension. Knowing future outcomes or events decreases anxiety. Fourth, knowing what will come makes one prepare for the encounter. Fifth, this knowledge may evoke gratitude in others who do not possess it themselves (e.g., palm readers are well rewarded for their service) so that the economic aspect of revealing the future is not irrelevant. Thus there is a constant duel between each individual's attempt to fight the unknown and the reality that unfolds according to its own "rules." Its unpredictability does not serve as a tranquilizer but, to the contrary, is a source of preoccupation, worry, or anxiety.

The future requires from psychologists a special approach for two reasons. In contrast to the past (known facts, unknown motives), what is known about the future is only one's intention, plan, or goal. The entire possible future is located in human cognitive-emotional processes. The future is potential and possible, not certain. Yet, this potentiality and cognitive reality does not make it psychologically less important than the facts of the past and the actions of today. It is so, because humans turn their minds toward and process the future. They live today but their guiding motor is always in front of them. The standards to be achieved by current actions are located in the future. Nuttin (1985) stresses one particular aspect related to future time perspective (FTP) and personality, suggesting that psychologically it is more important to us who we will be in the future than who we are now. As Nuttin stated, "it is far more important for man to know whom he wants to become than who he is now" (personal communication, 15 April 1986). This means that our existence, activities, decisions, and efforts are directed toward who we shall be in years ahead. Our cognitive capability makes us use the future time as our stage of activity. What we do now is often thought to be for those who will come after us. On a social level, past generations had the future welfare of society as their goal. This is how culture grew over the centuries. Therefore, we not only care for our own offspring but also for generations many years from now. Only humans can do this.

Is future orientation a self-standing dimension not dependent on other psychological factors and processes? The answer is no and our task is to show its connection with other individual and external characteristics. Most often FTP is described in psychology in positive terms and this is justified and constructive. We usually associate the term *future* with hope, goals, and success. However, there are other types of associations with the concept of future, such as fear, anxiety, preoccupation, and uncertainty. The negative aspects of FTP are often neglected and I would like to fill in this gap. In this chapter I concentrate on this negative aspect of future anxiety. The present era, full of global disasters, terrorism, attacks, corruption, and economic inequality, justifies this negative focus.

For further analysis we need to define the two concepts: future orientation (FTP) and future anxiety.

FUTURE TIME PERSPECTIVE

The concept of FTP or future orientation (see Zaleski, 1994) has earned its place in motivational psychology. Nuttin (1985) defines it as "the configuration of temporally localized objects that virtually occupy his mind in a certain situation" (p. 21). It is characterized by structure, length, and emotional attitude. The structural material consists of goals and the plans for their realization. These goals are set for different time periods, from immediate ones to life goals (sometimes even beyond the earthly life). This mentally conceived personal future evokes emotional attitudes that range from positive (hope, joy) to negative ones (distress, despair, anxiety). Recently Thiebaut (1998) added to this characteristic the dimension of density, how many strivings are set for a given time range.

The future horizon is used by people for their mental anticipatory activity of construing one's own self (Markus & Kitayama, 1991). The localization, in the future, of standards to be achieved has motivational properties as it makes one undertake effort and persist in activities leading to goal attainment. It not only activates an organism but also guides it to select effective strategies, and look for means and information (Zaleski, 1987). Locke and Latham (1990) report a large amount of research that indicates that goal-directed behavior is more effective than activities not guided by goals in many spheres (e.g., economic, scientific, education). Whether personal goals are of extrinsic or intrinsic character (Kasser & Ryan, 1996), they motivate one to pursue the desired standards set for the future. Goal attainment is rewarding, in part because intrinsic goals satisfy the needs of autonomy and personal growth (Ryan, 1995; Sheldon & Kasser, 1995) and are directly related to well-being (Kasser & Ryan, 2001). Thus the future horizon filled with goals that are gradually realized is an important basis for satisfying life. Such future is considered by some as a challenge full of strivings and achievements. For others, though, the future may appear as a terrain of unknown; one that is unpredictable, uncertain, and dangerous. Instead of creating hope and expectations of achievement, thinking of and imagining the future temporal space awakens fears and worries. In the next sections I concentrate on this negative side.

THE NOTION OF FUTURE ANXIETY

I presented a broader theoretical proposition on future anxiety (FA) in earlier work (Zaleski, 1994, 1996) and recommend these texts for the reader in order to avoid redundancy. There I define FA as "a state of apprehension, uncertainty, fear, worry and concern of unfavorable changes in a more remote personal future" (Zaleski, 1996, p. 165), one based on cognitive representations of negative future states/events. FA may have various facets ranging from the concern that one will not be a good husband to the fear that after reaching the age of 50 one will experience a heart attack. A mother may fear the future more as her son grows, when she realizes that his military service is coming up and he will serve in an area of political conflict (e.g., in Israel). The level of FA may increase in later stages of life. People's concern about later life (old age) tends to grow. They worry whether they will be able to maintain their independence, and their mental and physical capabilities and these concerns are very realistic. This may depend on receiving information that evokes either hope or fear, such as reading about statistics showing the high rate of deaths resulting from cancer or reading that the death index decreased due to new immune remedies. FA, as an anxiety, has a strong emotional component, namely worry. Eysenck (1992) argued that the increased anxiety is associated with worry. People worry about threatened values and this concerns potential frustrating, nonrewarded, personally important goals (see Borkovec, Metzger, & Pruzinsky, 1986). The intensity of worry is dependent on the subjective importance of value to be blocked, probability of occurrence, and controllability or self-efficacy.

The importance is very relevant as more important goals imply more investment of effort and as more personal existence depends on them. The chance of occurrence is of great relevance, too. This is a very subjective estimate, which may vary from one person to another. The more probable the event, the greater the FA. High-FA individuals may have elevated subjective probability because they do not spontaneously generate many reasons why threatening events will not happen. MacLeod, Williams, and Bekerian (1991) found in their experiment that high worriers, when asked to think of reasons why negative events would not happen to them, showed a substantial reduction in subjective probability. In contrast, the manipulation had no effect on low worriers.

Except for some cases of stable anxiety it is more justifiable to take the position that FA is dynamic and that its intensity fluctuates (whereas the mean value can characterize the person). However, the dynamics of anxious attitudes toward one's personal future is grounded in the new approach of dynamic social psychology proposed by Nowak and Vallacher (1998). As these authors state, the mechanism is stable but the thoughts, emotions, and actions are in constant interchange. Vallacher and Wegner (1985) discussed the processes by which people come to identify their past, present, and future actions. In anxious people, negative information or even the imagination of a harmful future event may almost automatically evoke anxiety, which, in a vicious circle, may maintain the informational processes. What is important is whether in this dynamic fluctuating process FA dominates over hope or neutral attitudes. If it does, it can be admitted that a person experiences FA and this is a characteristic feature of his or her general emotional state, although not static, but dynamic.

Bandura (1991) offers a cognitive approach to anxiety within his theory of self-efficacy. He defines anxiety as a state of anticipatory apprehension over possible deleterious happenings. According to him threat is a relational property reflecting the match between perceived coping capabilities and potentially harmful aspects of the environment. People who believe they can exercise control over potential threats do not engage in apprehensive thinking and are not perturbed by them. Similar conclusions were drawn by Lang, Shapiro, and Cobb (1991) from their research on biofeedback effects on heart rate control in response to the cold pressor test. They found that "belief in the given expectancy and perceived control were the major determinants of pain reports" (p. 176). Thus believing in ability to control an aversive situation relates to anxiety arousal.

Future and Anxiety—The Strong Links

The contextual neighborhood of these two concepts—future and anxiety—does not mean that they are always in a psychological duet. Under certain circumstances, following the attack of September 11, 2001, for example, every reader would agree that they are related. Important negative events may enhance anxiety and relate it to future possible events. Looking into or thinking of the future, people may perceive it through a more positive rather than negative periscope. The concepts of the unknown or uncertain as more cognitive characteristics of future perspective can be a very individual state.

Sorrentino, Roney, and Hanna (1992) propose a construct of uncertainty orientation. They state that "uncertainty oriented people are curious about the unknown, open to new and possible inconsistent information, and have high tolerance for ambiguity. They derive little satisfaction from knowing without working through the challenge of uncertainty" (p. 420). Furthermore they state that "Uncertainty orientation is assumed to be time-linked. The uncertainty oriented seek out new ideas and explore previously unknown possibilities. They are presumed to be future-oriented. The certainty-oriented cling to previously established ideas and are threatened by the uncertainty of the unknown" (p. 420). These formulations are clear within the cognitive approach. However, their idea of uncertainty conveys only positive, exploratory, and challenging meaning. Uncertainty is something unknown, which a rational and adventurous individual wants to clarify by undertaking some appropriate exploratory action. This is the uncertainty for scientists and explorers, filled with hope for discovery and clarifying answers. However, this uncertainty can be a source of concerns for anxiety-prone individuals, whose predictions of potential outcomes are more negative than positive. This means they anticipate negative events, outcomes or experiences concerning their country, their surroundings, or themselves personally.

This anticipated possibility has two features as anxiety sources. One is the unknown—in the sense that none can say for sure what will come tomorrow and it is still harder to preview the more distant future. The second is the negativity effect of past and present events/experiences that provides the basis for anticipating similar events to come. With the exception of the contrast effect showing that after negative events people expect positive ones (after a storm comes sunshine), a general biased expectation is that success gives way to another success and failure leads to another failure (Czapiński & Peters, 1991). Thus significant negative events sharpen our vigilance for similar ones to happen in the future and this seems to be a general phenomenon.

Different levels of future anxiety can characterize whole societies. Americans are regarded as optimists, Phillipinos as pessimists (Gandeza, 1986). The combination of political, economic, and social conditions gives rise to more hopeful or more fearful expectations concerning the future. On an individual level people differ in their length of FTP and the intensity of FA. The possible combinations can be seen in Table 7.1.

Which cell is the most functional? On a theoretical basis it can be assumed that individuals in Cell 3 are the most effective in coping with their future. This means that they have filled, with well-structured goals and plans, a long and stable future perspective and possess a positive attitude or set of hopes (Averill, Catlin, & Chon, 1990). Less functional is the FA to one's own future perspective in Cell 1—the longer the FTP the higher the FA. This can hinder progressive actions as fear paralyzes an actor in planning for the future and realizing personal projects. A person may want to and try to elaborate his future projects but the accompanying anxiety brings him or her back to inertia. The opposite can be said about the individuals with short FTP and high anxiety (Cell 2).

TABLE 7.1

FA	*FTP*	
	Long	*Short*
High	(1)	(2)
Low	(3)	(4)

They are crippled by fear of possible failures and mishaps and act within the limited boundaries of their social cognition and professional world. The last group (i.e., short FTP and low FA, Cell 4) can be described as happy-go-lucky actors focused on current issues, not extending their plans into the future. Life from day to day (carpe diem) is their credo. These are rough distinctions of various combinations of FTP and FA. Most often they are less clear-cut than in the figure as the dimensions are continuous. Yet, this theoretical classification may be useful for formulating hypotheses in future research.

DEVELOPMENT OF FUTURE ANXIETY

The origin of FA can be traced to personality predispositions along with personal experiences accumulated in the life course and in the current trends of human (world) history. Anxiety proneness, so much accentuated in psychology, and negative experience with the surrounding world, such as global conflicts emerging on the world's map will elicit anxious concerns. The list of possible harmful events is unlimited: ecological pollution, incurable disease, death of one's close family, social alienation, refined methods of manipulation, and control and dependency from sociopolitical systems are examples of aversive events. War is one of the most feared events.

Fears seem to increase historically, as reported by Nurmi (1987, 1988). His studies among youth in Finland showed that 11-year-old pupils expressed more fears in 1987 than 5 years earlier. The biggest proportion of the fears referred to future war. Fear of nuclear war is still one of the dominant fears among adolescents, although more recent research reports show this declining, at least among British youth (Gillies, 1989; Wilkins & Lewis, 1993). This decline is attributed by the authors to the détente atmosphere of the 1980s, inaugurated by Reagan and Gorbachev (see Lynn, 1975).

Children whose futures are naturally less secure would develop stronger FA. This refers, for example, to foster children (Molin, 1990). These children are perturbed by higher uncertainty and reduced permanency in their present life and also in the anticipated future.

Research by Susułowska (1985) focused on the contents and frequency of different fears in people aged 5–60. This research showed that fear of the future appeared at age 11–14 (2.8%), increased at age 15–19 (15.7%), has its highest frequency at age 20–29 (51.4%) and hardly appeared in older age, when the fear for their children became stronger. Her findings are relevant to our analysis for the reason that the age group 20–29 has many characteristics different from earlier stages. The personal record of previous successes and failures, the mature level of cognitive functioning, and the deeper understanding of the relationship between present actions and their future outcomes, constitute a fertile basis for incorporating FTP into one's psychological present. An effect of this developmental life stage is that the individual future horizon can contain a certain amount of anxiety.

SOCIAL AND POLITICAL DETERMINANTS OF FA

There is an open field for research on FA related to global changes in various zones of the globe. This suggestion is guided by the notion that political changes, local wars, disasters, and terrorist attacks make people look at their future with more fear and anxiety than before these events occurred. Is this anxiety related to political movements, party programs, major government plans? Until now we had at our disposal only some empirical indicators on these matters.

In this context I would like to underline an interesting contrast effect on a large social scale. Our research in the Ukraine in 1992 (Zaleski, Chlewiński, & Lens, 1994) and then the data obtained in the year 2000 (Niewiadomska, 2000) showed a relatively lower pessimism and future anxiety there than in other postcommunist countries. To interpret this finding in the first study we referred to a "jump from the bottom" effect. This nation had experienced all possible miseries in the past and when asked about the future and possible changes to come, people could think only of something more positive. Respondents believed that nothing worse than what had happened up to then could occur. Data from Bulgaria showed low FA among students (unpublished data from 2001). Bulgaria is also undergoing broad political, economic, and social transformations. Yet, there changes have less tragic character.

However, the fact that this positive inclination has been upheld over the years needs additional explanation. People who achieved independence for the first time in their history may have the feeling of a big gain (i.e., success). Despite their economic hardships, positive attitudes, and emotions, future expectations predominate and override the pessimism and negativity in other nations with different historical pasts.

At present more questions arise in this regard. Has the level of FA increased in the United States after the September 11, 2001, terrorist attack? Has the level of FA decreased in the Israeli population after Saddam Hussein's fall? Would the population of Spain show greater FA after the terrorist attack in Madrid on March 11, 2004? Are there changes in FA in countries joining the EU in 2004 and also some of them threatened by Al-Qaeda? Is the FA high or low among politicians observing the strong gap between the African postcolonial societies versus North American and European highly developed countries?

Should we be concerned about the growth of the world's population and experience FA related to increasing global population?

In some societies where there is a negative birth ratio, people are preoccupied with their retirement as there will be many old people economically dependent on a comparatively smaller number of citizens in their productive age. How can society cover the retirement pensions for so many? We need to add to this the cultural factor of family relationships. In many collectivist societies the elderly are traditionally taken care of by their children and grandchildren. But this family relationship model has changed. Nowadays, more and more often, the elderly are left to fend for themselves. They can no longer count on their offspring. When looking forward to their last period of life, the importance of this factor cannot be disregarded. Loneliness is one of the attributions that evoke anxiety, discomfort, and insecurity. It is no wonder that people with higher FA (in contrast to low) decide to bring up children in such a way that children should then care for their parents (Zaleski, 1989). If we risk making a causal link, it is plausible to assume that viewing one's own future without any help, care, assistance, or support from one's children, may enhance personal FA.

The concern for the word's and personal future may reflect factors affecting social consciousness, interests, and knowledge of what is happening in the local environment as well as on the world platform. Research showed the strong link between personal FA and (negative) expectations of solutions to global problems faced by humankind (Zaleski et al., 1994). The higher the personal FA the more pessimistic judgments of solutions to the global problems, of coping with threats and dangers.

In his well-known studies, Hofstede (1997) uses a concept of risk acceptance (vs. uncertainty avoidance) as a dimension for fruitful cross-cultural comparisons. The risk acceptance is, to some extent, a reversal concept of FA (see Zaleski, 1996). Those who do not accept a high level of risk or higher frequency of risky situations, act more cautiously, avoiding potential failures and their consequences. Uncertainty avoidance implies that people feel threatened by unclear situations and thus make them more predictable by setting up clear rules by which they live and demand others to follow.

Another question that arises is whether the cultures having lower risk acceptance are characterized by higher levels of FA. If so, would this be bound to (cultural) values, which according to Triandis (1995), constitute the structural ingredients of a given culture? Some cultures would then be described as looking at their future with more fear. Others are less fearful and more hopeful in anticipation of the future (see Averill et al., 1990). In his cross-cultural study, House (2000) reported on the basis of the GLOBE project that South African Blacks show a higher degree of future orientation than those in eastern countries, such as Zambia, Zimbabwe, and Namibia. European Americans (college students), as compared to South Korean Americans, reported greater positive affectivity and life satisfaction. Koreans reported having greater negative affectivity and depressive symptoms (Chang, Sanna, & Yang, 2003).

I would like to digress to the cultural aspect. Contrary to a commonly held stereotype about the Chinese, Chinese children have more positive general self-concepts than Dutch children (Peters, Ma, Mönks, & Ye, 1995), where self-concept is understood as not simply the sum of specific components of self but as a global image concerning the degree to which a person is satisfied with one's life and feels good about oneself (Rosenberg, 1979). It would be interesting to relate this state of life satisfaction to FA in this society.

THE IMPACT OF FA ON BEHAVIOR

As this stage of a more theoretical than elaborative discussion of the concept, some tenable assumptions can be formulated about the regulatory implications of FA. On a cognitive level we may have to block thinking about the future as a terrain of new accomplishments. More specifically, FA may lead to (a) a reduction in the subjective expectancy of positive outcomes of one's own actions, so lessening the probability of success, or (b) an attentional focus on the present time and events or escapism into the known past, both mechanisms limiting the temporal space of an individual. On the behavioral level FA can lead to (a) a passive awaiting of what might occur, (b) withdrawal from risky, open, and constructive activities, (c) keeping to routine ways and tested methods of dealing with situations encountered in life, (d) undertaking preventive activities in order to preserve the status quo rather than taking risks to increase present opportunities, (e) using different regressive-type defense mechanisms such as accusation, rationalization, or repression, in order to reduce the negative state, and (f) use of social relationships to help secure one's own future.

As in any other personality dimension, there is a question of the impact of FA on thinking, emotions, and overt behavior. Do people with high FA engage in different behavior than those with low FA? This is what is important from the psychological perspective. Does FA influence their decisions and social activity? An interesting question arises on the basis of anxiety psychology—namely, how intense should FA be to motivate people to act in a creative and preventive way, guarding and protecting themselves from possible harmful events (e.g., ecological, disease)? And the contrary, what degree of FA makes people adopt a dysfunctional mechanism of defense, of manipulation of available means in order to secure one's own future. There are some indices that shed light on this matter. The threatening future may lead to enactment of defense, as shown in experiments by Rappaport (1991). Subjects, when exposed to "threatening" lectures on world ecology, tend to constrict the future time zone. They may say to themselves "perhaps in my life these things will not happen—atomic war, petrol shortage, ecological disaster—so why should I invest my energy in preventing these calamities?" Better to be positive, even unrealistically optimistic (cf. Weinstein, 1980) and think that life is safe enough and let future generations worry about their own existence and take care of themselves. People's concerns are more limited to their present existence, to their own life standard, and Boniecki (1977) doubts whether people are concerned about the future in general. However, there are some who care for others (e.g., ecology activists). Perhaps they have developed nonego value systems and have acquired a substantially strong FA that leads them to fight for the future in ecological organizations and in different campaigns.

Using the standardized scale for measuring FA, Zaleski (1989) collected data from various groups. In one of the first studies (Zaleski, 1989), it was found that persons scoring high on FA (vs. those scoring low) declared stronger readiness to join clubs. These individuals postponed crucial and important actions for the future, and did not plan for the future in order to avoid disappointment. However, they prepare their children to take care of parents in older age and make new friends with persons who can be useful in the future. Also they are prone to foster the marriage of a daughter to someone for the sake of money rather than for love if this would provide more security for their personal future. In addition they have a high regard for power as the means to make life easier. The last results made us speculate on the particular influence that FA may have on social interactions within hierarchical settings, for example supervisor–subordinate relationships. Highly future-anxious persons may manipulate others and treat them instrumentally to secure their own, personal future, be that a relative, their own child, or a subordinate at work. Other people's welfare can be sacrificed in order to prepare a safer future for oneself. These people would use anything to decrease their uncomfortable anxiety.

FA appears to function neither according to Sarnoff and Zimbardo's (1961) hypothesis that anxiety leads to isolation from others and fear leads to affiliation, nor to the contradictory results obtained by Albas and Albas (1984). FA appears to follow another strategy for one's own welfare. This observation has been tested by the author in work settings. The guiding research question focused on supervisors' high or low FA as determinant of the power modes or influence styles used on their subordinates. Raven's (1992) conception of power, in which he proposed 11 power modes, was the theoretical basis for this research. Factor analysis grouped Raven's 11 modes into two styles of influence: a hard, and a cooperative/soft style. Data collected by Zaleski, Janson, and

Swietlicka, (1997; see also Zaleski & Janson, 2000) showed that high-anxious supervisors employ the hard style of management whereas low-anxious supervisors tend to use the cooperative, soft ways of influence.

In interpretation of these findings we assumed that supervisors who are uncertain of their personal future and anxious about future events try to make the actual stage activity more predictable. This means they demand from their subordinates well-defined actions by well-defined dates, in well-defined manners ordered by themselves. The subordinate is not given any degree of freedom as to how and when to carry out the work. If the decision were placed into the subordinate's hands, the supervisor would feel uncertain and anxious because an action conceived by the subordinate would only increase the supervisor's FA. Therefore, in order to raise the predictability of outcomes he or she will apply any means to keep the anxiety low or at least prevent its increase. One of the strategies is to arrange one's possible future according to personally imagined expectations by demanding of others concrete steps, making them act in expected ways and determining their choices to fit their own calculations.

In some cases FA may lead to cognitive and emotional blindness as a coping strategy. By this I mean "not seeing the possible danger." This blindness refers to the notion that people with strong FA may trivialize the danger, decrease its strength, cheat themselves by saying that it will not be so bad. This is a naive, wishful way of reducing the anxiety. Some analogies can be drawn from the studies on alexithymia. There are empirical reports that anxiety predicted alexithymia (Maruszewski & Scigala, 1998). If we extend this connection, it can be suggested that the inability to identify one's own negative emotion can be construed as a kind of avoidance of the trauma. Such findings support the suggestion that alexithymia may be a coping response to anxiety and negative affectivity and suggest that alexithymia might be construed as avoidance-based coping in survivors of trauma. Similarly, cutting the sight of future events (an "ostrich effect") is an erroneous strategy of protecting oneself from the consequences.

On the other hand, facing the uncertain future with "fully open eyes" may not be that comfortable. Some alleviating suggestions can refer to showing and realizing universal values common to all people (see Schwartz, 1994), to constructing goals by converging people's energy, to philosophizing the building of a secure human world. The present decade challenges such reflection particularly in economically differentiated societies.

SHOULD SMART PEOPLE FEAR THE FUTURE?

One of the particular questions refers to gifted children whose abilities, skills, and talents place them on a privileged track in school, family, and society. They stand out and are rewarded, but at the same time they are charged with others' expectations of living up to elevated standards. They should always win, score high, be the best and achieve in life, to a greater degree than the average ones or those not labeled *gifted*. If one is self-critical enough, would this social pressure not evoke relatively stronger preoccupations, as denoted by thoughts such as "Shall I reach these high standards?," "Shall I maintain the level of accomplishment that is expected of me?" The research conducted on sports figures, talented musicians, and so on, seems to support this assumption. They are not free from anxiety and tension and this is strongly bound to their motivation of being perfect. During the period of intense studies on gifted children in the years 1970–1980, Yadusky-Holahan and Holahan (1983) presented a view suggesting that gifted children may experience higher anxiety than children who are not gifted and this is attributable to self-imposed pressure and pressure to succeed imposed by parents and teachers. However their future anxiety can work in either direction—decreasing their outcomes or motivating them to still greater efforts.

Research by A. E. Gottfried (1985) led him to assume that gifted children (those who have higher achievements and higher IQ) may also be expected to evidence more positive perceptions of academic self-concept and lower academic anxiety. He also concluded "certainly, their greater academic success and mastery of cognitive tasks alone would predict better perception of competences and lower academic anxiety" (p. 14) and research confirms this (see Gottfried, Gottfried, Bathwist, & Guerin,1994, pp. 1002–1003). The literature reviewed by Schneider (1987) supported the notion that gifted children had a more positive self-concept in the cognitive and academic domains. When focusing on anxiety it is reasonable to expect a typical pattern of reversed relationship between anxiety and IQ/school achievements indicating that more capable children have lower levels of anxiety (see Hansen, 1977).

If we extend the perspective from a merely academic field to a broadened social context including future goals of becoming someone exceptional, then the assumption of enhanced FA is at least a challenging question for empirical research. The burden of talent is not light. In addition, the exposure of those gifted, via mass media, to the broader public may make the social pressure stronger.

An example, although remote, is the case of a Californian surgeon, whose fears were centered on whether he would be able to maintain his status, wealth, and reputation as a successful doctor in case he, for some reason, would not be able to perform operations anymore. "What if my hands start to tremble or shake and I cannot control the scalpel anymore," he stated (personal communication, Dec. 9 1994). This "what if" in people's minds may refer to various future conditions, events, and states that, when anticipated or considered now, do not disappear without leaving some trace of anxiety reactions related to them.

FUTURE ANXIETY AND WELL-BEING

Except for some preliminary indices, the literature lacks empirical data on the question of how future anxiety affects subjective well-being. Some indirect results can be considered in order to formulate research questions. The study by Salmela-Aro, Pennanen and Nurmi (2001) with young adolescents as well as the study by Lapierre, Leandre, Dube, Labelle, and Bastin (2001) with an elderly sample show that low subjective well-being is related to self-focused goals. This last research found negative correlations between state–trait anxiety and hope, which suggests that anxiety does affect the well-being components.

In one investigation, Zaleski, Cycoń, and Kurc (2001) looked at the links between FTP and well-being measured on the following criteria: emotional distress and hopelessness, life sense and emotional equilibrium, interpersonal intimacy, social contacts and competence, and realism and persistence. The data collected from high school students showed negative relationship between planning for the distant future and emotional distress and hopelessness and positive links with emotional equilibrium and feeling of sense of life. Also those with a longer temporal perspective experienced better interpersonal intimacy and social contacts. These results permit the assertion that those who do not fear the distant future, for whom the unknown future is not a source of anxiety, have a better life quality and are happier. This echoes Beck, Weissman, Lester, and Trexler's (1974) study showing that having goals acts as a buffer against depression that may accompany FA. As people strive for a happy life, FA is one of the handicaps along the way. More such aspects and questions appear as a challenge for further examination of this phenomenon (cf. Zaleski, 2002).

I close this chapter in a similar way as I began it, by saying that the current development of the political scene in various parts of the globe evokes future concern. The present time (year 2004) offers many such (vicious) "opportunities." A possible terrorist attack, tension between countries, new diseases (e.g., SARS) are the sources of anxiety. The future, however possibly good and happy, is, nevertheless, uncertain.

REFERENCES

Albas, D., & Albas, C. (1984). *Student life and exams: Stress and coping strategies*. Dubuque, IA: Kendall-Hunt.

Averill, J., Catlin, G., & Chon, K. (1990). *The rules of hope*. New York: Springer.

Bandura, A. (1991). Self-efficacy conception of anxiety. In R. Schwarzer & R. Wicklund (Eds.), *Anxiety and self-focused attention* (pp. 89–110). Chur, Switzerland: Harwood Academic Press.

Beck, A., Weissman, A., Lester, D., & Trexler, L. (1974). The measurement of pessimism: The Hopelessness Scale. *Journal of Clinical Psychology, 42,* 861–865.

Boniecki, G. (1977). Is man interested in his future? The psychological question of our times. *International Journal of Psychology, 12,* 59–64.

Borkovec, T. D., Metzger, R. L., & Pruzinsky, T. (1986). Anxiety, worry and the self. In L. M. Hartman & K. R. Blankstein (Eds.), *Perception of self in emotional disorder and psychotherapy.* New York: Plenum Press.

Chang, E., Sanna, L., & Yang, K. (2003). Optimism, pessimism, affectivity, and psychological adjustment in US and Korea: A test of meditation model. *Personality and Individual Differences, 34,* 1195–1208.

Czapiński, J., & Peters, G. (1991). The onion theory of happiness: basic concepts and cross-cultural test. In N. Bleichrot & P.-J.-D. Drenth (Eds.), *Contemporary issues in cross-cultural psychology* (pp. 196–216). Amsterdam: Swets & Zeitlinger.

Eysenck, M. (1992). *Anxiety: The cognitive perspective.* Hillsdale, NJ: Lawrence Erlbaum Associates.

Gandeza, L. (1986). The motivational significance of future time perspective among college students. *Saint Louis Research Journal, 17,* 19–32.

Gillies, P. A. (1989). A longitudinal study of the hopes and worries of adolescents. *Journal of Adolescence, 12,* 69–81.

Gottfried, A. E. (1985). Measures of socioeconomic status in child development research: Data and recommendations. *Merrill–Palmer Quarterly, 31,* 85–92.

Gottfried, A. W., Gottfried, A. E., Bathwist, K., & Guerin, D. W. (1994). *Gifted IQ: Longitudinal study.* New York: Plenum Press.

Hansen, R. A. (1977). Anxiety. In S. Ball (Ed.), *Motivation in education* (pp. 91–109). New York: Academic Press.

Heider, E. (1958). *The psychology of interpersonal relations.* New York: Wiley.

Hoffstede, G. (1997). *Cultures and organizations: Software of the mind.* New York: McGraw-Hill.

House, R. (Ed.). (2000). *Cultural influences on leadership and organizations.* Manuscript submitted for publication.

Kasser, T., & Ryan, R. M. (1996). Further examining the American dream: Differential correlates of intrinsic and extrinsic goals. *Personality and Social Psychology Bulletin, 22,* 280–287.

Kasser, T., & Ryan, R. M. (2001). Be careful what you wish for: Optimal functioning and the relative attainment of intrinsic and extrinsic goals. In P. Schmuck & K. Sheldon (Eds.), *Life goals and well-being: Towards a positive psychology of human striving* (pp. 116–131). Göttingen, Germany: Hogrefe & Huber.

Kruglanski, A. (1988). Contextual effects in hypothesis testing: The role of competing alternatives and epistemic motivations. *Social Cognition, 6,* 1–20.

Kruglanski, A., & Webster, D. (1996). Motivated closing of the mind: "Seizing" and "freezing." *Psychological Review, 103,* 263–283.

Lang, E., Shapiro, D., & Cobb, L. (1991). Modification of psychological and subjective responses to stress through heart rate and skin conductance biofeedback. In C. Spielberger, I. Sarason, Z. Kulcsar, & G. Van Heck (Eds.), *Stress and emotion* (Vol. 14, pp. 167–177). New York: Hemisphere.

Lapierre, S., Leandre, B., Dube, M., Labelle, R., & Bastin, E. (2001). Aspirations and well-being in old age. In P. Schmuck & K. Sheldon (Eds.), *Life goals and well-being: Towards a positive psychology of human striving* (102–115). Göttingen, Germany: Hogrefe.

Locke, E., & Latham, G. (1990). *A theory of goal setting and task performance.* Englewood Cliffs, NJ: Prentice-Hall.

Lynn, R. (1975). National differences in anxiety 1935–65. In I. Sarason & C. Spielberger (Eds.), *Stress and anxiety* (Vol. 2, pp. 257–274). Washington, DC: Hemisphere.

MacLeod, A., Williams, J. M., & Bekerian, D. A. (1991). Worry is reasonable: The role of explanations in pessimism about future personal events. *Journal of Abnormal Psychology, 100,* 478–486.

Markus, H., & Kitayama, S. (1991). Culture and the self: Implication for cognition, emotion and motivation. *Psychological Review, 98,* 224–253.

Maruszewski, T., & Scigala, E. (1998). *Emocje—aleksytymia—poznanie* [Emotion—alexithymia—cognition]. Poznan, Poland: Humaniora.

Molin, R. (1990). Future anxiety: Clinical issues of children in the latter phases of foster care. *Child and Adolescent Social Work Journal, 7,* 501–512.

Niewiadomska, I. (2000). Człowiek przełomu tysiąclecia. Problematyka leku przed przyszłoscią u młodzieży [Man at the brake of the millenium. Problem of future anxiety in youth]. In J. Makselon & B. Soiński (Eds.), *Człowiek przełomu tysiąclecia. Problemy psychologiczne* [Man at the brake of the millennia: Psychological problems.] (pp. 153-165). Kraków, Poland: Wydawnictwo Naukowe PAT.

Nowak, A., & Vallacher, R. (1998). *Dynamic social psychology.* New York: Guilford Press.

Nurmi, J.-E. (1987). Age, sex, social class, and quality of family interaction as determinants of adolescents' future: A development task interpretation. *Adolescence, 22,* 977–991.

Nurmi, J.-E. (1988). Experience of threat of war among Finnish adolescents: Effects of thinking about the future, and comparison of methods. *Medicine and War, 4,* 199–210.

Nuttin, J. (1985). *Future time perspective and motivation: Theory and research method.* Hillsdale, NJ: Lawrence Erlbaum Associates.

Peters, W., Ma, H., Mönks, F., & Ye, G. (1995). Self-concept of Chinese and Dutch gifted and nongifted children. In M. Katzako & F. Monks (Eds.), *Nurturing talent: Individual needs and social ability* (pp. 84–95). Proceedings of the Fourth Conference of the European Council for High Ability. Assen. The Netherlands: Van Gorcum.

Rappaport, H. (1991). Measuring defensiveness against future anxiety: Telepression. *Current Psychology Research and Reviews, 10,* 65–77.

Raven, B. (1992). Power interaction model of interpersonal influence: French & Raven thirty years later. *Journal of Social Behavior and Personality, 7,* 217–244.

Rosenberg, M. (1979). *Conceiving the self.* New York: The Free Press.

Ryan, R. M. (1995). Psychological needs and the facilitation of integrative processes. *Journal of Personality, 63,* 397–427.

Salmela-Aro, K., Pennanen, R., & Nurmi, J. E. (2001). Self-focused goals: What They Are, How They Function, And How They Relate To Well-Being. In P. Schmuck & K. Sheldon (Eds.), *Life Goals and well-being: Towards a positive psychology of human striving* (148–166). Göttingen, Germany: Hogrefe & Huber.

Sarnoff, L., & Zimbardo, P. (1961). Anxiety, fear and social affiliation. *Journal of Abnormal and Social Psychology, 62,* 356–363.

Schneider, B. H. (1987). *The gifted child in peer group perspective.* New York: Springer-Verlag.

Schwartz, B. (1994). *The cost of living: How market freedom erodes the best things in life.* New York: Norton.

Sheldon, K. M., & Kasser, T. (1995). Coherence and congruence: Two aspects of personality integration. *Journal of Personality and Social Psychology, 68,* 531–543.

Sorrentino, R. M., Roney, C. J., & Hanna, S. E. (1992). Uncertainty orientation. In C. Smith (Ed.), *Motivation and personality: Handbook of thematic content analysis* (pp. 419–427). Cambridge, England: Cambridge University Press.

Susułowska, M. (1985). Tresć lęków w przebiegu życia ludzkiego [Fear contents throughout people's life span]. *Przegląd Psychologiczny [Psychological Review], 28,* 1065–1085.

Thiebaut, E. (1998). La perspective temporelle, un concept a la recherché d'une definition operationnelle. [The temporal perspective, the concept in research for operational definition]. *L'Annee Psychologique, 98,* 101–125.

Triandis, H. (1995). *Individualism and collectivism.* Boulder, CO: Westview Press.

Vallacher, R. R., & Wegner, D. M. (1985). *A theory of action identification.* Hillsdale, NJ: Lawrence Erlbaum Associates.

Weinstein, N. (1980). Unrealistic optimism about future life events. *Journal of Personality and Social Psychology, 39,* 806–820.

Wilkins, R., & Lewis, C. (1993). Sex and drugs and nuclear war: Secular, developmental and Type A influences upon adolescents' fears of the nuclear threat, AIDS and drug addiction. *Journal of Adolescence, 16,* 23–41.

Yadusky-Holahan, M., & Holahan, W. (1983). The effect of academic stress upon the anxiety and depression levels of gifted high school students. *Gifted Child Quarterly, 27,* 42–46.

Zaleski, Z. (1987). Behavioral effects of self-set goals for different time ranges. *International Journal of Psychology, 22,* 17–38.

Zaleski, Z. (1989). Lęk przed przyszłoscią. Ramy teoretyczne i wstępne dane empiryczne. [The fear of the future: Theoretical frame and preliminary empirical data]. In A. Januszewski, Z. Uchnast, & T. Witkowski (Eds.), *Wykłady z psychologii w KUL* (pp. 167–184). Lublin, Poland: RW KUL.

Zaleski, Z. (Ed.). (1994). *Psychology of future orientation.* Lublin, Poland: TN KUL.

Zaleski, Z. (1996). Future anxiety: Concept, measurement and preliminary research. *Personality and Individual Differences, 21,* 165–174.

Zaleski, Z. (2002). Future horizon: A challenging concept for psychology. In J. Trempała & L.-E. Malmberg (Eds.), *Adolescents' future-orientation: Theory and research* (pp. 123–129). Frankfurt am Main, Germany: Peter Lang GmbH.

Zaleski, Z., Chlewiński, Z., & Lens, W. (1994). Importance of and optimism-pessimism in predicting solution to world problems: an intercultural study. In Z. Zaleski (Ed.) *Psychology of future orientation* (pp. 207–228). Lublin, Poland: TN KUL.

Zaleski, Z., Cycoń, A., & Kurc, A. (2001). Future time perspective and subjective well-being in adolescent samples. In P. Schmuck & K. Sheldon (Eds.), Life goals and well-being: Towards a positive psychology of human striving (pp. 58–67). Göttingen, Germany: Hogrefe & Huber.

Zaleski, Z., & Janson, M. (2000). Effect of future anxiety and locus of control on power strategies used by military and civilian supervisors. *Studia Psychologica, 42,* 87–95.

Zaleski, Z., Janson, M., & Swietlicka, D. (1997). Influence strategies used by military and civil supervisors: Empirical research. *Polish Psychological Bulletin, 28,* 325–332.

8

Personal Goals and Time Travel: How Are Future Places Visited, and Is It Worth It?

Kennon M. Sheldon
University of Missouri

Maarten Vansteenkiste
University of Leuven, Belgium

TIME TRAVEL, INTENTIONAL ACTIVITY, (NEG)ENTROPY, AND DISSIPATIVE SYSTEMS

Time Travel. We would like to start with a metaphor. The metaphor is: Personal goals offer a means of *traveling through time*. By this we do not exactly mean the science fiction plot device, in which time travel involves "jumping" to the chronological past or future then returning to the chronological present, without "doing the time" in between. Personal goals offer a means of steady forward motion through time, a means of *doing* the time, such that one arrives at a future place that is to one's liking. By investing and directing one's energy in stable and purposeful ways one may literally create the future, thereby building desirable life structures of one's choosing. We think that this metaphor can help illustrate two important ideas.

First is the notion that through goal-directed behavior, we can move toward greater personal coherence and complexity over time. In other words, personal goals offer a way to create new life configurations for ourselves, that is, elaborated cognitive, familial, social, and vocational networks that would not have happened otherwise. Because of the obvious importance of this intentional process, much of this chapter focuses on the factors that support personal goal attainment and thus successful time travel.

The second idea that the "traveling through time" metaphor conveys is that intentional behavior can make of life a *trip*—to future places that we have decided to try to visit, without always knowing what they will be like when we get there. To use a geographical metaphor, committing to a particular personal goal is like starting out on a journey, to a new country or continent that one has never seen. In terms of the time travel metaphor, it is like selecting a particular future (i.e., a target event or situation in the year 2015) that one will try to visit, hoping that this "place" will be as interesting and positive as one anticipated. However, not every goal commitment is likely to bring one to the most satisfactory possible future, even if one succeeds in attaining those goals. In other words, not all of the places we might succeed in visiting are nice places, once we get there! Because of the obvious importance of this issue, much of this chapter focuses on the factors that moderate the influence of goal attainment upon positive outcomes such as performance, adaptation, and psychological well-being.

(Neg)entropy and Intentional Activity. A common theme underlying both of these interpretations of time-travel is that of *making something happen*, while resisting the influences that might deter or detain one from the intended path. In more abstract terms, "traveling through time" involves both creating or approaching *negentropy*, that is, organized patterns of functioning or circumstances, as well as avoiding or negating *entropy*, that is, the disorganization that might be wrought by random or uncontrollable factors. Because of the (neg)entropy concept's potential relevance for understanding human behavior in its temporal context, and also because of its potential for helping psychology achieve greater consilience with the physical sciences (Wilson, 1998), it is worth considering the concept in more detail.

Entropy is the degree of disorder or randomness that exists within a system or environment. Entropy can be conceptualized and measured in many different ways, but the way most relevant for this chapter is to think about it in temporal and probabilistical terms—namely as the tendency for events to occur randomly and haphazardly, so that particular prespecified possible futures (i.e., double sixes in a dice throw) are unlikely to result just by chance. In contrast, temporal *neg*entropy exists when the sequence of events is highly structured and sequenced, such that a difficult or unlikely future condition is achieved after all.

To illustrate the concept of temporal negentropy in greater detail, consider the following branching corridor metaphor. Suppose that starting from any given present moment, there is a very large (and perhaps infinite) number of possible event chains branching off, leading to the set of "all possible futures." Any particular future or set of futures is only a small subset of the total set of potential futures—just one particular corridor that reality might or might not take. In stochastic or probabilistic terms, this means that particular possible futures tend to be particularly unlikely, without the intervention of some correspondingly unlikely set of intervening events or circumstances. To use a quantum mechanics analogy, reality is like a wave function, a manifold buzz of possibilities, awaiting final determination at each successive moment. According to the Copenhagen interpretation of the quantum mechanical equations (Feynman, Leighton, & Sands, 1995), the particular configuration of particles observed at time *t* depends, in large part, on when and how the observer makes the observation. The act of making the observation itself serves as the singular or unlikely event that collapses the wave function, elevating one possibility to actuality.

What are the specific events or circumstances that "collapse the wave function" at the level of the individual, affecting life events such that particular corridors of possible futures are entered, as life proceeds? This is doubtless determined by many factors, including personal history, geographic location, sociocultural norms, and the like (Sheldon, 2004), but herein, we suggest that one important factor determining final life courses is people's *personal goals and intentional activities*. In other words, one way to arrive at a particular (preferred) future is to try to intentionally create the new connections, contingencies, and coalitions that promise to lead one toward that future. In terms of the branching corridor metaphor, it is not easy to arrive at a particular possible future, because there are so few corridors or event chains that lead in that direction, relative to the set of all possible event chains. Still, if anything can preselect and then enact particular event chains, human intentional systems can.

Dissipative Systems. To further illustrate the nature of self-organizing intentional systems, it is useful to consider the concept of a *dissipative system* (Glansdorff & Prigogine, 1971). Dissipative systems were first discovered in the domain of chemical thermodynamics, and have since been demonstrated within the domains of biology, sociology, and culture as well (Jantsch, 1981).

They are functional systems consisting of many elements that maintain themselves far from equilibrium (i.e., in an advanced state of order) by neutralizing environmental perturbations to their structure (i.e., they dissipate entropy) through the application of coping mechanisms and compensatory procedures. In other words, dissipative systems develop the means to keep themselves at their current (high) level of order.

However, an important characteristic of dissipative systems is that they can not only counteract entropy, but also reduce it by "jumping" to a new, even more organized mode of functioning (i.e., they achieve negentropy), often in response to unusual threats or stresses. By coping with such challenges, dissipative systems may create new order within themselves, and within their circumstances. As might be obvious, living things are in many ways the *sine qua non* example of dissipative systems, maintaining themselves far from thermodynamic equilibrium, and at times also further elaborating and developing their structure. This can be seen at many levels of analysis regarding living things, ranging from the level of biological processes (i.e., self-maintaining and self-enhancing cellular systems) to the level of cognitive processes (i.e., self-maintaining and self-enhancing belief systems), and we believe that motivational goal systems also fall into this category.

To summarize, we believe that the dissipative systems idea can capture many characteristics of the active human striver, struggling to stay on course toward particular goals or "possible futures," in the face of influences and setbacks that might dis-entrain her from that course. By attempting to travel (rather than drift) through time, proactive humans may intentionally avoid temporal entropy or disorganization and may achieve greater temporal negentropy or structure, elaborating themselves and their skills in the process.

Having laid out the meta-theoretical foundation of our goal approach, we can now rephrase our two initial questions before tackling them in more detail. First, what factors enable the active human striver to dissipate entropy and create temporal negentropy? Or, in more common language, what factors enable strivers to overcome setbacks and generate creative new approaches to problems, so that they arrive at the intended possible futures? Second, if the human striver manages to attain the self-set possible futures, what factors determine whether the attained possible futures bring about the desired beneficial effects, or whether instead they turn out to be "mistakes," trips the person perhaps should not have taken? In other words, why does not every movement toward negentropy bring along the anticipated satisfaction, and which factors might moderate the relation between movement and satisfaction? In order to answer those two questions, it is helpful to distinguish between two somewhat different conceptualizations of optimal functioning.

Systemic Versus Organismic Perspectives on Optimal Functioning

Scrutiny of the literature on action, goals, and motivation reveals two major traditions, namely the systemic and organismic perspectives on optimal functioning (Sheldon & Kasser, 1995). Much can be learned concerning our questions by considering the difference between them. *Systemic perspectives* define optimal functioning in terms of the coherent and efficient operation of the person's behavioral or action systems. In other words, these theories focus on the structural coherence of action, such that a larger "*quantity*" of successful action is achieved. These perspectives include cognitive-behavioral, cybernetic, general systems, and functional theories. In contrast, *organismic perspectives* define optimal functioning in terms of the satisfaction of psychological needs and the attainment of personal growth and self-actualization. In other words, they focus on the congruence of the action and goal with deeper or more creative and resourceful parts of the self, such that action is of a higher "*quality*." These perspectives include humanistic, existential, spiritual, and some developmental and psychodynamic theories (Deci & Ryan, 2000; Sheldon, Williams, & Joiner, 2003). Next, we consider the systemic coherence perspective in detail, in order to demonstrate its importance.

Systemic Perspectives. Carver and Scheier's (1981, 1990, 1998) control theory model provides a good framework from which to consider systemic perspectives upon optimal goal functioning. This model depicts human behavior as underlain by a hierarchically ordered system of action units, in which lower level units exist, ideally, in functional relation to higher level units. Higher levels are more stable, abstract, and global, whereas lower levels are more transient, concrete, and local. Ideally, higher level goals set the standards for behavior, which is undertaken in order to reduce discrepancies between present situations and the desired future goals or preferences. Via processes of both negative feedback (using experience as information to adjust behavior to better reduce discrepancies) and positive feedback (setting new goals thereby creating new discrepancies to try to reduce), people move forward in time.

Let's consider an example. At a particular moment of time, we find a student driving in his car. What is happening? If we asked him what he is doing, he might say: "I'm going to the library, to get a book." According to Carver and Scheier, the person's behavior is being controlled by a goal ("get the book"). An elaborated sequence of more specific behaviors is being emitted in order to serve this goal (i.e., the person got in the car, backed out of the driveway, turned left, stopped at the stoplight, etc.). In the process, small discrepancies are steadily eliminated (i.e., the discrepancy between being on Elm Street and wanting to be on Oak Street), so that the larger discrepancy (between not having the book, and having the book) can be eliminated.

The aforementioned concerns the concrete means by which the goal is reached, that is, the "how" of retrieving the book. But we also want to know, *why* is the student retrieving the book? Control theory answers this question by looking further "up" in the action hierarchy. Thus, it may be that the person was getting the book in order to serve the higher level goal of "getting an A on my paper," which in turn serves the higher level goal of "earning a high undergraduate GPA," which in turn serves the higher level goal of "getting into medical school and becoming a doctor." In Carver and Scheier's model all of these levels of behavior can be operating at once, as expressed by every concrete behavior emitted within the sequential stream of behavior.

Notice that from the control theory perspective, the book-retrieving behavior is only intelligible within its broadest context, that is, when it is understood with relevance to the person's overarching life goals (i.e., of becoming a doctor). Also, control theory implies that people will be most motivated to engage in a particular action when it is seen as serving the most longer term or higher order goals (see Sheldon & Kasser, 1995, 1998, for supporting data). This assumption locates the theory within a family of theories that posit that the amount of motivation to emit a concrete behavior depends on the perceived instrumentality (Raynor, 1981) or the perceived utility value (Eccles & Wigfield, 2002) of the behavior for long-term goals. Finally, to return to our meta-theoretical suppositions, notice that successful functioning of the action system can serve to induce much temporal negentropy: It is very unlikely that the student will simply "find himself" in the branch of the future in which he is a doctor, unless he has invested long-term intentional effort into the project.

Starting from this foundation, one can identify a number of potential problems that might occur within an action system, which might prevent a person from making adequate forward progress toward desired futures. In terms of the first question raised in the introduction of this chapter, control theory does an especially good job in illuminating the factors that facilitate versus hinder successful time travel. First, the person may have *insufficient skills* or abilities relevant to goal pursuit. For example, a person seeking to go to the library may not be able to drive, and thus may have no way of getting there. In terms of Carver and Scheier's model, necessary lower level skills or action units may not be in place, so that the discrepancy between current state (no book) and the goal (get book) cannot be reduced. As one illustration of the importance of this factor, Sheldon and Kasser (1998) showed that those with stronger life skills (such as the abilities to delay gratification, to perceive and follow social norms, and to use emotions as information) better attained their personal goals over a 2-month period.

Second, it is also possible to *have* a relevant skill or ability, but *not apply* that skill. For example, the aforementioned nondriver may have the ability to find and follow a bus schedule, or to call up an acquaintance to ask for a favor, or to ride her bicycle, but she may not think of these action alternatives. In Sheldon and Kasser's (1995) terminology, a person may not have "vertical coherence" in his or her action system, such that lower and higher level action units fail to contact each other, even though a functional relation is possible. Vertical incoherence may exist not only at lower levels of the action system, such that relatively simple and concrete daily goals are not achieved, but also at higher levels of the action system, such that the person's entire life plan lacks coherence. For example, a person's daily goals and behaviors might be irrelevant to, or even conflict with, his or her broader ambitions. To return to the aspiring doctor example, the person might spend much of his time trying to beat a particular game on his computer, an activity unlikely to serve any of his longer term goals.

Third, besides potential inconsistency or conflict between goals at different levels of the action system (vertical incoherence), there can also be inconsistency or conflict between goals at the *same* level of the action system (Emmons & King, 1988; termed "horizontal incoherence" by Sheldon & Kasser, 1995). For example, the person driving to the library might also need to complete a calculus assignment that day, finish moving out of his old dorm room, discuss and resolve an issue with his girlfriend, and coordinate a fraternity project, goals that directly conflict with each other because of time and other resource limitations (Wilensky, 1983). He may have to give up one or more of these ambitions.

In sum, systemic perspectives upon optimal goal functioning can make a number of suggestions and prescriptions for how people may best "travel through time." First, people should develop as many skills as possible that can be used to achieve their self-set goals and to attain greater negentropy over time. Second, they should practice recognizing when their skills are relevant to problems at hand, so that discrepancies can be reduced whenever possible. Hopefully, concrete skills and short-term goals will directly serve longer term goals, such that the person can make maximal progress and thus maximally influence which possible futures come to be (vertical coherence). Third, it is good when one's various projects are consistent with each other, so that working on one does not take away from another. Better yet is when one's projects mutually reinforce each other—as when one can simultaneously pursue the goals of obtaining the library book, visiting a specialty store, and renewing one's driver's license, all in one trip to the other side of town. Here, much is accomplished, via a relatively small overall investment of time and energy (horizontal coherence).

See Heckhausen (2002) or Shah and Kruglanski (2000) for other summaries of contemporary systemic approaches, and also for further consideration of the many positive effects of systemic integration upon life outcomes.

In addition to indicating how goal attainment can be facilitated, Carver and Scheier's control theory also has relevance for the second question raised in the introduction of this chapter, namely, whether goal attainment yields well-being effects. Carver and Scheier assert that people are happier to the extent that they are achieving a larger quantity of goals or, to put it in temporal terms, if they are "moving more quickly" through time. More specifically, people are happiest when their actual goal progress most exceeds their own "meta-goal" regarding the desired *rate* of progress (Carver & Scheier, 1990). To illustrate, if the student expects to spend 5 years studying to become a doctor, but he finishes his studies in 4 years, he will be happier than if he finishes in the expected 5 years. From this perspective, the key to happiness is to get as much done as quickly as possible; goal attainment will lead to well-being, *if* it occurs faster than the person expected it to.

Readers may have observed that this analysis suggests that the optimal striver is like an "achievement freak," working constantly to make huge numbers of things happen. In terms of the metaphors developed earlier, such a highly active person travels the farthest distance in time, visiting the most new and/or difficult temporal countries, while also creating the most negentropy in her wake. But is this really the ideal model? What if the person is hyperactive, a slave to her compulsions, who seems to lack all balance in life, or who seems to have cockeyed values and priorities? This is where the organismic perspective comes in.

Organismic Perspectives. Again, systemic perspectives on optimal goal functioning primarily concern efficient and effective functioning by the action system, such that many goals are achieved in short order. However, intuition suggests that other factors matter, also. For example, *what* are the goals that the person is striving for? Are some types or contents of goals (such as for personal growth, emotional intimacy, or community contribution) more healthy or beneficial than other goal types or contents (such as goals for material success, social popularity, or an attractive image)? Also, *why*, or for what personal reasons, is the person striving for the goals? Does he feel a sense of pressure or a sense of being controlled by external forces, or does he instead feel that the goals are intrinsically interesting, valuable, and important to him? To return to the time travel metaphor, *what* is the nature of the future place the person is trying to visit, and, *why* is the person trying to visit it? Sheldon and Kasser (1995) defined these factors as relevant for two different types of "organismic congruence" (see also

Deci & Ryan, 2000, or Sheldon, Ryan, Kasser, & Deci, 2004). As we hope to show, both the content and the motive factors impact both people's goal attainment (i.e., successful time travel), and the degree to which goal attainment produces well-being (i.e., whether the "trip" has been worthwhile). We first consider the content of goals, in the following subsections.

The "What" of Goals. Considerable research now demonstrates that the particular types of goals a person pursues have implications for her happiness or well-being (Emmons, 1991; Pohlman, 2001). For example, Kasser and Ryan (1993) showed that there may be a "dark side" to the American dream, or at least to one version of the dream—the ideal of attaining wealth and luxury. Specifically, participants who pursued financial success (an "extrinsic" goal) more strongly than they pursued emotional intimacy, community feeling, and personal growth (three "intrinsic" goals) reported lower well-being, and also showed more evidence of psychopathology during clinical interviews. Notably, in the studies discussed here and later, well-being is most often measured in terms of high positive affect, high life satisfaction, and low negative affect.

Kasser and Ryan (1996) extended the extrinsic-goal concept, showing that those who strongly value social popularity and physical attractiveness (two more purported extrinsic goals) also evidenced lower well-being. Notably, factor analyses supported the idea that money, beauty, and popularity goals tend to go together, whereas intimacy, growth, and community goals tend to go together (Kasser & Ryan, 1996). Thus, Kasser and Ryan's conceptual distinction between the two kinds of goal content is also echoed in the empirical data.

Sheldon and Kasser (1998) extended these cross-sectional results by conducting a 2-month-long study of goal attainment and its effect on changes in well-being. They showed that attaining goals strongly linked to the three extrinsic "possible futures" listed earlier had no effect upon increasing well-being, whereas attaining goals that were helpfully linked to the three intrinsic futures *did* have a positive effect upon longitudinal well-being. These findings suggest that some types of vertical coherence may be more beneficial than others—it depends on to *what* higher level goals the person's lower level goals are functionally linked. Again, this issue is typically not represented within purely systemic perspectives upon optimal functioning, which tend to assume that "all goals are created equal" (Ryan, Sheldon, Kasser, & Deci, 1996).

Notably, the association between strong extrinsic valuing and negative well-being is not large, and thus, people who pursue money, fame, and/or beauty harder than they pursue growth, intimacy, and community are not necessarily unhappy. Still, it appears that such orientations serve as risk factors for unhappiness, or for less than complete happiness.

What is the source of the content-to-well-being effects? Deci and Ryan (1985, 2000) theorize that the overpursuit of extrinsic goals may fail to meet universal organismic needs, such as the three needs proposed by self-determination theory: for autonomy, competence, and relatedness (Deci & Ryan, 2000; Ryan & Deci, 2000; Sheldon, Elliot, Kim, & Kasser, 2001). For example, a person who is fixated on acquiring wealth or fame may ignore deeper interests and passions, and often feel pressure and tension to perform behaviors he or she does not really want to do (reduced autonomy); also, he or she may often focus on the possibility of negative evaluations by others, or may objectify others, perceiving them as means to self-centered ends (reduced relatedness); and finally, he or she may feel that wealth or fame outcomes are difficult to attain, and may feel little control over such outcomes (reduced competence; see Kasser, 2002, for a more detailed discussion of these dynamics). According to self-determination theory and considerable empirical data (Sheldon & Elliot, 1999; Sheldon, Ryan, & Reis, 1996; Sheldon et al., 2001), autonomy, competence, and relatedness are all important for well-being, and the evidence suggests that the process of pursuing extrinsic goals may yield fewer such experiences, on average. This claim was again confirmed in a recent set of empirical studies, showing that employees who value extrinsic work goals over intrinsic work goals display lower positive well-being and more symptoms of burnout due to a lack of experienced need satisfaction on the work floor (Vansteenkiste, De Witte, & Deci, 2003).

The "Why" of Goals. The second issue relevant to organismic congruence concerns the "why" of goals—that is, the quality of the reasons for which people pursue goals (Sheldon & Kasser, 1995). Is the person acting because the actions express his or her developing interests and core values, or is he or she instead acting with a sense of being controlled by external constraints or internal compulsions? As we show later, this question also has considerable relevance for psychological need satisfaction (Sheldon & Elliot, 1999; Sheldon et al., 2004).

To illustrate the many possible kinds of dynamic motives for pursuing a goal, let us return to the "going to get a book at the library" example. The question is, *why*, personally, does the person in our example want to become a doctor? There are a variety of possibilities: for example, because he sees this goal as a route to external rewards, or as something that his parents expect or demand, as something that will prove his worth and value to others, or as something that will express his personal values and ideals. Or, perhaps, he simply likes learning about medical issues!

The latter reason for acting is not well handled by systemic perspectives, because such perspectives assume that behaviour is always initiated and controlled by some higher level or more abstract goal. But as in the "pleasure of learning" example just mentioned, many behaviors are engaged in simply because they yield enjoyable experiences in the present, not because they are immediately instrumental to some longer term goal. Self-determination theory calls enjoyment-based reasons for acting "intrinsic motivation," and has demonstrated that intrinsic motivation has strong positive effects on people's mood and state of mind, even though they may be largely irrelevant to higher level goals (Deci & Ryan, 2000). Notably, intrinsic motivations for acting are conceptualized as distinct from intrinsic types of values, although intrinsic values and intrinsic motivations typically covary.

Are all instrumental or extrinsic motivations problematic, to be avoided, according to self-determination theory? No—again, it depends on the type of motivation represented. In addition to focusing on intrinsically motivated reasons for acting, self-determination theory also distinguishes between three types of extrinsic reasons for acting: external, introjected, and identified. These reasons vary in their degree of felt self-determination.

External motivation (striving because external factors seem to "force" one to) is least self-determined, because the person feels no ownership at all of the action. Introjected motivation (striving because one would feel ashamed or guilty if one did not) is slightly more self-determined, because the compulsion is now inside the person, although it is not yet inside the self. Identified motivation (striving because it expresses one's values and ideals, even if the behavior is not enjoyable) is mostly self-determined, because the internal force is now self-congruent. External, introjected, and identified motivations are all classified as extrinsic motivations, but identified motivation (like intrinsic motivation) is nevertheless viewed as an autonomous or self-determined form of motivation because the person feels a full sense of choice and volition in the behavior. In contrast, external and introjected motivations are defined as controlled forms of motivation, because the person feels a sense of compulsion in the behavior.

Although considerable research demonstrates that controlled (external and introjected) forms of motivation tend to have negative effects upon performance and well-being outcomes, whereas autonomous (identified and intrinsic) motives tend to have positive effects (see Sheldon, 2002), this fact is not easily represented within systemic and "quantity"-oriented theories. Whereas organismic perspectives consider the why of motivation in terms of the quality of felt volition for a goal, systemic perspectives address the why question by analyzing the degree of helpful linkage between a goal and the higher level goals in the person's system. We suggest that these are two equally important, but quite different, ways of considering "why" a person behaves.

The Self-Concordance Concept. Sheldon and colleagues (Sheldon & Elliot, 1999; Sheldon & Houser-Marko, 2001; Sheldon, 2002) have referred to the feeling that one's goals are caused by the self as "self-concordance" and have operationally defined self-concordance as identified motivation plus intrinsic motivation minus external motivation minus introjected motivation. Thus, self-concordance is, in one sense, just another label for autonomous or self-determined. However, Sheldon (2002) has argued that the concept has an additional or extended meaning in the case of self-generated personal goals and initiatives. Specifically, pursuing one's proactive goals in a self-determined manner implies that one has likely chosen the goals based on one's deeper values, interests, needs, and personality preferences rather than on external forces or nonintegrated parts of the self. In other words, self-concordance is thought to measure the "fit" between self and personality, that is, between people's conscious objectives and their underlying dispositions and organismic needs.

From this perspective, the self's ability to correctly intuit the interests, needs, and values of its underlying organism, and also the ability to select and act upon goals that are consistent with these, are important skills that are highly relevant to successful "time travel." Self-concordance is a developmental achievement, in which the person has learned to observe his or her proactive growth and health impulses, and to give them voice. The idea is that people have native gifts and inclinations that they might be advised to try to recognize and develop, rather than blindly following parental, societal, or outgrown self-expectations.

For example, suppose that the student, discussed previously, is trying to become a doctor primarily for the money, or because that is what his parents insist upon, or because he hopes that this will finally prove his intelligence, or because he feels strong shame at the thought of failing. In other words, he is striving for non-self-concordant reasons. What is the problem with this? First, he may never get there; the data show that, on average, goal attainment (and thus successful time travel) is less likely to occur when motivation is non-self-concordant (see Sheldon & Elliot, 1999). Indeed, it seems logical that people would be more resistant to investing sustained effort in goal pursuits that they feel pressured to pursue or that are not experienced as fitting with their underlying needs and preferences; instead they would tend to give them up, just as people usually give up New Year's resolutions. Second, even if the person manages to reach the goal, he may end up regretting his choice (or, from the existential perspective, regretting his failure to make a more *authentic* choice; see Sheldon & Elliot, 1999, and Sheldon & Kasser, 1998, for supportive data). In terms of the time travel metaphor, he may arrive at a place in the future from which it is difficult to backtrack, but that, again, is not as satisfying as would be desired. Perhaps he should have pursued some other career besides medicine, such as his recurring idea of attending graduate school in cultural anthropology?

In sum, the organismic perspective on optimal goal functioning suggests that it is very important to consider not just *how effectively* or *quickly* a person is striving, as suggested by control theory and utility theories more generally, but also, toward *what* a person is striving, and *why* he or she is striving for it (King, Richards, & Stemmerich, 1998). By considering these questions, we propose that the organismic perspective usefully extends the systemic perspective. Again, however, this does not mean that it *supplants* it; systemic and organismic perspectives are ultimately complementary, and both are needed to fully understand the nature of optimal human functioning (Sheldon & Kasser, 1995).

In this summary it is also worth further discussing the fact that control theory and self-determination theory think about the "why" of motivation in two different ways, with different implications for optimal functioning. The control theory perspective implies that one has a good answer to the question "why are you doing that?" if one can say "because it is helping me move quickly toward one or more of my broader goals." In terms of the first question of this chapter, one might be traveling rapidly through time; however, it is not certain that one is moving toward a desirable "place." In contrast, the self-determination theory perspective would say that a good answer to the "why" question is "because I enjoy doing the behavior, or because I fully identify with the longer term goal that the behavior is serving." In terms of the first question of this chapter, one is likely to travel more quickly to one's desired place, but one is also more likely to experience the place as more satisfying and desirable, once arrived at.

Using Future Time Perspective Theory to Further Compare Systemic and Organismic Approaches

So far, we have tried to clarify how both systemic and organismic theoretical approaches shed light on the time travel metaphor that we developed in the introduction of this chapter. However, these two perspectives only implicitly deal with the notion of time. Other researchers (Lens, Simons, & Dewitte, 2002; Strathman, Gleicher, Boninger, & Edwards, 1994; Zimbardo & Boyd, 1999) have more directly examined future time perspective (FTP), and its impact upon people's degree of optimal functioning. It seems important to discuss this research and theory for three reasons. First, goal striving implies an anticipation of future events, and thus seems to require an FTP. In other words, it is likely that FTP theory can supply fruitful tools for further analysis of goal-striving processes. Second, FTP theory will help us to draw further links between organismic and systemic theoretical perspectives. As we see later, FTP theory has traditionally been more like a systemic theory, but has more recently begun introducing organismic concepts into the mix. Third, consideration of FTP theory will allow us to discuss a seeming conflict between organismic (present-oriented) and systemic (future-oriented) theories, and to once again show that organismic and systemic perspectives are more complementary than conflicting.

Typically, the concept of FTP refers to a personality disposition to be preoccupied with the future. Kastenbaum (1961) defined FTP as "a general concern for future events" (p. 204). Wallace (1956) went further and focused on the length or breadth of the temporal vision, defining FTP as "the length of the future time span over which one conceptualises personalised future events" (p. 240). Similarly, Nuttin (1984) defined FTP as the aggregate temporal distance to a person's self-listed goals. The primary question for all of these researchers is: To what extent is the individual not only anticipating the immediate consequences, but also the long-term consequences, of his or her present behavior (De Volder & Lens, 1982; Simons, Vansteenkiste, Lens, & Lacante, 2004)? Past research using a variety of FTP measures demonstrated that having a long FTP is correlated with many positive outcomes. These include internal locus of control, the ability to delay gratification, higher grades, taking fewer health and safety risks, and greater psychological well-being (Strathman et al., 1994; Zimbardo & Boyd, 1999; Zimbardo, Keough, & Boyd, 1997). It appears that a long-term or extended future orientation is a good thing.

Such results seem logical when the concept of FTP is placed in the context of systemic perspectives. Again, systemic perspectives focus on the functional coherence of behavior, such that discrepancies between current states and desired future states are steadily or speedily reduced. Similarly, FTP theory focuses on the perceived degree of instrumentality of current behavior for desired future states, so that a person with a long FTP achieves many desirable advantages and adaptations.

However, when considered from an organismic perspective, it seems that being future oriented might in some cases yield less positive consequences, in the same way that being systemically but not organismically integrated might sometimes create problems. When future oriented, people are almost by definition extrinsically motivated for their present activities, because their actions are, to a large degree, undertaken only as means to some future end. This raises an intriguing question. Are future and instrumental orientations compatible with experiencing immediate pleasure and enjoyment in the moment? That is, can one have both a long FTP and also experience enough satisfaction in the moment to make life worthwhile? This question arises because past research has shown that intrinsic motivation is typically reduced or undermined when people become focused on instrumental rewards that are separable from the task itself (see Deci, Koestner, & Ryan, 1999, for a comprehensive meta-analysis of this literature). Analogously, we can wonder whether a similar effect occurs for future-oriented individuals, so that their enjoyment of the present is necessarily undermined.

Simons, Dewitte, and Lens (2000) recently made an important contribution to resolving this possible conflict between FTP and intrinsic motivation. They showed that when people consider their activities as leading to a future goal that they personally value, their intrinsic motivation for the activity is not undermined. Thus, being oriented toward the future seems quite compatible with experiencing enjoyment in the present, given that the future goal is personally meaningful. These results can be easily interpreted from a self-determination theory perspective. Extrinsically motivated goals that have been internalized into the self will not undermine intrinsic motivation, because the person has autonomously chosen to engage in them and has thus fully accepted the less-appealing tasks necessary to achieve them (Deci, Eghrari, Patrick, & Leone, 1994; Deci & Ryan, 2000). Indeed, identified motivation (the internalized form of extrinsic motivation) and intrinsic motivation are typically positively correlated, indicating that these two autonomous forms of motivation are mutually supportive (Sheldon & Elliot, 1998).

More recent FTP research, also exploring organismic issues, has investigated the possibility that the "what" of goals matters, just as the "why" of goals matters. For example, Vansteenkiste, Simons, Soenens, and Lens (2004) recently demonstrated in an experimental study that when participants were told that their present physical exercise would serve the long-term goal of enhancing their physical attractiveness (an extrinsic goal content) their enjoyment of the exercises was undermined compared to participants who were not given any future goal for engaging in the exercises. Conversely, when participants' present exercising was linked to the intrinsic future goal of physical fitness, they experienced the exercises as more enjoyable than did those who were not given any future goal. Thus, these results indicate that the type or content of future goals contained within the person's FTP helps determine whether they experience their current behavior as enjoyable or not.

In sum, contemporary FTP research is doing precisely what we believe is necessary to extend systemic perspectives upon optimal functioning; namely, consider the specific content of, and the dynamic motives underlying, the "futures" that are sought. Once again, it appears that organismic and systemic perspectives are not necessarily antithetical—instead, both perspectives are needed to fully understand optimal functioning (Sheldon & Kasser, 1995).

FTP and Present Orientation. However, systemic and organismic perspectives might still be viewed as incompatible, from yet another angle. It might be argued that organismic theories, because of their dominant focus on intrinsic motivation and present need satisfaction, are inherently incompatible with systemic theories, because the latter focus on instrumental motivation and future goal attainment. In line with this, Strathman and colleagues (1994) suggested that future orientation and present orientation form two extremes of a single continuum. In other words, present-oriented individuals cannot also be future oriented; the two constructs are antithetical to one another. Similarly, organismic congruence (which involves harmonious present experience) and systemic coherence (which involves subordinating present concerns to future concerns) might be incompatible.

We believe this would be an incorrect conclusion, or at least, would be put too strongly. Whether or not present and future orientations are compatible (or not) depends, in large part, on the *type* of present orientation involved. At least three different types of present orientation have emerged in recent time-related research: a present-fatalistic orientation, a present-hedonistic orientation (Zimbardo & Boyd, 1999), and mindfulness (Brown & Ryan, 2003). These three concepts all share at least two elements: First, they are said to be quite stable dispositions over time, and second, they are said to reflect a similar time orientation or time focus in which the present is dominating. However, these three present-oriented constructs also differ considerably.

First, a *present-fatalistic* orientation reflects a negative time attitude and thus yields a rather pessimistic and helpless outlook toward the present and the future. It reveals a belief that "the future is predestined and can not be uninfluenced by individual actions, whereas the present must be borne with resignation because humans are at the whimsical mercy of fate" (Zimbardo & Boyd, 1999, p. 1278). Because planning ahead in the future is worthless for such people, the negative correlation of present-fatalistic orientation with future orientation ($r = -.26$) reported by Zimbardo and Boyd does not come as a surprise. A present-fatalistic orientation was also a strong positive predictor of depression and aggression, whereas it was negatively related to self-esteem.

Second, a *present-hedonistic* orientation reflects a risk-taking, devil-may-care attitude toward time (Zimbardo & Boyd, 1999). Hedonism, which dates back to the philosophical writings of the Greek philosopher Aristippus and utilitarian thinkers such as Bentham, equates well-being with experiencing a maximum amount of pleasure, so that happiness consists of the totality of these hedonic moments (Ryan & Deci, 2001). As a consequence, a present-hedonistic orientation is characterized by a focus on immediate pleasure, sensation, and excitement, and a readiness to sacrifice today for tomorrow. The expression "carpe diem" might well define such a present orientation. In line with this definition, Zimbardo and Boyd found strong positive relations between this type of present orientation and sensation seeking/novelty seeking. Further consistent with Strathman et al.'s (1994) single-continuum assumption, Zimbardo and Boyd also reported that hedonistic present orientation was negatively correlated with future orientation.

Notably, the label "hedonistic" suggests that this type of present orientation should positively predict well-being, and be negatively related to ill-being. In other words, a person who pursues frequent enjoyments and pleasures should be happy, shouldn't she? Surprisingly, however, Zimbardo and Boyd (1999) found hedonism to be positively correlated with depression and aggression, whereas it was unrelated to self-esteem. Thus, perhaps, being focused only on immediate enjoyment is not so beneficial after all. Why might hedonism be maladaptive? We suggest that hedonistic present orientations are often adopted in order to cope with or to compensate for a lack of deeper, meaning-based satisfactions in life, and/or to compensate for psychological problems that the person wishes to escape. In order to distract themselves from inner conflicting feelings and frustrations, people might start to focus on seeking the next momentary pleasure, thereby continually avoiding the stable, unpleasant aspects of their lives.

In short, both the fatalistic and the hedonistic forms of present orientation are associated with problems and both seem to be incompatible with being future oriented. What kind of present orientation, if any, is positive and beneficial, and might not be antithetical to adopting a future orientation? We suggest that *mindfulness* (Brown & Ryan, 2003) is a valuable candidate. Mindfulness implies a conscious awareness of and a receptive attention to one's inner psychological experiences and states (Brown & Ryan, 2003; Martin, 1997). Deci and Ryan (1985) argued that such awareness facilitates more self-regulated or autonomous functioning, because it permits introspective access to one's needs, desires, and potentials in any given situation, and thereby facilitates the selection of behaviors that are fully congruent with those inner promptings and needs (Brown & Ryan, 2003). Furthermore, mindfulness does not entail escaping from life difficulties by hiding in the short-term present, but rather, facilitates the resolution of these difficulties through direct confrontation with the anxieties and frustrations that can accompany them.

When contrasted with a hedonic present orientation, we suggest that mindfulness reflects an eudaimonic present orientation. Eudaimonism is an ethical theory rooted in Aristotelian philosophy that calls people to recognize and live in accordance with the "daimon" or "true self" (Ryan & Deci, 2001; Waterman, 1993). The daimon is an ideal in the sense of being an excellent state of self-expression, and any efforts to live in accordance with this perfection give raise to a condition of "eudaimonism." Mindfulness might well represent one prerequisite for obtaining such a condition, because it allows people to engage in personally expressive activities and to experience higher well-being. The latter supposition was strongly confirmed by Brown and Ryan (2003), who reported that mindfulness predicted well-being of both the hedonic sort (i.e., life satisfaction, positive and negative affect) and eudaimonic sort (i.e., vitality, self-actualization) (Ryan & Deci, 2001; Vansteenkiste, Lens, Dewitte, De Witte, & Deci, 2004), and that it was negatively related to ill-being (i.e., depression, anxiety). In sum, there appears to be at least one form of present orientation that is compatible with a future orientation, although future research will have to establish the positive association of mindfulness with FTP.

How are mindful present and future orientations compatible, from a functional perspective? Going even further, are there some ways in which a mindful present orientation may actually *help* a person's future-oriented efforts? We believe that there are. This belief is based on two assumptions: first, that people automatically evaluate the significance of the present moment for their future goals and objectives, and second, that they can access this information, if they are alert to it. Presumably, mindful individuals *are* alert in this way; that is, they are sensitive to internal information and intuitions bearing upon their future goals and ambitions. In contrast, less mindful people are more likely to lose sight of their future goals, or to fail to notice the relevance of current occurrences for those goals. However, it will take further research to bear out this supposition.

CONCLUSION

In this chapter we have developed the metaphor of "traveling through time," not via a magical time machine, but rather, by sustained intentional action. A vision of the active human being was developed, in which people can to some extent create the future through the intentional pursuit of goals. In the process people not only dissipate entropy, they also create new order over time, thereby achieving temporal negentropy. Both systemic and organismic perspectives can make recommendations for effective time travel; that is, they can show how people propel themselves through time, rather than merely drifting passively through time. However, whereas systemic perspectives emphasize the hierarchal structure of one's goal pursuits and the *quantity* of one's motivation, organismic perspectives raise the question of where such goal hierarchies and quantities come from, and whether the reasons and the content of one's goal striving are compatible with one's underlying needs, personality, and talents. In other words, the *quality* of motivation also counts, both as one determinant of motivational "quantity" and as a moderator of goal attainment's effects upon well-being (Sheldon & Elliot, 1999).

In the last part of the chapter we used contemporary FTP theory to consider the potential conflicts that might arise between systemic models and organismic models. Again, the potential conflict is that being strongly oriented toward attaining higher level future goals (systemic coherence) might well cut people off from the present, hindering their ongoing enjoyment and satisfaction in life (organismic congruence); the well-known stereotype of the person who never "stops and smells the roses" comes to mind. However, we tried to show that such a conflict can be avoided if people manage to *identify* with their nonenjoyable future goals, and/or if they focus on future goals with *intrinsic* (growth, community, intimacy) more so than extrinsic (money, beauty, popularity) content. As another way of demonstrating the ultimate compatibility of present-oriented and future-oriented personality styles, we discussed mindful present orientation. We tried to show that, in contrast to fatalistic and hedonistic present orientation, mindfulness promotes both positive experience in the moment and timely movement toward desired possible futures.

REFERENCES

Brown, K. W., & Ryan, R. M. (2003). The benefits of being present: Mindfulness and its role in psychological well-being. *Journal of Personality and Social Psychology, 84,* 822–848.

Carver, C. S., & Scheier, M. F. (1981). *Attention and self-regulation: A control theory approach to human behavior.* New York: Springer-Verlag.

Carver, C. S., & Scheier, M. F. (1990). Origins and functions of positive and negative affect: A control-process view. *Psychological Review, 97,* 19–35.

Carver, C., & Scheier, M. (1998). *On the self-regulation of behavior.* Cambridge, England: Cambridge University Press.

De Volder, M., & Lens, W. (1982). Academic achievement and future time perspective as a cognitive-motivational concept. *Journal of Personality and Social Psychology, 42,* 566–571.

Deci, E. L., Eghrari, H., Patrick, B. C., & Leone, D. R. (1994). Facilitating internalization: The self-determination perspective. *Journal of Personality, 62,* 119–142.

Deci, E. L., Koestner, R., & Ryan, R. M. (1999). A meta-analytic review of experiments examining the effects of extrinsic rewards on intrinsic motivation. *Psychological Bulletin, 25,* 627–668.

Deci, E. L., & Ryan, R. M. (1985). *Intrinsic motivation and self-determination in human behavior.* New York: Plenum Press.

Deci, E. L., & Ryan, R. M. (2000). The "what" and the "why" of goal pursuits: Human needs and the self-determination of behavior. *Psychological Inquiry, 11,* 227–268.

Eccles, J. S., & Wigfield, A. (2002). Motivational beliefs, values, and goals. *Annual Review of Psychology, 53,* 109–132.

Emmons, R. A. (1991). Personal strivings, daily life events, and psychological and physical well-being. *Journal of Personality, 59,* 453–472.

Emmons, R. A., & King, L. (1988). Conflict among personal strivings: Immediate and long-term implications for psychological and physical well-being. *Journal of Personality and Social Psychology, 54,* 1040–1048.

Feynman, R., Leighton, M., & Sands, M. (1995). *Six easy pieces: Essentials of physics, explained by its most brilliant teacher.* Reading, MA: Addison-Wesley.

Glansdorff, P., & Prigogine, I. (1971). *Thermodynamic theory of structure, stability and fluctuations.* New York: Wiley.

Heckhausen, J. (2002). Developmental regulation of life-course transitions: A control theory approach. In L. Pulkkinen & A. Caspi (Eds), *Paths to successful development: Personality in the life course.* (pp. 257–280). New York: Cambridge University Press.

Jantsch, E. (1981). *The evolutionary vision: Toward a unifying paradigm of physical, biological, and sociocultural evolution.* Boulder, CO: Westview Press.

Kasser, T. (2002). *The high price of materialism.* London: MIT Press.

Kasser, T., & Ryan, R. M. (1993). A dark side of the American dream: Correlates of financial success as a central life aspiration. *Journal of Personality and Social Psychology, 65,* 410–422.

Kasser, T., & Ryan, R. M. (1996). Further examining the American dream: Differential correlates of intrinsic and extrinsic goals. *Personality and Social Psychology Bulletin, 22,* 280–287.

Kastenbaum, R. (1961). The dimensions of future time perspective: An experimental analysis. *Journal of General Psychology, 65,* 203–218.

King, L. A., Richards, J. H., & Stemmerich, E. (1998). Daily goals, life goals, and worst fears: Means, ends, and subjective well-being. *Journal of Personality, 66,* 713–744.

Lens, W., Simons, J., & Dewitte, S. (2002). From duty to desire: The role of students' future time perspective and instrumentality perceptions for study motivation and self-regulation. In F. Pajares & T. Urdan (Eds.), *Academic motivation of adolescents* (pp. 221–245). Greenwich, CT: Information Age.

Martin, J. R. (1997). Mindfulness: A proposed common factor. *Journal of Psychotherapy Integration, 7,* 291–312.

Nuttin, J. R. (1984). *Motivation, planning and action: A relational theory of behavior.* Hillsdale, NJ: Lawrence Erlbaum Associates.

Pohlman, K. (2001). Agency and communion-orientation in life goals: Impacts on goal pursuit strategies and psychological well-being. In P. Schmuck & K. Sheldon (Eds.), *Life goals and well-being: Towards a positive psychology of human striving* (pp. 68–84). Göttingen, Germany: Hogrefe & Huber.

Raynor, J. O. (1981). Future orientation and achievement motivation: Toward a theory of personality functioning and change. In G. d'Ydewalle & W. Lens (Eds.), *Cognition in human motivation and learning* (pp. 199–231). Hillsdale, NJ: Lawrence Erlbaum Associates.

Ryan, R. M., & Deci, E. L. (2000). Self-determination theory and the facilitation of intrinsic motivation, social development and well-being. *American Psychologist, 55,* 68–78.

Ryan, R. M., & Deci, E. L. (2001). On happiness and human potentials: A review of research on hedonic and eudaimonic well-being. *Annual Review of Psychology, 52,* 141–166.

Ryan, R. M., Sheldon, K. M., Kasser, T., & Deci, E. L. (1996). All goals were not created equal: An organismic perspective on the nature of goals and their regulation. In P. M. Gollwitzer & J. A. Bargh (Eds.), *The psychology of action: Linking motivation and cognition to behavior* (pp. 7–26). New York: Guilford Press.

Shah, J. Y., & Kruglanski, A. W. (2000). Aspects of goal networks: Implications for self-regulation. In M. Boekaerts, P. Pintrich, & M. Zeidner (Eds.), *Handbook of self-regulation* (pp. 85–110). San Diego: Academic Press.

Sheldon, K. M. (2002). The self-concordance model of healthy goal-striving: When personal goals correctly represent the person. In E. L. Deci & R. M. Ryan (Eds.), *Handbook of self-determination research* (pp. 65–86). Rochester, NY: University of Rochester Press.

Sheldon, K. M. (2004). *Optimal human being: An integrated multi-level perspective.* Mahwah, NJ: Lawrence Erlbaum Associates.

Sheldon, K. M., & Elliot, A. J. (1998). Not all personal goals are personal: Comparing autonomous and controlled reasons for goals as predictors of effort and attainment. *Personality and Social Psychology Bulletin, 5,* 546–557.

Sheldon, K. M., & Elliot, A. J. (1999). Goal-striving, need satisfaction, and well-being: The self-concordance model. *Journal of Personality and Social Psychology, 76,* 482–497.

Sheldon, K. M., Elliot, A. J., Kim, Y., & Kasser, T. (2001). What is satisfying about satisfying events? Testing 10 candidate psychological needs. *Journal of Personality and Social Psychology, 80,* 325–339.

Sheldon, K. M., & Houser-Marko, L. (2001). Self-concordance, goal-attainment, and the pursuit of happiness: Can there be an upward spiral? *Journal of Personality and Social Psychology, 80,* 152–165.

Sheldon, K. M., & Kasser, T. (1995). Coherence and congruence: Two aspects of personality integration. *Journal of Personality and Social Psychology, 68,* 531–543.

Sheldon, K. M., & Kasser, T. (1998). Pursuing personal goals: Skills enable progress, but not all progress is beneficial. *Personality and Social Psychology Bulletin, 24,* 1319–1331.

Sheldon, K. M., Kasser, T., Smith, K., & Share, T. (2002). Personal goals and psychological growth: Testing an intervention to enhance goal-attainment and personality integration. *Journal of Personality, 70,* 5–31.

Sheldon, K. M., Ryan, R. M., Kasser, T., & Deci, E. L. (2004). The independent effects of goal contents and motives on well-being: It's both what you pursue and why you pursue it. *Personality and Social Psychology Bulletin, 30,* 475–486.

Sheldon, K. M., Ryan, R. M., & Reis, H. R. (1996). What makes for a good day? Competence and autonomy in the day and in the person. *Personality and Social Psychology Bulletin, 22,* 1270–1279.

Sheldon, K. M., Williams, G., & Joiner, T. (2003). *Motivating health: Applying self-determination theory in the clinic.* New Haven, CT: Yale University Press.

Simons, J., Dewitte, S., & Lens, W. (2000). Wanting to have versus wanting to be: The effect of perceived instrumentality on goal orientation. *British Journal of Psychology, 91,* 335–351.

Simons, J., Vansteenkiste, M., Lens, W., & Lacante, M. (2004). Placing future time perspective theory in a temporal perspective. *Educational Psychology Review, 16,* 121–139.

Strathman, A., Gleicher, F., Boninger, D. S., & Edwards, C. S. (1994). The consideration of future consequences: Weighting immediate and distant outcomes of behavior. *Journal of Personality and Social Psychology, 86,* 742–752.

Vansteenkiste, M., De Witte, H., & Deci, E. L. (2003). *Explaining the well-being and behavioral benefits of adopting an intrinsic work orientation: The role of basic need satisfaction.* Internal Research Report, University of Leuven, Belgium.

Vansteenkiste, M., Lens, W., Dewitte, S., De Witte, H., & Deci, E. L. (2004). The "why" and "why not" of job search behavior: Their relation to searching, unemployment experience and well-being. *European Journal of Social Psychology, 34*(3), 345–363.

Vansteenkiste, M., Simons, J., Soenens, B., & Lens, W. (2004). How to become a persevering exerciser: The importance of providing a clear, future goal in an autonomy-supportive way. *Journal of Sport and Exercise Psychology, 26,* 232–249.

Wallace, M. (1956). Future time perspective in schizophrenia. *Journal of Abnormal and Social Psychology, 52,* 240–245.

Waterman, A. S. (1993). Two conceptions of happiness: Contrast of personal expressiveness (eudaimonia) and hedonic enjoyment. *Journal of Personality and Social Psychology, 64,* 678–691.

Wilensky, R. (1983). *Planning and understanding: A computational approach to human reasoning.* Reading, MA: Addison-Wesley.

Wilson, E. O. (1998). *Consilience: The unity of knowledge.* New York: Knopf.

Zimbardo, P. G., & Boyd, J. N. (1999). Putting time in perspective: A valid, reliable individual-difference metric. *Journal of Personality and Social Psychology, 77,* 1271–1288.

Zimbardo, P. G., Keough, K. A., & Boyd, J. N. (1997). Present time perspective as a predictor of risky driving. *Personality and Individual Differences, 23,* 1007–1023.

9

Yesterday, Today, and Tomorrow: Counterfactual Thinking and Beyond

Lawrence J. Sanna
Seth E. Carter
Edward Burkley
University of North Carolina at Chapel Hill

Time it was,
And what a time it was,
It was …

—Paul Simon (*Bookends Theme*, 1968)

It was. It is. It might have been. *Counterfactual thinking* involves mentally simulating "what might have been" in contrast to what was and what is. Such thoughts focus on alternatives to the *past* that did not actually happen but that could be imagined having happened instead, typified by "if only," "at least," or "what if" (see Miller, Turnbull, & McFarland, 1990; Roese, 1997; Roese & Olson, 1995, for reviews). They can occur spontaneously (Sanna, Parks et al., 2003; Sanna & Turley, 1996) and vary by intensity (Sanna & Turley-Ames, 2000). Like Simon and Garfunkel waxing about earlier times, people's thoughts also often center on how the present might otherwise have been different. Counterfactuals have both theoretical and applied implications due to relations with diverse topics such as affective reactions (Gleicher et al., 1990; Landman, 1987), accident and victim compensation (Macrae & Milne, 1992; Turley, Sanna, & Reiter, 1995), coping and blame assignment (Holman & Silver, 1998; Miller & Gunasegaram, 1990), and causal ascriptions (Lipe, 1991; Spellman & Mandel, 1999), among others.

In keeping with the theme of this book, not only do we describe theory and research on counterfactuals, but we do so while simultaneously considering the relationship of these thoughts to other temporal orientations. We do this by placing counterfactuals within a broader context of mental simulations (Sanna, 2000; Sanna, Stocker, & Clarke, 2003; Taylor, Pham, Rivkin, & Armor, 1998) in this chapter, including anticipations of the future, assessments of the present, and retrospections of the past. We also outline an integrative model of imaginative mental simulations that summarizes and synthesizes several areas of theory and research, and moves beyond them. Our approach not only may enable a greater awareness of conceptual relations between existing findings, but also suggests novel and unique predictions for future research. Finally, we end by offering conclusions and by suggesting how these proposals might have further applications by informing an understanding of other phenomena, such as temporal biases, personality processes and strategies, and automaticity and coping with life events.

THEORY AND RESEARCH

Theory and research on counterfactual thinking has been ongoing for several decades, and has been of interest to philosophers, mathematicians, and psychologists, among others. We first provide a brief historical background on counterfactual thinking from a psychological perspective, focusing on some major antecedents and consequences. We then present an integrative framework that conceptualizes counterfactuals within a broader context of mental simulations that considers implications for past, present, and future temporal orientations.

THINKING ABOUT WHAT MIGHT HAVE BEEN: A BRIEF HISTORICAL CONTEXT

Counterfactuals are classically represented by *if–then* conditional propositions (Johnson-Laird & Byrne, 2002; Roese, Sanna, & Galinsky, 2004), whereby factual realities are contrasted with imagined alternatives. They can run from personal (e.g., "If only I married Norma my life might be immeasurably happier") to historical (e.g., "At least the Nazis did not develop the atomic bomb first"; Cowley, 1999; Tetlock & Belkin, 1996) or whimsically scientific ("What if the moon did not exist?"; Comins, 1993). The ability to think counterfactually may be so critical that it is an essential component to consciousness and intelligence (Hofstadter, 1979).

Simulation Heuristic and Norm Theory

Kahneman and Tversky's (1982) discussion of the simulation heuristic forms the genesis for most psychological interest in counterfactual thinking. They proposed that people "run through" or imagine various possible alternatives when determining responses. Einhorn and Hogarth (1981) suggested they are "based on a mental simulation in which 'what might be,' or 'what might have been,' is combined with 'what is'" (p. 456). The simulation heuristic plays an influential role in at least five major tasks, including counterfactual thinking: (a) generating predictions (G. P. Brown, MacLeod, Tata, & Goddard, 2002), (b) assessing event probabilities (Sanna, Schwarz, & Stocker, 2002), (c) assessing conditioned probabilities (McKenzie, Ferreira, Mikkelsen, McDermott, & Skrable, 2001), (d) assessing causality (McGill & Tenbrunsel, 2000); and (e) generating counterfactuals (Galinsky & Moskowitz, 2000). In each case, the imagined alternatives can have a wide variety of implications, and also can play a significant role in people's experienced affective states (Kahneman & Tversky, 1982).

But not all mental simulations are equally likely. Easily accessible (Schwarz, 1998; Tversky & Kahneman, 1973) alternatives are judged most probable (Kahneman & Tversky, 1982). When thinking counterfactually, people are more apt to change *(mutate)* surprising—*abnormal*—events by returning them to more typical—*normal*—states than vice versa (Kahneman & Miller, 1986; Kahneman & Tversky, 1982). People are more likely to change exceptional, committed, and unconstrained behaviors, rather than routine, omitted, or constrained behaviors (Miller et al., 1990). For example, a person involved in a car accident driving home from work via an atypical (exceptional) route at a typical (routine) time of day would be more likely to think "if only" about sticking to a typical route rather than about leaving at an atypical time (Kahneman & Tversky, 1982; Wells & Gavanski, 1989). In this way, counterfactual thoughts take the abnormal and recapitulate normality (Kahneman & Miller, 1986).

Kahneman and Miller's (1986) norm theory provides the most detailed and influential discussion of this idea. Although counterfactuals may be elicited by violated expectations, what is seen as normal can also be completely *postcomputed* in an ad hoc fashion (Kahneman & Miller, 1986). Specific evoking events activate sets of related exemplars from memory that combine to create *norms*. Thus, counterfactuals can occur independent of prior expectations (Byrne, 2002; Medvec, Madey, & Gilovich, 1995). Consider the classic example of two people missing plane flights, one by 30 minutes and one by 5 minutes (Kahneman & Tversky, 1982). In the latter case, counterfactuals are more readily available, as it is easier to imagine having shaved off only 5 minutes than a full 30 minutes. Prior expectations and objective situations are identical (both people expected to make their flights but did not), but the relative availability of counterfactual alternatives accounts for differing affect—people say they experience greater regret when a flight is missed by 5 minutes than by 30 minutes (see also Miller et al., 1990; Sanna & Turley-Ames, 2000).

Affect, Content, and Goals

There are several other antecedents and consequences of counterfactual thinking. For example, Roese (1997; Roese & Olson, 1997) proposed a two-stage model. The key feature of the model is a separation of determinants of counterfactual thinking into those influencing: (a) *activation* and (b) *content*. Negative affect is presumed to activate counterfactual thinking, that is, to "turn it on" or begin the process. Once activated, normality is then presumed to dictate the content of counterfactual thoughts, that is, what people think about. The primary divergence between the two-stage model and norm theory is that negative affect (not norm violation) is presumed to initiate counterfactual generation (Roese et al., 2004). For instance, the negative affect produced by an accident or a romantic breakup may then instigate a search for unusual causal features that might have been performed differently. Thus, although counterfactuals are directed at the past, they can also provide prescriptions for future behaviors.

This follows from functionalist views of counterfactuals (Markman, Gavanski, Sherman, & McMullen, 1993; Roese, 1994; Sanna, 1996). As derived from social comparison theory (Collins, 1996; Taylor & Lobel, 1989; Wood, 1989), the notion of counterfactual direction has been indispensable to views of people's goals and motives. *Upward* counterfactuals contrast reality with better alternatives (e.g., "If only I studied harder, I might have been admitted to medical school"); *downward* counterfactuals compare reality with worse alternatives (e.g., "At least I was wearing my seat belt, or I might have been more seriously injured"). Generally, upward counterfactuals are useful for future preparation and *self-improvement,* and downward counterfactuals are useful for self-enhancement or *mood repair* (Roese, 1994; Sanna, Turley-Ames, & Meier, 1999; Taylor & Schneider, 1989). This is related to ideas about problem-focused versus emotion-focused coping (Lazarus & Folkham, 1984).

However, research has shown that people can mentally simulate alternatives, *both* retrospectively and prospectively, in the service of several diverse goals or motives. Sanna, Chang, and Meier (2001) demonstrated that people can retrospectively engage in *mood-repair* or *mood-maintenance*, and can prospectively engage in *self-improvement* or *self-protection*. Each of these self-motives can be cued by positive and negative life events or outcomes and accompanying positive and negative affect, respectively. We elaborate on each of these motives in our discussion later. Valenced outcomes and affect are connected directly because numerous life events, like failures and successes, exert influences through moods (J. D. Brown & Mankowski, 1993). Mood repair can be achieved by downward simulations after failure; mood maintenance can be achieved by downward simulations after success; and self-improvement and self-protection can be achieved by upward simulations when anticipating performances (Sanna, Chang et al., 2001; see also Sirois, 2004). The ensuing mental simulations, including counterfactual thoughts, can be either automatic or effortful (Sanna, Chang et al., 2001).

Summary

Research on counterfactual thinking has been ongoing for several decades and has been of focus in divergent areas. Psychological interest in counterfactuals has its basis in seminal discussions of the simulation heuristic and norm theory. The two-stage model and ideas about diverse self-motives and goals suggest conditions under which counterfactuals may be activated, what their content might be, and what motivational purposes they may serve.

BEYOND COUNTERFACTUALS: THINKING ABOUT THE PAST, PRESENT, AND FUTURE

The imagination, goals, and affect (IGoA, "ego") model depicted in Fig. 9.1 places counterfactuals within a broader integrative framework of mental simulations, including thoughts about the past, present, and future (see also Sanna, 2000; Sanna, Small, & Cook, 2004; Sanna, Stocker et al., 2003). *Ego* also refers to the degree to which self-motives or goals are activated. The model incorporates prior theorizing (e.g., comparisons over time, simulation direction) while also moving beyond it. Three underlying mental simulation dimensions are proposed. Positive and negative life events, and accompanying good and bad moods, elicit downward and upward simulations. Mental contrasting and assimilation further determine the nature and consequences of the simulations. We provide examples of mental simulations within our framework.

Underlying Mental Simulation Dimensions

We propose that mental simulations can be classified and better understood when viewed in accordance with three underlying dimensions.

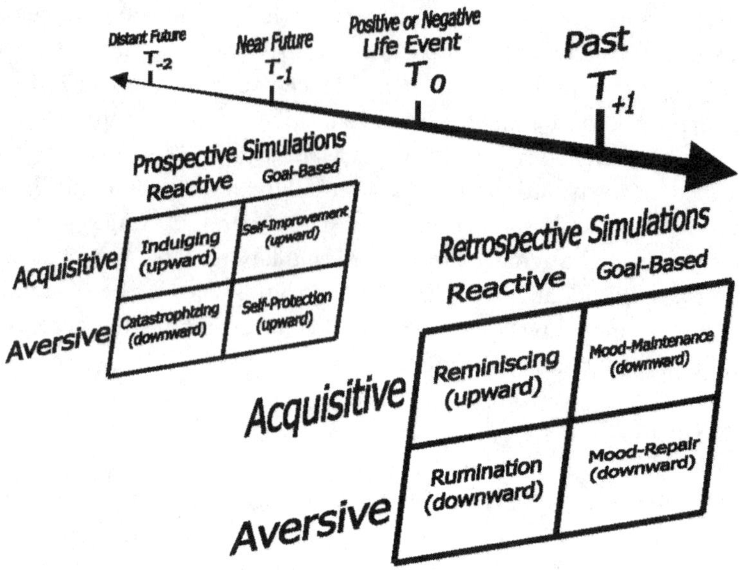

FIG. 9.1. The imagination, goals, and affect (IGoA) model over time. Mental contrasting underlies goal-based simulations, whereas mental assimilation underlies reactive simulations.

Reactive Versus Goal Based. Mental simulations can be *reactive* or *goal based*, referring to whether there is a clear underlying goal, or whether there is not. For instance, Klinger (1975, 1977) divided thoughts into two basic types: respondent and operant. The former occur without premeditation or purpose, and may in fact involve shifts away from goal-directed tasks (see also Rachman, 1981). Other mental simulations are more directly goal based (Sanna, Chang et al., 2001; Taylor & Schneider, 1989). What is meant by goal based is a matter of degree. This is not to say that what we have classified as reactive simulations can never be goal based, just that they are normally less so than clear goal-based ones (see also Klinger, 1977).

Retrospective Versus Prospective. Mental simulations can differ by temporal perspective, *retrospective* or *prospective*. Retrospective simulations focus on the past or previously obtained outcomes. Counterfactual thoughts (Miller et al., 1990; Roese, 1997) exemplify this. Focusing on just the past, however, would be a lot like trying to drive a car while looking only through the rearview mirror. Prospective simulations focus on the future or expected outcomes. Prefactual thoughts of "what may be" (Sanna, 1996, 1999) exemplify this. One could forecast losing retirement savings in a stock market crash, or being ecstatic while playing shortstop for the New York Yankees (see also Gilbert, Pinel, Wilson, Blumberg, & Wheatley, 1998).

Acquisitive Versus Aversive. Finally, mental simulations can be *acquisitive* or *aversive*. Acquisitive refers to obtaining or retaining something good or positive (Arkin & Shepperd, 1989); aversive refers to something negative or avoiding or protecting against negative possibilities (Baumeister, Tice, & Hamilton, 1989). A person might, for example, reminisce about the glory days of old when he or she was a star high school athlete, or a student might think about how an exam score might have been worse if only he or she had not looked at the study guide. When it comes to goal-based simulations, acquisitive might also be viewed as promotion focused, whereas aversive might be viewed as prevention focused (Higgins, 1998).

Mental Contrasting and Assimilation: Mechanisms Comparing "Today"

Mental contrasting and assimilation can provide mechanisms for how present life events and moods may serve as comparison referents or standards for mental simulations of the past and future (Oettingen, Pak, & Schnetter, 2001; Sanna, 2000; Schwarz & Bless, 1992). Research on counterfactual thinking had mainly emphasized mental contrasting, whereby reality is compared with better or worse alternatives (Miller et al., 1990; Roese, 1997). Research on prefactual thinking also emphasized mental contrasting (Sanna, 1998, 1999). Oettingen et al. (2001) argued that comparing desired futures with present realities underlies goal setting, a mental contrasting. The IGoA model extends this idea to include all goal-based mental simulations, both prospective and retrospective, depicted in Fig. 9.1. Each involves a contrast with present realities.

To the contrary, reactive simulations depicted in Fig. 9.1 do not involve contrasts with present realities, at least not to the same degree. For example, Oettingen et al. (2001) found that people may "indulge" in free fantasies by focusing only on the future without regard to reality (see also McMullen, 1997; Oettingen, 1996), a mental-assimilation process. The possibility of mental assimilation has recently begun to receive increased attention among counterfactual researchers (Markman & McMullen, 2003; Sanna, 2000; see also Mussweiler, 2003). The IGoA model extends this idea to all reactive mental simulations, both prospective and retrospective, focusing on only the future or the past without any specific contrast with present realities.

Retrospective Mental Simulations: Thinking About "Yesterday"

Retrospective mental simulations depicted on the right in Fig. 9.1 can be either goal based or reactive, and they can be either acquisitive or aversive.

Mood Maintenance. People in happy moods, or those who have experienced positive life events or outcomes such as successes, behave in ways that preserve positive affect. The goal or motive underlying this is *mood maintenance* (Isen, 1987). For example, good moods make accessible pleasant self-thoughts (Bower, 1991) or induce people to select only information that prolongs positive affective states (Wegener & Petty, 1994). With regard to counterfactuals, evidence for mood maintenance comes from the finding that people in happy moods generate larger numbers of downward counterfactuals, and they report high enjoyment in doing so (Sanna, Meier, & Wegner, 2001). We propose that mood maintenance is retrospective, goal based, and involves a contrast of pleasant reality with even worse alternatives to perpetuate positive affect.

Mood Repair. People in bad moods, or who have experienced negative life events or outcomes such as failures, work to reinstate positive affect. The underlying goal or motive is *mood repair* (Erber & Erber, 1994; Sedikides & Strube, 1997). For example, when confronted with poor outcomes or bad moods, people recall positive information about the self (Parrott & Sabini, 1990), denigrate outgroups (Fein & Spencer, 1997), or perform helpful acts (Schaller & Cialdini, 1990) to feel better. With regard to counterfactuals, people have been shown to generate downward counterfactuals to repair mood (Sanna, Meier, & Turley-Ames, 1998; Sanna et al., 1999). We propose that mood repair is retrospective, goal based, and involves contrasting currently negative realities with worse alternatives to regain positive affect.

Reminiscing. People reminisce by thinking about positive aspects of their lives with a focus on past accomplishments, achievements, or successes (Fivush, Haden, & Reese, 1996; Strack, Schwarz, & Gschneidinger, 1985). Reminiscing increases life satisfaction, extends cognitive functioning, alleviates depression, and sustains a positive self-image (Rybarczyk, 1995). Self-reported frequency of reminiscing is quite high (Webster, 1993), especially among older adults. People who reminisce about past glories and assimilate these thoughts to their current state experience an increased sense of well-being (Strack et al., 1985; see also McMullen, 1997, and Sanna, 1997). We propose that reminiscing is retrospective, reactive, and involves a mental assimilation of upward, or positive, thoughts with a focus only on the past.

Rumination. Ruminations are persistent or reoccurring thoughts about past events that focus on negatives or are associated with bad moods (Nolen-Hoeksema, 1996; Nolen-Hoeksema & Morrow, 1991). They can have an irrevocable quality (Gold & Wegner, 1995), are respondent or intrusive (Klinger, 1977), and may have little to do with goal setting in this case. Ruminations can intensify reactions to negative life events. For example, men who ruminated about losing a partner to AIDS showed increased negative emotions and risk for prolonged periods of depression (Nolen-Hoeksema, McBride, & Larson, 1997; see also Holman & Silver, 1998). When viewed in this way, we propose that ruminations are retrospective and aversive and do not entail a contrast with a present state—that is, they involve mental assimilation.

Prospective Mental Simulations: Thinking About "Tomorrow"

Prospective mental simulations depicted on the left in Fig. 9.1 can also be either goal based or reactive, and they can be either acquisitive or aversive.

Self-Improvement. People can be motivated to improve traits, abilities, health status, or well-being (Collins, 1996; Taylor & Lobel, 1989). Generating alternatives are useful for goal attainment (L. L. Martin & Tesser, 1989, 1996; L. L. Martin, Tesser, & McIntosh, 1993). Feelings of threat, inadequacy, or negative affect often instigate motives to get better (Taylor & Schneider, 1989). For example, cancer patients focus on better alternatives to improve coping (Wood, Taylor, & Lichtman, 1985), and people can mentally simulate better possibilities when preparing for the future (Markman et al., 1993; Roese, 1994; Sanna, 1996). We propose that self-improvement is prospective, acquisitive, and goal based. It involves comparing present realities with better futures, as does the original view of mental contrasting (Oettingen et al., 2001).

Self-Protection. People can protect themselves from negative possibilities by "bracing for loss" (Shepperd, Findley-Klein, Kwavnick, Walker, & Perez, 2000; Shepperd, Ouelette, & Fernandez, 1996). Mentally simulating how the worst may transpire can mitigate the sting of failure should it occur. This is especially likely when tests of important attributes are close at hand. Defensive pessimism (Norem & Cantor, 1986; Sanna, 1996) and self-handicapping (Berglas & Jones, 1978; Sanna & Mark, 1995) are both strategies that, in part, involve self-protection. Generating upward prefactuals has been similarly shown to help people protect from possible failure (Sanna, 1999; Sanna & Meier, 2000). We propose that self-protection is prospective, aversive, goal based, and involves a mental contrast with the present.

Indulging. Oettingen et al. (2001) argued that people can look to the future without specific goals in mind, called *indulging.* People think about desired futures without a contrast with the present (or only a weak contrast with the present). Indulging is a free fantasy in which people "enjoy the desired future in the here and now" (Oettingen et al., 2001, p. 737; see also Oettingen, 1996). Evidence comes from work on goal setting (Oettingen et al., 2001) and from counterfactual work indicating that people assimilate upward simulations (McMullen, 1997; Sanna, 1997, 2000). For example, a person buying a lottery ticket may envision him or herself vacationing on sandy beaches while enjoying fancy feasts and fine champagne. We propose that indulging is prospective, acquisitive, reactive, and occurs by mental assimilation.

Catastrophizing. Catastrophizing entails simulating negative "what if" scenarios (Kendall & Ingram, 1987) and worrying about bad things that may transpire (Vasey & Borkovec, 1992), resulting in progressive gloom. Poor problem-solving confidence is a contributing factor (Davey, Jubb, & Cameron, 1996). A passenger at an airport gate waiting for her or his departure might think about all the bad things that may happen (e.g., "what if my plane crashes," "how will my family get along without me"). There is an internal dialogue characterized by problem-specific pessimism, feelings of inadequacy, despair, and hopelessness. Catastrophizing is linked through lifestyles to negative events (Peterson, Seligman, Yurko, L. R. Martin, & Friedman, 1998). We view catastrophizing as prospective, aversive, reactive, and involving mental assimilation.

Summary

The IGoA model goes beyond counterfactuals by placing these thoughts within a broader integrative framework of mental simulations about the past, present, and future. People can retrospectively maintain good moods (mood maintenance), repair bad moods (mood repair), or reflect on positive (reminiscing) or negative (rumination) pasts. They can also prepare for future events (self-improvement), brace for loss (self-protection), or reflect on positive (indulging) or negative (catastrophizing) futures. This occurs via mental contrasting or assimilation to present states, and we provided several examples of mental simulations that fit within this framework.

APPLICATIONS AND FUTURE DIRECTIONS

As we have illustrated, researchers and theorists working in many areas are interested not only in counterfactuals but also in other conceptually related mental simulations, although they have approached each topic in slightly different ways. This has had an unfortunate consequence of making domains somewhat insular, to the detriment of all involved. Our chapter begins to fill this gap by taking a more integrative perspective. Considered in a more consolidated light, not only may a greater understanding of existing findings be achieved, but also novel and unique propositions can be formulated and new avenues for further research opened. We conclude by offering a few other possible applications and implications of this approach.

Temporal Biases and Mentally Simulating Events

Figure 9.1 also depicts patterns related to several temporal biases: confidence changes, planning fallacy, affective forecasting, and hindsight bias. T_0 indicates an event outcome has occurred or is expected. T_{-2} and T_{-1} represent the distant and near future. T_{+1} indicates a time after event outcomes have occurred or are known, representing the past. *Confidence change* research compares distant and near futures. Nisan's (1972, 1973) participants anticipated taking a test immediately (T_{-1}) or in 4 weeks (T_{-2}); those anticipating an immediate test were less confident than those anticipating a distant test (see also Gilovich, Kerr, & Medvec, 1993). Shepperd et al. (1996) suggested that people brace for the worst or possible loss by reducing confidence over time (see also Shepperd et al., 2000). Confidence changes may be related to increases in self-protective mental simulations at T_{-1} (Sanna, 1999; Sanna & Meier, 2000).

Planning fallacy research (Buehler, Griffin, & Ross, 1994; Kahneman & Tversky, 1979) shows that people underestimate task completion in the distant future (e.g., T_{-2}) compared with actual completion (T_0). People adopt an optimistic "narrow focus on successful future plans" (Buehler, Griffin, & MacDonald, 1997, p. 239) at a distance. *Affective forecasting* research (Gilbert et al., 1998; Wilson, Wheatley, Meyers, Gilbert, & Axom, 2000) compares people's predicted feelings (e.g., T_{-2}) with actual feelings after events (T_{+1}); people "think too much about the focal event" (Wilson et al., 2000, p. 833) when predicting. Planning fallacy may be due in part to more indulging at T_{-2}; that is, people "enjoy the desired future in the here and now" (Oettingen et al., 2001, p. 737), resulting in undue optimism. Biases in affective forecasting may be similarly due to indulging (when predicting success reactions) and catastrophizing (when predicting failure reactions) at T_{-2}; people may also inadequately account for other processes like mood repair (Gilbert et al., 1998) at T_{+1} when predicting future reactions at T_{-2}.

Hindsight bias research indicates that once event outcomes are known (T_{+1}), people believe they "knew all along" what would happen even though their preevent predictions (T_{-2}) indicated otherwise (Christensen-Szalanski & Willham, 1991; Fischhoff, 1975). The past is also viewed as particularly inevitable after experiencing positive versus negative events (Louie, 1999; Mark & Mellor, 1991), suggesting the operation of mood maintenance and mood repair motives at T_{+1}, respectively. There may be at least one other process that all temporal biases share: whether people's thoughts emphasize focal or alternative events (Sanna & Schwarz, 2004; Sanna, Schwarz, & Small, 2002; Sanna, Schwarz, & Stocker, 2002). Future research examining counterfactuals and other mental simulations on temporal biases may need to attend not only to temporal, but also to social and hypothetical comparisons (Mark & Folger, 1984; Markman & McMullen, 2003). That is, these biases may be influenced not only by a person's judgments over time (e.g., "Am I better off now than in the past?") but also by social (e.g., "Am I better off than most other people?") and purely imagined (e.g., "Am I as good as I can be?") comparisons. Taking these various perspectives into account can provide fuller insights into people's decision making.

Personality Processes and Strategies

Personality characteristics, processes, or strategies may also be related to counterfactuals and other mental simulations. For example, one individual difference studied by counterfactual researchers is people's propensity to consider the immediate versus distant consequences of their behaviors (Boninger, Gleicher, & Strathman, 1994; Strathman, Gleicher, Boninger, & Edwards, 1994). Within the IGoA model, this line of research suggests that anticipated reactions might be extended beyond T_{+1} to T_{+2} or T_{+n}, recent versus more distant pasts. Some people think about more remote consequences of their actions, and this can have a variety of implications for future behaviors and emotions in diverse contexts (Strathman et al., 1994). Other researchers have also proposed individual differences in rumination (Nolen-Hoeksema et al., 1997), reminiscence (Webster, 1993), catastrophizing (Peterson et al., 1998), and so on. Each may be associated with various mental simulations, and thus they seem certainly deserving of future research.

Other well-known personality characteristics may also have interesting relationships with particular mental simulations. Sanna (2000) already has connected several individual differences to such thoughts. Within the IGoA model, personality types, traits, or strategies may represent more global macrogoals or macroreactions within the model. For example, *defensive pessimism, pessimism,* and *optimism* present a comparison of how people might use prospective (T_{-2}, T_{-1}) versus retrospective (T_{+1}) mental simulations (Sanna, 1996, 1998; Sanna & Chang, 2003): Defensive pessimists use upward prefactual simulations prospectively, but optimists use downward counterfactual ones retrospectively. Defensive pessimists' strategy involves a combination of self-improvement and self-protective motives, whereas that of optimists involves a mood repair motive. Both strategies make use of mental contrasting.

Self-esteem, another example, refers to self-worth (Baumeister et al., 1989; J. D. Brown & Mankowski, 1993). Low-self-esteem persons are governed by self-protective motives; high-self-esteem persons are governed by acquisitive motives. Similar to optimists, high-self-esteem persons use downward counterfactuals (T_{+1}) to repair moods (Sanna et al., 1999); low-self-esteem persons do not. Although both use upward prefactuals (T_{-2} vs. T_{-1}), low-self-esteem persons contrast them in a self-protective fashion (Sanna & Meier, 2000), similar to defensive pessimists. High-self-esteem persons assimilate upward prefactuals. Findings for self-efficacy, naive optimism, true pessimism, and others may also fit the model (Sanna, 2000). Future research on counterfactuals and other mental simulations might thus inform a myriad of individual differences, as mental simulations may be one distinguishing underlying process. Other motives (Helgeson & Mickelson, 1995) may also underlie mental simulations.

Automaticity and Coping With Life Events

Positive and negative life events (T_0), and accompanying good and bad moods, can trigger mental simulations of a particular direction, and this may enlighten what is known about coping and well-being. Research has shown that negative events or bad moods induce upward simulations, whereas positive events or good moods induce downward simulations (Sanna, 2000; Sanna et al., 1999; Sanna, Chang et al., 2001). Life events can exert strong influences through affect (J. D. Brown & Mankowski, 1993), although it should be obvious that we are not arguing that valenced events are the only antecedent. Moods produce an identical pattern, and existing evidence suggests that this is because they inform people's current states (Sanna et al., 1998; Schwarz & Clore, 1996). That is, people interpret their lives negatively in bad moods (e.g., "I am a failure" or "There is a problem") but positively in good moods (e.g., "I am a success" or "Things are fine").

Mental assimilation involves a focus on only the past or the future, as described. The IGoA model proposes that affect here influences simulations in a mood-congruent manner. That is, just as assimilating upward simulations leads to good moods (McMullen, 1997), good moods reciprocally lead to upward simulations (Sanna, 2000). In other words, affect influences the mental simulation directly, and reactive simulations are assimilated to current states. Mental contrasting, however, involves a comparison of future (Oettingen et al., 2001) or past with present realties or current states. The IGoA model proposes, and evidence indicates, that affect here influences perceptions of the present directly (Sanna, 2000; Sanna et al., 1999). Goal-based simulations are contrasted with this present. Mood maintenance occurs for positive affect, and self-improvement, self-protection, and mood repair occur for negative affect (Sanna, Chang et al., 2001).

The broader implication is that when simulation direction activated by affect (upward for negative and downward for positive) matches that listed in Fig. 9.1, responses are automatic. When there is a mismatch, responses are effortful. To date, moods and only the four goal-based simulations have been varied (Sanna, Chang et al., 2001). Half of participants were put under time pressure, getting at automatic responses (Bargh, 1994; Smith & DeCoster, 2000). In bad moods, with no time pressure, quick reactions arose for upward simulations for self-improvement or self-protection (matches); slow reactions for downward simulations for mood repair (mismatch). In good moods with no time pressure, quick reactions arose for downward simulations for mood maintenance (match). Under time pressure, responses were quick for upward and downward simulations in bad and good moods, consistent with that activated by affect. Future research is needed to test whether this match versus mismatch idea extends to reactive simulations.

Summary

Counterfactuals and other mental simulations might have further applications and implications across diverse areas. They may help to explain temporal biases such as confidence changes, planning fallacy, affective forecasting, and hindsight bias; they may help conceptualize a myriad of individual differences by identifying common underlying processes; and they may help to inform the nature and consequences of people's coping and responses to life events.

CODA

Much may be gained by pursuing a more inclusive conception of counterfactual thinking within the context of other mental simulations. We attempted to broaden perspectives by articulating relations between diverse areas in this chapter. The IGoA model incorporates several previous findings and coalesces them in a way that organizes current developments so that all areas can benefit while affording new insights. Researchers and theorists addressing divergent topics such as self-motives, and temporal biases, among others, may be better able to identify similar underpinnings. Of course, no framework can profess to be a sole way to look at the field. However, bringing together ideas may help to define and promote the state of scientific knowledge in this area, moving theory and research along in ways that are even more productive, exponentially increasing an understanding of counterfactual thinking ... and beyond.

REFERENCES

Arkin, R. M., & Shepperd, J. A. (1989). Self-presentation styles in organizations. In R. A. Giacalone & P. Rosenfeld (Eds.), *Impression management in the organization* (pp. 125–139). Hillsdale, NJ: Lawrence Erlbaum Associates.

Bargh, J. A. (1994). The four horsemen of automaticity: Awareness, intention, efficiency, and control in social cognition. In R. S. Wyer, Jr. & T. K. Srull (Eds.), *Handbook of social cognition* (2nd ed., Vol. 1, pp. 1–40). Hillsdale, NJ: Lawrence Erlbaum Associates.

Baumeister, R. F., Tice, D. M., & Hamilton, D. G. (1989). Self-presentational motivations and personality differences in self-esteem. *Journal of Personality, 57,* 547–579.

Berglas, S., & Jones, E. E. (1978). Drug choice as a self-handicapping strategy in response to noncontingent success. *Journal of Personality and Social Psychology, 36,* 405–417.

Boninger, D. S., Gleicher, F., & Strathman, A. (1994). Counterfactual thinking: From what might have been to what may be. *Journal of Personality and Social Psychology, 67,* 297–307.

Bower, G. H. (1991). Mood congruity of social judgments. In J. P. Forgas (Ed.), *Emotion and social judgments* (pp. 31–53). Elmsford, NY: Pergamon Press.

Brown, G. P., MacLeod, A. K., Tata, P., & Goddard, L. (2002). Worry and the simulation of future outcomes. *Anxiety, Stress, and Coping: An International Journal, 15,* 1–17.

Brown, J. D., & Mankowski, T. A. (1993). Self-esteem, mood, and self-evaluation: Changes in mood and the way you see you. *Journal of Personality and Social Psychology, 64,* 421–430.

Buehler, R., Griffin, D., & MacDonald, H. (1997). The role of motivated reasoning in optimistic time predictions. *Personality and Social Psychology Bulletin, 23,* 238–247.

Buehler, R., Griffin, D., & Ross, M. (1994). Exploring the "planning fallacy": Why people underestimate their task completion times. *Journal of Personality and Social Psychology, 67,* 366–381.

Byrne, R. M. J. (2002). Mental models and counterfactual thoughts about what might have been. *Trends in Cognitive Science, 6,* 426–431.

Christensen-Szalanski, J. J., & Willham, C. F. (1991). The hindsight bias: A meta-analysis. *Organizational Behavior and Human Decision Processes, 48,* 147–168.

Collins, R. L. (1996). For better or worse: The impact of upward social comparison on self-evaluations. *Psychological Bulletin, 119,* 51–69.

Comins, N. F. (1993). *What if the moon didn't exist? Voyages to Earths that might have been.* New York: HarperCollins.

Cowley, R. (Ed.). (1999). *What if? The world's foremost historians imagine what might have been.* New York: Putnam.

Davey, G. C. L., Jubb, M., & Cameron, C. (1996). Catastrophic worrying as a function of changes in problem-solving confidence. *Cognitive Therapy and Research, 20,* 333–344.

Einhorn, H. J., & Hogarth, R. M. (1981). Behavioral decision theory: Processes of judgment and choice. *Annual Review of Psychology, 32*, 53–88.

Erber, R., & Erber, M. W. (1994). Beyond mood and social judgment: Mood incongruent recall and regulation. *European Journal of Social Psychology, 24*, 79–88.

Fein, S., & Spencer, S. J. (1997). Prejudice as self-image maintenance: Affirming the self through derogating others. *Journal of Personality and Social Psychology, 73*, 31–44.

Fischhoff, B. (1975). Hindsight ≠ foresight: The effect of outcome knowledge on judgment under uncertainty. *Journal of Experimental Psychology: Human Perception and Performance, 1*, 288–299.

Fivush, R., Haden, C., & Reese, E. (1996). Remembering, recounting, and reminiscing: The development of autobiographical memory in social context. In D. Rubin (Ed.), *Remembering our past: Studies in autobiographical memory* (pp. 341–359). New York: Cambridge University Press.

Galinsky, A. D., & Moskowitz, G. B. (2000). Counterfactuals as behavioral primes: Priming the simulation heuristic and consideration of alternatives. *Journal of Experimental Social Psychology, 36*, 257–383.

Gilbert, D. T., Pinel, E. C., Wilson, T. D., Blumberg, S. J., & Wheatley, T. P. (1998). Immune neglect: A source of durability bias in affective forecasting. *Journal of Personality and Social Psychology, 59*, 617–638.

Gilovich, T., Kerr, M., & Medvec, V. H. (1993). Effect of temporal perspective on subjective confidence. *Journal of Personality and Social Psychology, 64*, 552–560.

Gleicher, F., Kost, K. A., Baker, S. M., Strathman, A. J., Richman, S. A., & Sherman, S. J. (1990). The role of counterfactual thinking in judgments of affect. *Personality and Social Psychology Bulletin, 16*, 284–295.

Gold, D. B., & Wegner, D. M. (1995). Origins of ruminative thought: Trauma, incompleteness, nondisclosure, and suppression. *Journal of Applied Social Psychology, 25*, 1245–1261.

Helgeson, V. S., & Mickelson, K. D. (1995). Motives for social comparison. *Personality and Social Psychology Bulletin, 21*, 1200–1209.

Higgins, E. T. (1998). Promotion and prevention: Regulatory focus as a motivational principle. In M. P. Zanna (Ed.), *Advances in experimental social psychology* (Vol. 23, pp. 305–331). New York: Academic Press.

Hofstadter, D. R. (1979). *Gödel, Escher, Bach: An eternal golden braid*. New York: Vintage Books.

Holman, E. A., & Silver, R. C. (1998). Getting "stuck" in the past: Temporal orientation and coping with trauma. *Journal of Personality and Social Psychology, 74*, 1146–1163.

Isen, A. M. (1987). Affect, cognition, and social behavior. In L. Berkowitz (Ed.), *Advances in experimental social psychology* (Vol. 20, pp. 203–253). San Diego: Academic Press.

Johnson-Laird, P. N., & Byrne, R. M. J. (2002). Conditionals: A theory of meaning, pragmatics, and inference. *Psychological Review, 109*, 646–678.

Kahneman, D., & Miller, D. T. (1986). Norm theory: Comparing reality to its alternatives. *Psychological Review, 93*, 136–153.

Kahneman, D., & Tversky, A. (1979). Intuitive prediction: Biases and corrective procedures. *Management Science, 12*, 313–327.

Kahneman, D., & Tversky, A. (1982). The simulation heuristic. In D. Kahneman, P. Slovic, & A. Tversky (Eds.), *Judgment under uncertainty: Heuristics and biases* (pp. 201–208). New York: Cambridge University Press.

Kendall, P. C., & Ingram, R. E. (1987). The future for cognitive assessment of anxiety: Let's get specific. In L. Michaelson & L. M. Ascher (Eds.), *Anxiety and stress disorders: Cognitive-behavioral assessment and treatment* (pp. 89–104). New York: Guilford Press.

Klinger, E. (1975). Consequences to commitment to and disengagement from incentives. *Psychological Review, 82*, 223–231.

Klinger, E. (1977). *Meaning and void: Inner experience and the incentives in people's lives*. Minneapolis: University of Minnesota Press.

Landman, J. (1987). Regret and elation following action and inaction: Affective responses to positive versus negative outcomes. *Personality and Social Psychology Bulletin, 13,* 524–536.

Lazarus, R. S., & Folkham, S. (1984). *Stress, appraisal, and coping*. New York: Springer-Verlag.

Lipe, M. G. (1991). Counterfactual thinking as a framework for attribution theories. *Psychological Bulletin, 109,* 456–471.

Louie, T. A. (1999). Decision makers' hindsight bias after receiving favorable and unfavorable feedback. *Journal of Applied Psychology, 84,* 29–41.

Macrae, C. N., & Milne, A. B. (1992). A curry for your thoughts: Empathic effects on counterfactual thinking. *Personality and Social Psychology Bulletin, 18,* 625–630.

Mark, M. M., & Folger, R. (1984). Responses to relative deprivation: A conceptual framework. In P. Shaver (Ed.), *Review of personality and social psychology: Emotions, relationships, and health* (Vol. 5, pp. 192–218). Beverly Hills, CA: Sage.

Mark, M. M., & Mellor, S. (1991). Effect of self-relevance of an event on hindsight bias: The foreseeability of a layoff. *Journal of Applied Psychology, 76,* 569–577.

Markman, K. D., Gavanski, I., Sherman, S. J., & McMullen, M. N. (1993). The mental simulation of better and worse possible worlds. *Journal of Experimental Social Psychology, 29,* 87–109.

Markman, K. D., & McMullen, M. N. (2003). A reflection and evaluation model of comparative thinking. *Personality and Social Psychology Review, 7,* 244–267.

Martin, L. L., & Tesser, A. (1989). Toward a motivational and structural theory of ruminative thought. In J. S. Uleman & J. A. Bargh (Eds.), *Unintended thought* (pp. 306–326). New York: Guilford Press.

Martin, L. L., & Tesser, A. (1996). Some ruminative thoughts. In R. S. Wyer (Ed.), *Ruminative thoughts: Advances in social cognition* (Vol. 9, pp. 1–47). Mahwah, NJ: Lawrence Erlbaum Associates.

Martin, L. L., Tesser, A., & McIntosh, W. D. (1993). Wanting but not having: The effects of unattained goals on thoughts and feelings. In D. M. Wegner & J. W. Pennebaker (Eds.), *The handbook of mental control* (pp. 552–572). New York: Prentice-Hall.

McGill, A. L., & Tenbrunsel, A. E. (2000). Mutability and propensity in causal selection. *Journal of Personality and Social Psychology, 79,* 677–689.

McKenzie, C. R. M., Ferreira, V. S., Mikkelsen, L. A., McDermott, K. J., & Skrable, R. P. (2001). Do conditional hypotheses target rare events? *Organizational Behavior and Human Decision Processes, 85,* 291–309.

McMullen, M. N. (1997). Affective assimilation and contrast in counterfactual thinking. *Journal of Experimental Social Psychology, 33,* 77–100.

Medvec, V. H., Madey, S. F., & Gilovich, T. (1995). When less is more: Counterfactual thinking and satisfaction among Olympic athletes. *Journal of Personality and Social Psychology, 69,* 603–610.

Miller, D. T., & Gunasegaram, S. (1990). Temporal order and the perceived mutability of events: Implications for blame assignment. *Journal of Personality and Social Psychology, 59,* 1111–1118.

Miller, D. T., Turnbull, W., & McFarland, C. (1990). Counterfactual thinking and social perception: Thinking about what might have been. In M. P. Zanna (Ed.), *Advances in experimental social psychology* (Vol. 23, pp. 305–331). New York: Academic Press.

Mussweiler, T. (2003). Comparative processes in social judgment: Mechanisms and consequences. *Psychological Review, 110,* 472–489.

Nisan, M. (1972). Dimension of time in relation to choice behavior and achievement orientation. *Journal of Personality and Social Psychology, 21,* 175–182.

Nisan, M. (1973). Perception of time in lower-class black students. *International Journal of Psychology, 8,* 109–116.

Nolen-Hoeksema, S. (1996). Chewing the cud and other ruminations. In R. S. Wyer (Ed.), *Ruminative thoughts: Advances in social cognition* (Vol. 9, pp. 135–144). Mahwah, NJ: Lawrence Erlbaum Associates.

Nolen-Hoeksema, S., McBride, A., & Larson, J. (1997). Rumination and psychological distress among bereaved partners. *Journal of Personality and Social Psychology, 72,* 855–862.

Nolen-Hoeksema, S., & Morrow, J. (1991). A prospective study of depression and posttraumatic stress symptoms after a natural disaster: The 1989 Loma Prieta earthquake. *Journal of Personality and Social Psychology, 61,* 155–171.

Norem, J. K., & Cantor, N. (1986). Defensive pessimism: "Harnessing" anxiety as motivation. *Journal of Personality and Social Psychology, 51,* 1208–1217.

Oettingen, G. (1996). Positive fantasy and motivation. In P. M. Gollwitzer & J. A. Bargh (Eds.), *The psychology of action: Linking cognition and motivation to behavior* (pp. 236–259). New York: Guilford Press.

Oettingen, G., Pak, H., & Schnetter, K. (2001). Self-regulation and goal-setting: Turning free fantasies about the future into binding goals. *Journal of Personality and Social Psychology, 80,* 736–753.

Parrott, W. G., & Sabini, J. (1990). Mood and memory under natural conditions: Evidence for mood incongruent recall. *Journal of Personality and Social Psychology, 59,* 321–336.

Peterson, C., Seligman, M. E. P., Yurko, K. H., Martin, L. R., & Friedman, H. S. (1998). Catastrophizing and untimely death. *Psychological Science, 9,* 127–130.

Rachman, S. (1981). Unwanted intrusive cognitions. *Advances in Behavioral Research and Therapy, 3,* 89–99.

Roese, N. J. (1994). The functional basis of counterfactual thinking. *Journal of Personality and Social Psychology, 66,* 805–818.

Roese, N. J. (1997). Counterfactual thinking. *Psychological Bulletin, 121,* 133–148.

Roese, N. J., & Olson, J. M. (Eds.). (1995). *What might have been: The social psychology of counterfactual thinking.* Hillsdale, NJ: Lawrence Erlbaum Associates.

Roese, N. J., & Olson, J. M. (1997). Counterfactual thinking: The intersection of affect and function. In M. P. Zanna (Ed.), *Advances in experimental social psychology* (Vol. 29, pp. 1–59). San Diego: Academic Press.

Roese, N. J., Sanna, L. J., & Galinsky, A. D. (2004). The mechanics of imagination: Automaticity and control in counterfactual thinking. In J. A. Bargh, J. S. Uleman, & R. R. Hassin (Eds.), *The new unconscious* (pp. 138–170). New York: Oxford University Press.

Rybarczyk, B. D. (1995). *The reminiscence interview: A new approach to stress intervention in the medical setting.* New York: Springer-Verlag.

Sanna, L. J. (1996). Defensive pessimism, optimism, and simulating alternatives: Some ups and downs of prefactual and counterfactual thinking. *Journal of Personality and Social Psychology, 71,* 1020–1036.

Sanna, L. J. (1997). Self-efficacy and counterfactual thinking: Up a creek with and without a paddle. *Personality and Social Psychology Bulletin, 23,* 654–666.

Sanna, L. J. (1998). Defensive pessimism and optimism: The bitter-sweet influence of mood on performance and prefactual and counterfactual thinking. *Cognition and Emotion, 12,* 635–665.

Sanna, L. J. (1999). Mental simulations, affect, and subjective confidence: Timing is everything. *Psychological Science, 10,* 339–345.

Sanna, L. J. (2000). Mental simulation, affect, and personality: A conceptual framework. *Current Directions in Psychological Science, 9,* 168–173.

Sanna, L. J., & Chang, E. C. (2003). The past is not what it used to be: Optimists' use of retroactive pessimism to diminish the sting of failure. *Journal of Research in Personality, 37,* 388–404.

Sanna, L. J., Chang, E. C., & Meier, S. (2001). Counterfactual thinking and self-motives. *Personality and Social Psychology Bulletin, 27,* 1023–1034.

Sanna, L. J., & Mark, M. M. (1995). Self-handicapping, expected evaluation, and performance: Accentuating the positive and attenuating the negative. *Organizational Behavior and Human Decision Processes, 64,* 84–102.

Sanna, L. J., & Meier, S. (2000). Looking for clouds in a silver lining: Self-esteem, mental simulations, and temporal confidence changes. *Journal of Research in Personality, 34,* 236–251.

Sanna, L. J., Meier, S., & Turley-Ames, K. J. (1998). Mood, self-esteem, and counterfactuals: Externally attributed moods limit self-enhancement strategies. *Social Cognition, 16,* 267–286.

Sanna, L. J., Meier, S., & Wegner, E. C. (2001). Counterfactuals and motivation: Mood as input to affective enjoyment and preparation. *British Journal of Social Psychology, 40,* 235–256.

Sanna, L. J., Parks, C. D., Meier, S., Chang, E. C., Kassin, B. R., Lechter, J. L., Turley-Ames, K. J., & Miyake, T. M. (2003). A game of inches: Spontaneous use of counterfactuals by broadcasters during Major League Baseball playoffs. *Journal of Applied Social Psychology, 33,* 455–475.

Sanna, L. J., & Schwarz, N. (2004). Integrating temporal biases: The interplay of focal thoughts and accessibility experiences. *Psychological Science, 15,* 474–481.

Sanna, L. J., Schwarz, N., & Small, E. M. (2002). Accessibility experiences and the hindsight bias: I knew it all along versus it could never have happened. *Memory & Cognition, 30,* 1288–1296.

Sanna, L. J., Schwarz, N., & Stocker, S. L. (2002). When debiasing backfires: Accessible content and accessibility experiences in debiasing hindsight. *Journal of Experimental Psychology: Learning, Memory, and Cognition, 28,* 497–502.

Sanna, L. J., Small, E. M., & Cook, L. M. (2004). Social problem solving and mental simulation: Heuristics and biases on the route to effective decision making. In E. C. Chang, T. J. D'Zurilla, & L. J. Sanna (Eds.), *Social problem solving: Theory, research, and training* (pp. 135–149). Washington, DC: American Psychological Association.

Sanna, L. J., Stocker, S. L., & Clarke, J. A. (2003). Rumination, imagination, and personality: Specters of the past and future in the present. In E. C. Chang & L. J. Sanna (Eds.), *Virtue, vice, and personality: The complexity of behavior* (pp. 105–124). Washington, DC: American Psychological Association.

Sanna, L. J., & Turley, K. J. (1996). Antecedents to spontaneous counterfactual thinking: Effects of expectancy violation and outcome valence. *Personality and Social Psychology Bulletin, 22,* 906–919.

Sanna, L. J., & Turley-Ames, K. J. (2000). Counterfactual intensity. *European Journal of Social Psychology, 30,* 273–296.

Sanna, L. J., Turley-Ames, K. J., & Meier, S. (1999). Mood, self-esteem, and simulated alternatives: Thought-provoking affective influences on counterfactual direction. *Journal of Personality and Social Psychology, 76,* 543–558.

Schaller, M., & Cialdini, R. B. (1990). Happiness, sadness, and helping: A motivational integration. In E. T. Higgins & R. M. Sorrentino (Eds.), *Handbook of motivation and cognition: Foundations of social behavior* (Vol. 2, pp. 265–296). New York: Guilford Press.

Schwarz, N. (1998). Accessible content and accessibility experiences: The interplay of declarative and experiential information in judgment. *Personality and Social Psychology Review, 2,* 87–99.

Schwarz, N., & Bless, H. (1992). Constructing reality and its alternatives: An inclusion/exclusion model of assimilation and contrast in social judgment. In L. L. Martin & A. Tesser (Eds.), *The construction of social judgments* (pp. 217–245). Hillsdale, NJ: Lawrence Erlbaum Associates.

Schwarz, N., & Clore, G. L. (1996). Feelings as phenomenal experiences. In E. T. Higgins & A. W. Kruglanski (Eds.), *Social psychology: Handbook of basic principles* (pp. 433–465). New York: Guilford Press.

Sedikides, C., & Strube, M. J. (1997). Self-evaluation: To thine own self be good, to thine own self be sure, to thine own self be true, and to thine own self be better. In M. P. Zanna (Ed.), *Advances in experimental social psychology* (Vol. 29, pp. 209–269). San Diego: Academic Press.

Shepperd, J. A., Findley-Klein, C., Kwavnick, K. D., Walker, D., & Perez, S. (2000). Bracing for loss. *Journal of Personality and Social Psychology, 78,* 620–634.

Shepperd, J. A., Ouellette, J. A., & Fernandez, J. K. (1996). Abandoning unrealistic optimism: Performance estimates and the temporal proximity of self-relevant feedback. *Journal of Personality and Social Psychology, 70,* 844–855.

Simon, P. (1968). Bookends theme [Recorded by P. Simon & A. Garfunkel]. On *Bookends* [record]. Los Angeles: Columbia Records. (April 3, 1968).

Sirois, F. M. (2004). Procrastination and counterfactual thinking: Avoiding what might have been. *British Journal of Social Psychology, 43,* 269–286.

Smith, E. R., & DeCoster, J. (2000). Dual-process models in social and cognitive psychology: Conceptual integration and links to underlying memory systems. *Personality and Social Psychology Review, 4,* 108–131.

Spellman, B. A., & Mandel, D. R. (1999). When possibility informs reality: Counterfactual thinking as a cue to causality. *Current Directions in Psychological Science, 8,* 120–123.

Strack, F., Schwarz, N., & Gschneidinger, E. (1985). Happiness and reminiscing: The role of time perspective, affect, and mode of thinking. *Journal of Personality and Social Psychology, 49,* 1460–1469.

Strathman, A., Gleicher, F., Boninger, D. S., & Edwards, C. S. (1994). The consideration of future consequences: Weighing immediate and distant outcomes of behavior. *Journal of Personality and Social Psychology, 66,* 742–752.

Taylor, S. E., & Lobel, M. (1989). Social comparison activity under threat: Downward evaluation and upward contact. *Psychological Review, 96,* 569–575.

Taylor, S. E., Pham, L. B., Rivkin, I. D., & Armor, D. A. (1998). Harnessing the imagination: Mental simulation, self-regulation, and coping. *American Psychologist, 53,* 429–439.

Taylor, S. E., & Schneider, S. K. (1989). Coping and the simulation of events. *Social Cognition, 7,* 174–194.

Tetlock, P. E., & Belkin, A. (Eds.). (1996). *Counterfactual thought experiments in world politics: Logical, methodological, and psychological perspectives.* Princeton, NJ: Princeton University Press.

Turley, K. J., Sanna, L. J., & Reiter, R. L. (1995). Counterfactual thinking and perceptions of rape. *Basic and Applied Social Psychology, 17,* 285–303.

Tversky, A., & Kahneman, D. (1973). Availability: A heuristic for judging probability and frequency. *Cognitive Psychology, 5,* 207–232.

Vasey, M. W., & Borkovec, T. D. (1992). A catastrophizing assessment of worrisome thoughts. *Cognitive Therapy and Research, 16,* 505–520.

Webster, J. D. (1993). Construction and validation of the reminiscence functions scale. *Journal of Gerontology: Psychological Sciences, 48,* 256–262.

Wegener, D. T., & Petty, R. E. (1994). Mood management across affective states: The hedonic contingency hypothesis. *Journal of Personality and Social Psychology, 66,* 1034–1048.

Wells, G. L., & Gavanski, I. (1989). Mental simulation of causality. *Journal of Personality and Social Psychology, 56,* 161–169.

Wilson, T. D., Wheatley, T., Meyers, J. M., Gilbert, D. T., & Axom, D. (2000). Focalism: A source of durability bias in affective forecasting. *Journal of Personality and Social Psychology, 78,* 821–836.

Wood, J. V. (1989). Theory and research concerning social comparison of personal attributes. *Psychological Bulletin, 106,* 231–248.

Wood, J. V., Taylor, S. E., & Lichtman, R. R. (1985). Social comparison in adjustment to breast cancer. *Journal of Personality and Social Psychology, 49,* 1169–1183.

10

Attitudes Over Time: Attitude Judgment and Change

Rick D. Brown
Dolores Albarracín
University of Florida

Years ago, Sherif and Hovland (1961) suggested that people who hold an attitude about an object are likely to perceive new information about the object in relation to the attitude. For example, one's prior attitude toward condom use will partially determine the effect of a strong external endorsement for condom use on one's subsequent attitude. If recipients favor condom use only slightly, the campaign will offer a relatively stronger endorsement than the recipients'. The size of the difference, however, should matter a great deal in predicting the actual effects of the campaign. If the distance between the two positions is small, recipients may perceive the campaign as congruent with their prior attitudes and shift in the direction of increased support for condom use. In contrast, as the distance between the two positions increases, recipients will be likely to distinguish or contrast the two positions. Under these conditions, they may shift their attitude in the opposite direction of the campaign, resulting in decreased rather than increased support for condom use. Following this logic, prior attitudes, other accessible evaluative information, and the comparison of these sources of information jointly determine the stability of attitudes over time.

On the heels of Sherif and Hovland's (1961) seminal work, however, the majority of researchers have concentrated on the simpler influence that prior attitudes exert on subsequent evaluative judgments (for a review, see Eagly & Chaiken, 1993; Johnson, Maio, & Smith-McLallen, in press; Petty, Wheeler, & Tormala, 2003). As part of this attempt, the representation of attitudes in memory and the mechanisms through which these attitudes influence subsequent judgments have been a matter of considerable attention. For example, Fazio (1995) described attitude representations in permanent memory as a concept node (flower) linked to an evaluative node (e.g., pleasant). When this link is strong, the prior evaluation is more accessible in memory and, thus, it is more likely that the evaluative node will be activated whenever one encounters or recalls the attitude object. Important to note, these accessible attitudes do not rest in memory inconsequentially (e.g., Fazio, 1990). They influence not only future evaluative judgments, but also the future processing of information about the attitude object and the behaviors in which people engage. For example, this impact is so important that sometimes having a prior attitude blinds people to changes in the real object (Fazio, Ledbetter, & Towles-Schwen, 2000). Thus, for Sherif and Hovland as well as Fazio, evaluative judgments that are based on existing attitudes are fundamentally different from judgments that require novel evaluations (see also Albarracín, Wallace, & Glasman, 2004).

In contrast to Sherif and Hovland's (1961) assumptions, other researchers have downplayed the role of attitudes stored in permanent memory. As the most visible proponents of this perspective, Schwarz and colleagues (Schwarz & Bohner, 2001; Schwarz & Strack, 1991) have persuasively argued that attitudes are momentary evaluative judgments constructed on the basis of whatever information happens to be accessible at the time. Thus, even when a prior evaluation serves as an input for this construction, the old evaluation is translated and typically integrated with other information rather than just applied automatically after its recall. After all, the argument goes, if people cannot even report their attitude toward a political party without the response being biased by subtle features of the questions, why should researchers dedicate themselves to explicating the structure of enduring attitudes stored in permanent memory? For these reasons, Schwarz and colleagues have modeled attitude judgment while minimizing the importance of representations of attitudes in memory and the study of attitude change.

Although attitudes have long been considered central to social psychology (see Allport, 1935; Eagly & Chaiken, 1993), adequate integration between representational and constructionist perspectives has been scarce. The lack of integration is unfortunate because, although a disparity exists between these perspectives, both views are likely to capture aspects of the true picture. For instance, most theorists would probably agree that the enduring attitudes people hold are highly diagnostic for making evaluative judgments. At the same time, most would also agree that judgments involve the selection of an informational basis as well as response generation. Therefore, although attitudes may be highly diagnostic, other types of available information can also contribute to judgments and constraints in response formats can force individuals to further reconstruct their judgments. In this context, a comprehensive model of attitudes must acknowledge the contributions of both memory-based and online evaluations, and capture both the durability and the fluidity of evaluative judgments. As we see later, some of the most recent models of attitudes (e.g., Bassili & Brown, in press) and models of attitude change (e.g., Albarracín et al., 2004) have embraced this logic, and thus appear to have embraced the implications of Sherif and Hovland's (1961) seminal analysis.

Moreover, in keeping with the central theme of this volume we highlight how the consideration of future orientation has contributed to our understanding of attitudes. Research in other domains has already shown that one's temporal perspective is likely to influence how one elaborates upon novel information and ultimately one's judgments (see chaps. 2 and 15, this volume). Perhaps surprisingly, little effort has been made to directly explore the role of this temporal factor in attitude judgment and change (see Strathman, Gleicher, Boninger, & Edwards, 1994). Nonetheless, on the basis of the existing literature we speculate how the consideration of future orientation may cast further light on the contributions of memory-based and online evaluative processes to the dynamics of attitudes over time.

REPRESENTATIONAL MODELS OF ATTITUDES

Fazio's Model of the Attitude Representation

Fazio (1986, 1990, 1995; Fazio & Towles-Schwen, 1999) offers what is arguably the most prominent contemporary representational model of attitudes. According to him, attitudes are represented in memory as summary evaluations associated with the attitude object. Although the object–evaluation associations are presumed to be part of much broader networks that may include other information about the object, the model concentrates on the strength of the association between an evaluation and an attitude object. Attitudes are thought to fall on a continuum defined, at one end, by representations of attitude objects that are not associated with a summary evaluation (i.e., nonattitudes; see Converse, 1964, 1974) and, at the other end, by representations of attitude objects that are strongly associated with a summary evaluation.

According to the model, attitude accessibility is determined by the strength of the association between an attitude object and its evaluation. When the object–evaluation link is strong, the attitude is highly accessible and exposure to the attitude object will activate the evaluation. The process is thought to be automatic and important to the attitude–behavior relationship because activated evaluations can guide thought and behavior in the presence of the attitude object (Fazio, Powell, & Herr, 1983; for a review, see Ajzen & Fishbein, in press). For example, highly accessible attitudes exert strong influences on behavior (Fazio, 1990) and can bias perceptions of attitude objects (Fazio et al., 2000). To this extent, information about an object is likely to have a different impact when people possess a prior attitude and when they do not.

Perhaps ironically, even though Fazio's model recognizes that existing attitudes often influence the impact of novel information, it does not attempt to describe specific processes that allow these representations to be incorporated with other information at the time of evaluative judgment, or how representations change in the face of compelling contradictory information. Other models, however, have attempted to explicate these processes.

MODELS OF ONLINE INFORMATION PROCESSING

Constructionist Models

In stark contrast to traditional representational models of attitudes, constructionist models emphasize the contribution of online evaluations derived on the basis of whatever information happens to be accessible at the time. The strong form of this argument implies that evaluative judgment is exclusively guided by information present in the external context rather than on elements that are represented in memory (Schwarz & Bohner, 2001). For example, individuals may use the affective reactions (e.g., Schwarz & Clore, 1983) or the physiological arousal (e.g., Valins, 1966; Wells & Petty, 1980) they momentarily experience to determine their evaluations of objects they encounter, without ever bothering to recall a prior attitude about these objects. A more tempered application of Schwarz and Bohner's construal model suggests that memory-based evaluative information about an attitude object plays a role in judgment, but this role is often no more important than that of external inputs. Moreover, even when a prior judgment serves as a basis for a subsequent judgment, the judgment is still constructed anew, just constructed using old information from memory.

On the basis of their inclusion/exclusion model, Schwarz and colleagues (Schwarz, 1999; Schwarz & Bless, 1992) described how momentarily accessible information can contribute to evaluative assimilation and contrast effects and corresponding instability in attitude judgment. For example, in one study Stapel and Schwarz (1998) drew participants' attention to Colin Powell's (a highly popular military leader at the time) decision to join the Republican Party or to his decision to reject an offer to run as a presidential candidate for the Republican Party before they were asked to evaluate the party. When participants were asked what party Colin Powell recently joined, evaluations of the party were more favorable presumably because the question led participants to think of well-liked Colin Powell as being part of the Republican Party. Under these conditions, participants' extremely positive regard for Powell may have been used as an input for the evaluation of the party. When participants were asked about Powell's refusal to run as a presidential candidate for the Republican Party, evaluations of the party were less favorable, presumably because the question led participants to distance Powell from the party. Under these conditions, participants' extremely positive regard for Powell may be used as a standard of comparison, rendering the judgment of the party less favorable by contrast. Assimilation and contrast effects like these are often difficult to explain if one assumes that evaluative judgments are driven by the retrieval of stable attitudes represented in memory.

By incorporating evaluative inputs associated with momentarily accessible evaluative information and concentrating on the role of processes underlying judgment construction, the construal model signifies a great step forward in addressing the context-sensitivity of attitude judgment. Despite its strengths, however, it has garnered criticism due to its limited attention to evaluative structures encoded in memory. Contrary to the assumption that evaluations stored in memory are not important, the literature and daily experience are filled with examples that reveal the dramatic influence of some attitudes stored in memory (for a review, see Bassili & Brown, in press; Petty & Krosnick, 1995; Petty et al., 2003). Hence, insofar as the scope of the construal model does not adequately address the representational aspects of attitudes and the issue of attitude change, it presents a partial picture of the processes involved in attitude judgment.

Reception Models

Models of attitude change have emerged largely from an interest in persuasion and from attempts to account for enduring changes in attitudes. Initially, social psychologists simply applied learning theory and assumed that people change their attitudes when they receive the "right" information (for a recent review see Johnson et al., in press). Thus, incentive models prescribed associating a particular message recommendation or idea with an adequately reinforcing stimulus. Associating the idea with the proper reward, such as approval from a communicator, should in turn elicit attitude change.

As research on persuasion evolved and as psychology shifted from learning to cognition, theories developed an understanding of the processes triggered by a persuasive communication. As the most prominent example, McGuire (1968a) conceptualized the impact of persuasive messages as contingent on the stages of exposure to the communication, attention, comprehension, yielding or acceptance of the message's position, retention of the new attitude, and behavior. However, research following McGuire's analysis was slow to deepen understanding of the various stages of processing he identified. In fact, McGuire (1968b) himself abandoned the distinction between exposure, attention, and comprehension and proposed a single stage comprising all aspects of message reception, though retaining yielding as the second stage.

In spite of the simplification the reception-yielding model represents, the model is useful to generate a number of interesting predictions concerning the relationship between factors of the recipients (personality, intelligence) or the environment (distraction) and attitude change. According to the model, individual difference variables can exert opposing effects on reception and yielding. For example, McGuire predicted that self-esteem and intelligence should relate positively to reception but negatively to yielding. Presumably, persons with higher intelligence or self-esteem are better able to attend to and comprehend information (increased reception) but also better able to defend their initial attitudes and be critical of new information (depressed yielding). As a result of the play of these two antagonistic influences, the overall impact of intelligence and self-esteem on persuasion should be curvilinear, with persons on the midscale positions being persuaded more than those at higher or lower positions.

The predictions derived from the reception-yielding model have received some support. For example, Eagly and Warren (1976) explored the influence of intelligence on reception and yielding by exposing participants to persuasive messages containing complex or simple arguments. Compared to their less intelligent counterparts, intelligent participants were expected to better comprehend the complex arguments and thus demonstrate greater attitude change when such arguments were included in the message. At the same time, intelligent participants were expected to defend their own attitudes better than their less intelligent counterparts, and thus demonstrate less attitude change when simple arguments were presented. As predicted, when the messages were supported with complex arguments there was a positive correlation between intelligence and attitude change, whereas when the messages were supported with simple arguments, there was a negative correlation between intelligence and attitude change. Overall, the empirical support for McGuire's predictions has, however, been relatively weak (see Eagly & Chaiken, 1993, for a review). Perhaps the streamlined model fails to address processes that contribute to the integration of prior attitudes and the evaluative implications of other available information, thus producing inadequate predictions of communication outcomes.

MODELS OF THE ROLE OF ATTITUDE REPRESENTATIONS AND ONLINE INFORMATION PROCESSING

Dual-Process Models

Other models have emphasized the processes that account for the selection of specific information at the time of judgment, recognizing that factors that decrease ability to comprehend or think about a message determine the way in which people express and ultimately change their attitudes. Specifically, dual-process theories like the elaboration likelihood model (ELM; Petty & Cacioppo, 1981, 1986; Petty & Wegener, 1999) and the heuristic-systematic model (HSM; Chaiken, 1987; Chaiken, Lieberman, & Eagly, 1989) both propose that recipients of a persuasive communication scrutinize the arguments the communication presents if and only if they are able and motivated to do so. Cognitive resources and motivation, however, are limited. Therefore, when ability and motivation are scarce, recipients of a communication are influenced by information other than the arguments contained in the communication. For example, the communication might be more persuasive if it contains three arguments instead of one, regardless of the quality of the arguments being considered. Although the ELM and HSM are distinct models that stem from different traditions, the models share many fundamental features and can accommodate the same findings (Petty et al., 2003). For our present purposes, we simply focus on the ELM.

At the heart of the ELM is the elaboration likelihood continuum. An individual's position along the continuum, determined by their motivation and ability to carefully think about the issues at hand, has qualitative and quantitative implications. The qualitative and quantitative implications of elaboration likelihood were demonstrated by Petty, Cacioppo, and Goldman (1981). In this study, student participants were told that the university was currently considering policy changes to academic programs, including the implementation of comprehensive exams prior to college graduation. Participants were instructed to evaluate the broadcast quality of arguments in favor of the implementation of the exams for possible use on the university radio station. The researchers manipulated personal involvement, message argument quality, and message source expertise to see if argument quality and source expertise were contingent on the level of personal involvement with the issue, which presumably affects recipients' motivation to think about the message.

Personal involvement was manipulated by telling half the participants that the comprehensive exam policy would be implemented during the following year, thereby affecting them personally. The remaining participants were told that the policy would be implemented 10 years down the line, thus never affecting them directly. To manipulate argument quality, half of the participants received messages with strong arguments that contained statistics and evidence (e.g., "the institution of the exams has led to a reversal in the declining scores on standardized achievement tests at other universities"). The remaining participants received messages with weak arguments (e.g., "a friend of the author's had to take a comprehensive exam and now has a prestigious academic position"). Finally, to manipulate the expertise of the source of the message, half of the participants were led to believe that the policy report had been prepared by a class at a local high school (low expertise), whereas the remaining half of the participants were told that the report had been prepared by a professor of education at Princeton University (high expertise).

As predicted by the ELM, argument quality exerted a significant impact on attitudes toward comprehensive exams when personal involvement was high. Independent of source expertise, strong arguments produced more agreement than did weak arguments. Under low involvement, however, the researchers observed the opposite pattern. Independent of argument quality, participants agreed with the message more when the source was an expert than when it was not. In terms of the ELM, highly involved participants believed that the institution of the senior comprehensive exam would directly affect them and were, therefore, motivated to process the issue-relevant merits of the message. Hence, strong arguments exerted a greater impact on these participants' attitudes than did weak ones. In contrast, low-involvement participants had little motivation to elaborate on the arguments of the message. Implementation of the exam was 10 years away and by then they would have long since graduated. Having little motivation to elaborate on the message, these participants primarily focused on the peripheral cue—the expertise of the source.

The Role of Future Orientation. One of the most remarkable characteristics of dual-process models like the ELM is their ability to address the influence of a variety of individual and environmental factors. As noted earlier, research in other domains has already shown that one's future orientation is likely to influence how one elaborates upon novel information and ultimately one's judgments (see chaps. 2 and 15, this volume). To our knowledge, however, only Strathman et al. (1994) have directly explored the influence of this temporal factor in the domain of attitudes. In particular, Strathman and colleagues demonstrated that chronic differences in consideration of future consequences (CFC) influenced attitude change toward issues that involve immediate and distant consequences of events such as offshore oil drilling. Specifically, among low-CFC participants, who tend to focus on the immediate implications of their current actions, attitudes about drilling were more favorable when the advantages were characterized as immediate and the disadvantages as far off. In contrast, high-CFC participants, who tend to focus on the future implications of their current actions, were more persuaded when the advantages were distant and the disadvantages immediate.

These findings can be interpreted as a reflection of individual differences in the processing of information concerning future versus immediate outcomes, leading to increased elaboration of messages that match the recipients' temporal perspective. Interestingly, this bias remains unaltered when one introduces instructions that might be expected to correct it. Subsequent research demonstrated that even when low-CFC individuals are prompted to think about the future, they do not seem to weigh possible futures much at all (Boninger, Gleicher, & Strathman, 1994).

Attitude Stability. One of the most important conclusions offered by process models is that attitudes formed on the basis of careful processing of information will show greater temporal stability, greater impact on behavior, and greater resistance to counterpersuasion than attitudes formed on the basis of peripheral processing. Petty and Cacioppo (1986) outlined various mediating mechanisms that are presumably responsible for the effects of elaboration on attitude strength. Generally speaking, elaboration involves greater thinking about the attitude object and contributes to heightened accessibility of the attitude. Heightened accessibility, in turn, increases the probability that the same attitude will be expressed at two points in time and that the attitude will be available to guide behavior in the future (Fazio, 1990, 1995). In some cases, however, thinking about the attitude object even superficially, as induced by repeated peripheral processing of peripheral cues (Petty & Cacioppo, 1986) or by rehearsal of the message content without elaboration (Zanna, Fazio, & Ross, 1994), can also contribute to attitude strength through increased accessibility.

The Role of Prior Attitudes. Although we have classified dual-process models like the ELM as addressing both prior attitudes and online information processing, these models only superficially acknowledge the influence and role of prior attitudes. In part, the relative lack of importance assigned to prior attitudes derives from the fact that the model assumes the same processes apply for the case of attitude formation as of change (Petty & Cacioppo, 1986). Past research, however, has not clearly established whether people who possess prior attitudes utilize the same types of heuristics as people who lack prior attitudes. To examine this issue Kumkale and Albarracín (2003) conducted a meta-analytic review of longitudinal research on the effects of source credibility and distinguished between cases in which participants possessed an initial attitude toward the target issue and those in which they did not. The results revealed that the effects of source credibility were small when participants possessed an initial attitude or had sufficient ability or motivation to form a new attitude online on the basis of the arguments contained in the message. In contrast, when participants could neither retrieve an attitude from memory nor form a new attitude online, the impact of the credibility of the source of the communication strengthened significantly.

The Potentiated Recruitment Framework

Models inspired by connectionism (see Smith, 1996) offer an alternative means to account for the influences of both enduring attitudes and the evaluative implications of momentarily accessible information. For example, according to the potentiated recruitment framework (PRF; Bassili & Brown, in press) attitudes are represented as molecular elements that have the potential to be recruited in various mixes depending on the eliciting context and chronic potentiating factors. Evaluation emerges in response to the activity of microconceptual networks that are activated by contextually situated attitude objects, goals, and task demands. Therefore, evaluations are as fluid and context dependent as the combined activation of chronic and temporary relevant patterns allows.

According to the model, evaluations emerge as a result of four primary sources of potentiation. One source is recent cognitive experiences that prime particular microconcepts in memory. Another source is the current information available about the attitude object and the context in which it is situated. This source of potentiation is particularly important because it comprises activation resulting from both enduring evaluations toward the attitude object and the specific eliciting conditions. Thus, even subtle features of the context can exert considerable influence on the emergent evaluation. The third source of potentiation consists of the flow of activation between linked microconcepts and accounts for the potential influence of general knowledge and culture on attitudes. Finally, cognitive activity in working memory is an important source of potentiation, particularly in marking the distinction between implicit and explicit evaluative judgments. By explicating all these sources of variability in the potentiation of attitudes, this framework increases the possibility of making predictions about evaluations at different points in time. Hopefully, we will see developments from this model in the area of attitude change in the years to come.

The Activation and Comparison Model

In contrast to previous theories, the activation and comparison model of attitude survival and change (Albarracín et al., 2004) attempts to incorporate the key aspects of both representational and information processing models by embracing the distinction between attitude formation and attitude change. Naturally, the model attempts to take a broader look at processes that take place at the time of attitude judgment and account for the possible simultaneous contributions of memory-based and online evaluations.

As discussed at the outset of this chapter, the attitudes people hold are likely to be diagnostic whenever an evaluative judgment is necessary. Extant research, however, indicates that judgment processes involve the selection of an informational basis as well as response generation (Wyer & Srull, 1989; see also Albarracín, 2002). As diagnostic as attitudes may be, other types of information that are chronically or momentarily available can also be selected and contribute to responses. Embracing this logic, Albarracín and colleagues attempt to provide a parsimonious yet comprehensive account of how the old attitude and the present information collectively contribute to the dynamics of change in evaluative judgments over time.

The activation and comparison model is simple to the point of obviousness. It emphasizes that understanding and predicting attitude change requires examination of three processes: (a) activating the prior attitude (retrieving it from memory), (b) activating information related to the prior attitude (which can come from memory or an external source), and (c) comparing the prior attitude with the related information. None of the processes is inevitable, and each process can have different implications for attitude change and maintenance. On the one hand, the sole activation of either attitude-consistent information or the prior attitude itself will lead to attitude maintenance. On the other hand, online reconstruction of an attitude based on the sole activation of attitude-inconsistent information, as well as comparison of the prior attitude with attitude-consistent or -inconsistent information should generally produce attitude change. Nevertheless, these two processes do not always occur independently of each other, and better understanding of attitude change emerges from a joint consideration of the two. Although the processes themselves are not counterintuitive, their joint implications, as elaborated in the model, often contrast with prior assumptions and predictions. The outcomes of each process in isolation and in combination are described next.

Activation. The model allows for independent effects resulting from the activation of existing attitudes and other accessible evaluative information. Activation of an existing attitude in the absence of comparison with other available information typically results in judgment stability. In contrast, when other information is accessible and an existing attitude is not activated, judgments should be based primarily on the online evaluation of this information. Under these conditions, attitude stability should occur when the evaluative implications of the accessible information are congruent with the prior attitude, but judgments should change when these implications are incongruent with the prior attitude.

Comparison. Individuals increase the confidence or extremity of an existing attitude when the evaluative implication of other accessible information corroborates their attitude. Correspondingly, individuals maintain the confidence or extremity of an existing attitude when the attitude and novel information are evaluatively congruent but only one is valid. Furthermore, individuals increase their confidence in, or extremity of, attitude judgment when they perceive their evaluation as valid and comparative processes have resulted in the invalidation of other evaluatively incongruent information. In contrast, comparing a prior attitude with incongruent but equally valid novel information results in moderation of the prior attitude. As implied by these possibilities, generally speaking, comparative processes will contribute to attitude change.

Reciprocal Influence of Activation and Comparison. Although the activation of a prior attitude will increase attitude maintenance in the absence of comparative processes, attitude activation will contribute to change when comparative processes are active. Simply put, in order to compare an existing attitude with other available information, the initial attitude must first be brought to mind. When a prior attitude is relatively inaccessible, the motivation to compare these two elements will facilitate attitude activation but may not be sufficient to produce the actual comparison. Under these conditions, the evaluative implication of the other available information is not likely to serve as an input for judgment construction, leading to stability in attitude judgments.

Empirical Support. Wallace and Albarracín (2003) conducted three studies exploring how factors associated with information selection and comparative processes influence attitude survival and change. As part of the procedures, participants were induced to form an initial positive attitude toward a proposal to implement comprehensive exams. After a period of time, participants received additional information about the proposal before reporting their attitude toward comprehensive exams for a second time.

As part of the procedures of Studies 1 and 2, the initial message advocating comprehensive exams was followed by another message that contained new arguments also in support of the exams. In Study 1, activation and comparison were elicited by presenting the first message, which served as a basis for the initial attitude, at the time of presenting the second message. In Study 2, the comparison manipulation entailed explicit instructions to compare the implications of the second message with participants' attitudes based on the first message. The results of Studies 1 and 2 indicated that, as predicted by the model, judgments polarized when participants were induced to compare the new message with their earlier attitude, but showed greater stability when no comparison induction was in place.

Wallace and Albarracín's (2003) Study 3 was particularly helpful in clarifying the role of the accessibility of a prior attitude in subsequent evaluative judgments. In Studies 1 and 2, individual differences in need to evaluate (NE; Jarvis & Petty, 1996)—which in an independent sample correlated negatively with attitude response latencies (lower NE = slower attitude responses)—were used to estimate prior attitude accessibility (low NE = low attitude accessibility). To complement these findings, Study 3 assessed initial attitude accessibility more directly by measuring initial attitude response latencies (speedy attitude judgments were interpreted as an indicator of subsequent accessibility). Furthermore, Study 3 manipulated prior attitude accessibility by reminding half of the participants of their initial attitude judgment before they read the second information set, which contained information contradictory to the first set.

As expected, attitudes were generally stable when the initial attitude was highly accessible. Hence, akin to previous findings by Fazio (1989), these findings show that highly accessible attitudes can impede the online formation of new inconsistent attitudes. Of greater interest, however, are the findings from conditions in which comparative processes were active and the initial attitude was inaccessible. When participants' initial attitudes were inaccessible, evaluative judgments changed to a greater extent when the comparison instructions were presented with a reminder of the prior attitude rather than alone. These results, therefore, support the central tenet of the activation and comparison model that when people are motivated to compare their prior attitudes with new information, having a highly accessible prior attitude can actually promote attitude change. In contrast, when participants' initial attitudes were accessible, attitudes changed in response to comparison instructions regardless of whether or not an attitude reminder was present.

Wallace and Albarracín's (2003) findings are particularly interesting because they are consistent the model's prediction that both attitude stability and change can occur due to memory-based and online attitudinal processes. According to the model, when asked to report their attitude at Time 2, participants who were not induced to compare the second set of information with their initial attitude should have constructed a judgment primarily on the basis of their initial attitude if their initial attitude was accessible. In contrast, participants should construct a judgment primarily on the basis of the evaluative implications of the second information set if their initial attitude was not accessible. In Studies 1 and 2, the second information set was consistent with the first information set. Under these conditions, it is difficult to distinguish between judgments constructed on the basis of memory-based or online evaluations because both can (and did) yield attitude judgments consistent with the initial attitude. More informative, however, are the results of Study 3 in which the second information set was inconsistent with the first. According to the model, participants whose initial attitude is inaccessible should construct their judgment at Time 2 on the basis of the second evaluatively inconsistent message and thus change their attitude judgments. In contrast, those participants who spontaneously activated or were induced to activate their initial attitudes should maintain their initial judgments when no comparison instructions are present. Again, the results were in line with the predictions of the model.

Summary. In light of these findings, Albarracín and colleagues provide compelling evidence in support of their hypotheses that the activation of prior attitudes, the activation of attitude-related information, and the comparison of the attitude with the other information jointly determine the survival and change of prior attitudes. As a result, researchers may now have a better understanding of parts of the many different processes that govern the evolution of attitudes over time. Nevertheless, important aspects of these processes remain outside of the model. One such area is highlighted by Wilson, Lindsey, and Schooler (2000), who suggest that an attitude that changes does not perish. According to Wilson and colleagues, when people change a prior attitude, the prior attitude can persist at the implicit level and reemerge under some conditions. Because the scope of the activation and comparison model does not extend to storage processes per se, readers should consult Wilson et al.'s work for a treatment of how different attitudes may coexist in memory (but see Bassili & Brown, in press; Fazio & Olson, 2003).

In presenting their conceptualization of attitude survival and change, Albarracín and colleagues (2004) considered the possibility that people can activate up to two cognitive elements (i.e., the prior attitude and novel attitude relevant information) at a time. Without a doubt, however, individuals spend their lives in environments with large amounts of information. As a result, they must often make decisions after considering multiple elements that have the potential to guide their future attitudes. Hence, Albarracín and colleagues acknowledge that the presence of multiple prior attitudes or multiple pieces of novel information should have important implications for the processes they postulate. Future research should, therefore, address the processes elicited by information of greater complexity.

GENERAL DISCUSSION AND FINAL COMMENTS

For some time, the literature on attitudes has lacked a comprehensive interpretation of the mechanisms underlying judgment survival and change. In particular, we emphasized the inadequate integration between representational and constructionist perspectives. We also suggested that a comprehensive model of attitudes must capture both the durability and the fluidity of evaluative judgments over time, and thus must acknowledge the contributions of prior attitudes and of the evaluative implications of other available information. As part of our review, we presented two recently introduced models that have embraced this logic: the potentiated recruitment framework of attitudes (Bassili & Brown, in press) and the activation and comparison model of attitude survival and change (Albarracín et al., 2004). We believe that these integrative models offer researchers exciting new tools with which to shape future research.

Earlier we noted that a recent consideration in attitude change is that one's concern for future consequences is likely to influence how one elaborates upon novel information and ultimately one's evaluative judgments (e.g., Strathman et al., 1994). It interesting to contemplate the extent to which individual differences along this dimension may also cast further light on the contributions of memory-based and online evaluative processes in response to persuasive messages. On one hand, low-CFC individuals who are grounded "in the moment" may be more likely to base their evaluations on information available in the here and now. Moreover, these online evaluations may be primarily derived from available information that concerns the immediate consequences of one's actions. Following this logic, online information that contradicts previous information about the immediate consequences of one's actions should prompt low-CFC individuals to engage in comparative and elaborative processes, thereby provoking changes in attitude judgments.

On the other hand, high-CFC individuals, who have a chronic tendency to relate present actions to future consequences, may be more likely to rely on evaluations stored in memory, which may provide a more efficient means to relate current actions to long-standing distal goals. Moreover, these memory-based evaluations may be primed by available cues that concern future consequences. Following this logic, novel information that challenges one's prior evaluations of the future consequences should prompt high-CFC individuals to engage in comparative processes, contributing to changes in attitude judgments. Clearly, this line of reasoning is highly speculative and calls for future research. The CFC construct, however, may offer researchers a promising tool with which to examine the contribution of online and memory-based evaluative processes to attitude judgment and change.

In closing, many researchers have investigated the predictors and consequences of attitude survival and change. As a result, contemporary researchers have a greater understanding of the many different processes that govern the dynamics of attitudes over time. Unfortunately, the existing literature addressing attitude survival and change has suffered from a lack of comprehensive, theoretical integration, which is not surprising considering the difficulties inherent in achieving such integration. In view of this situation, we have called for an enhanced integration and theoretical development in attitude research and highlight two recent advancements toward this goal. The integrative models we have reviewed may offer researchers a new means to guide the development of programs to improve judgments and behaviors that are important for individuals in society. Given the important societal consequences of attitude survival and change we hope that these models and further consideration of the future orientation construct will stimulate such efforts in the future.

REFERENCES

Ajzen, I., & Fishbein, M. (in press). The influence of attitudes on behavior. In D. Albarracín, B. T. Johnson, & M. P. Zanna (Eds.), *Handbook of attitudes and attitude change: Basic Principles*. Mahwah, NJ: Lawrence Erlbaum Associates.

Albarracín, D. (2002). Cognition in persuasion: An analysis of information processing in response to communications. In M. P. Zanna (Ed.), *Advances in experimental social psychology*, (Vol. 34, pp. 61–130). San Diego: Academic Press.

Albarracín, D., Wallace, H. M., & Glasman, L. R. (2004). Survival and change in judgments: A model of activation and comparison. In M. P. Zanna (Ed.), *Advances in experimental social psychology* (Vol. 36, pp. 251–315). San Diego: Academic Press.

Allport, G. W. (1935). Attitudes. In C. Murchison (Ed.), *Handbook of social psychology* (pp. 798–844). Worcester, MA: Clark University Press.

Bassili, J. N., & Brown, R. D. (in press). Implicit and explicit attitudes: Research, challenges, and theory. In D. Albarracín, B. T. Johnson, & M. P. Zanna (Eds.), *Handbook of attitudes and attitude change: Basic principles*. Mahwah, NJ: Lawrence Erlbaum Associates.

Boninger, D. S., Gleicher, F., & Strathman, A. (1994). Counterfactual thinking: From what might have been to what may be. *Journal of Personality & Social Psychology, 67*, 297–307.

Chaiken, S. (1987). The heuristic model of persuasion. In M. P. Zanna, J. M. Olson, & C. P. Herman (Eds.), *Social influence: The Ontario Symposium* (Vol. 5, pp. 3–39). Hillsdale, NJ: Lawrence Erlbaum Associates.

Chaiken, S., Liberman, A., & Eagly, A. H. (1989). Heuristic and systematic processing within and beyond the persuasion context. In J. S. Uleman & J. A. Bargh (Eds.), *Unintended thought* (pp. 212–252). New York: Guilford Press.

Converse, P. E. (1964). The nature of beliefs systems in mass publics. In D. Apter (Ed.), *Ideology and discontent* (pp. 206–261). New York: The Free Press.

Converse, P. E. (1974). Comment: The status of nonattitudes. *American Political Science Review, 68,* 650–660.

Eagly, A. H., & Chaiken, S. (1993). *The psychology of attitudes.* New York: Harcourt Brace.

Eagly, A. H., & Warren, R. (1976). Intelligence, comprehension, and opinion change. *Journal of Personality, 44,* 226–242.

Fazio, R. H. (1986). How do attitudes guide behavior? In R. M. Sorrentino & T. E. Higgins (Eds.), *Handbook of motivation and cognition: Foundations of social behavior* (pp. 204–243). New York: Guilford Press.

Fazio, R. H. (1989). On the power and functionality of attitudes: The role of attitude accessibility. In S. J. Breckler & A. G. Greenwald (Eds.), *Attitude structure and function* (pp. 153–179). Hillsdale, NJ: Lawrence Erlbaum Associates.

Fazio, R. H. (1990). Multiple processes by which attitudes guide behavior: The MODE model as an integrated framework. In M. P. Zanna (Ed.), *Advances in experimental social psychology* (Vol. 23, pp. 75–109). New York: Academic Press.

Fazio, R. H. (1995). Attitudes as object–evaluation associations: Determinants, consequences, and correlates of attitude accessibility. In R. E. Petty & J. A. Krosnick (Eds.), *Attitude strength: Antecedents and consequences* (pp. 247–282). Hillsdale, NJ: Lawrence Erlbaum Associates.

Fazio, R. H., Ledbetter, J. E., & Towles-Schwen, T. (2000). On the costs of accessible attitudes: Detecting that the attitude object has changed. *Journal of Personality & Social Psychology, 78,* 197–210.

Fazio, R. H., & Olson, M. A. (2003). Implicit measures in social cognition research: Their meaning and use. *Annual Review of Psychology, 54,* 297–327.

Fazio, R. H., Powell, M. C., & Herr, P. M. (1983). Toward a process model of the attitude-behavior relation: Accessing one's attitude upon mere observation of the attitude object. *Journal of Personality & Social Psychology, 44,* 723–735.

Fazio, R. H., & Towles-Schwen, T. (1999). The MODE model of attitude-behavior processes. In S. Chaiken & Y. Trope (Eds.), *Dual-process theories in social psychology* (pp. 97–116). New York: Guilford Press.

Johnson, B. T., Maio, G. R., & Smith-McLallen, A. (in press). In D. Albarracín, B. T. Johnson, & M. P. Zanna (Eds.), *Handbook of attitudes and attitude change: Basic principles.* Mahwah, NJ: Lawrence Erlbaum Associates.

Kumkale, T. G., & Albarracín, D. A. (2003). *Attitude formation and change: A meta-analytic review of longitudinal research on the effects of source credibility.* Unpublished manuscript, University of Florida, Gainesville.

McGuire, W. J. (1968a). Personality and attitude change: An information-processing theory. In A. G. Greenwald, T. C. Brock, & T. M. Ostrom (Eds.), *Psychological foundations of attitudes* (pp. 171–196). San Diego: Academic Press.

McGuire, W. J. (1968b). Personality and susceptibility to social influence. In E. F. Borgatta & W. W. Lambert (Eds.), *Handbook of personality theory and research* (pp. 1130–1187). Chicago: Rand McNally.

Petty, R. E., & Cacioppo, J. T. (1981). *Attitudes and persuasion: Classic and contemporary approaches.* Dubuque, IA: Brown.

Petty, R. E., & Cacioppo, J. T. (1986). *Communication and persuasion: Central and peripheral routes to attitude change.* New York: Springer/Verlag.

Petty, R. E., Cacioppo, J. T., & Goldman, R. (1981). Personal involvement as a determinant of argument-based persuasion. *Journal of Personality & Social Psychology, 41*, 847–855.

Petty, R. E., & Krosnick, J. A. (1995). *Attitude strength: Antecedents and consequences.* Hillsdale, NJ: Lawrence Erlbaum Associates.

Petty, R. E., & Wegener, D. T. (1999). The elaboration likelihood model: Current status and controversies. In S. Chaiken & Y. Trope (Eds.), *Dual-process theories in social psychology* (pp. 41–72). New York: Guilford Press.

Petty, R. E., Wheeler, S. C., & Tormala, Z. L. (2003). Persuasion and attitude change. In T. Mellon & M. J. Lerner (Eds.), *Handbook of psychology: Vol. 5. Personality and social psychology* (pp. 353–382). Hoboken, NJ: Wiley.

Schwarz, N. (1999). Self-reports: How the questions shape the answers. *American Psychologist, 54*, 93–105.

Schwarz, N., & Bless, H. (1992). Scandals and the public's trust in politicians: Assimilation and contrast effects. *Personality & Social Psychology Bulletin, 18*, 574–579.

Schwarz, N., & Bohner, G. (2001). The construction of attitudes. In A. Tesser & N. Schwarz (Eds.), *Blackwell handbook of social psychology: Intraindividual processes* (pp. 436–457). Malden, MA: Blackwell.

Schwarz, N., & Clore, G. L. (1983). Mood, misattribution, and judgments of well-being: Informative and directive functions of affective states. *Journal of Personality & Social Psychology, 45*, 513–523.

Schwarz, N., & Strack, F. (1991). Context effects in attitude surveys: Applying cognitive theory to social research. In W. Stroebe & M. Hewstone (Eds.), *European review of social psychology* (Vol. 2, pp. 31–50). Chichester, England: Wiley.

Sherif, M., & Hovland, C. I. (1961). Placement of items on controversial issues. In M. Sherif & C. I. Hovland (Eds.), *Social judgment* (pp. 99–126). New Haven, CT: Yale University Press.

Smith, E. R. (1996). What do connectionism and social psychology offer each other? *Journal of Personality and Social Psychology, 70*, 893–912.

Stapel, D. A., & Schwarz, N. (1998). The Republican who did not want to become president: Colin Powell's impact on evaluations of the Republican Party and Bob Dole. *Personality & Social Psychology Bulletin, 24*, 690–698.

Strathman, A., Gleicher, F., Boninger, D. S., & Edwards, C. S. (1994). The consideration of future consequences: Weighing immediate and distant outcomes of behavior. *Journal of Personality & Social Psychology, 66*, 742–752.

Valins, S. (1966). Cognitive effects of false heart-rate feedback. *Journal of Personality & Social Psychology, 4*, 400–408.

Wallace, H. W., & Albarracín, D. (2003). *Maintenance and change of attitudes as a function of activation and comparison.* Unpublished manuscript, University of Florida, Gainesville.

Wells, G. L., & Petty, R. E. (1980). The effects of overt head movement on persuasion: Compatibility and incompatibility of responses. *Basic and Applied Social Psychology, 1*, 219–230.

Wilson, T. D., Lindsey, S., & Schooler, T. Y. (2000). A model of dual attitudes. *Psychological Review, 107*, 101–126.

Wyer, R. S., & Srull, T. K. (1989). Memory cognition in its social context. Hillsdale, NJ: Lawrence Erlbaum Associates.

Zanna, M. P., Fazio, R. H., & Ross, M. (1994). The persistence of persuasion. In R. C. Schank & E. Langer (Eds.), *Beliefs, reasoning, and decision making: Psychology in honor of Bob Abelson* (pp. 347–362). Hillsdale, NJ: Lawrence Erlbaum Associates.

III

Groups and Interpersonal-Level Processes

III

11

The Dimension of Time in Interdependence Theory

Ann C. Rumble
The Ohio State University

Robert has been married to Mary for 5 years. Recently Robert and Mary have experienced greater levels of dissatisfaction with their relationship than in the past. Robert is now faced with an opportunity to start a new relationship with an attractive woman from work or choose to work on his current relationship with Mary.

Israeli officials are negotiating with the Palestinian Authority on a peace accord. At the same time, Israeli politicians decide to build a wall between Israeli and Palestinian territories. Palestinian officials now must decide how to respond to the wall being built and how it will shape the current peace accord negotiations.

Each of these situations is an example of an *interdependent* situation or relationship. Interdependence can be defined as any situation in which two or more interacting individuals have some degree of control over the other's outcomes; the structure of the situation encourages awareness of this control; and the nature of the relationship affects how this control is expressed (Kelley et al., 2003; Kelley & Thibaut, 1978; Thibaut & Kelley, 1959). Interdependence theory is an extension of Lewin's (1946) basic assertion that behavior is a function of the person and the situation [in Lewin's terms, $B = f(S, P)$]

(Holmes, 2002; Kelley et al., 2003; Lewin, 1946; Rusbult & Van Lange, 2003). Interdependence theorists assert that social interactions are a function of each individual and the particular situation in which they are interacting [I = *f*(Situation, Person A, Person B)], and these situations may include features that encourage cooperative or competitive behavior (Kelley et al., 2003; Kelley & Thibaut, 1978; Thibaut & Kelley, 1959). In the Israeli/Palestinian negotiation example, the current situation involves competition for scarce land resources, which heightens awareness that the two groups must coexist in the same physical space, and there is additionally a history of distrust, which could easily influence which of the available behaviors each party will select during the next interactions.

But in addition to the influence of the individuals, the situation, and the relationship, time also has an effect on the course of human interactions (Kelley, 1984, 1997a). In this chapter, I examine how time, as well as cognitions about time, influence interdependent interactions. Specifically, I want to consider how interactions evolve within a continuing relationship, as well as how time affects whether people choose to remain in an existing relationship or move to a new one. I first briefly review interdependence theory and related research on close relationships and negotiation, including a discussion of how different aspects of the situation, the interaction partners, and their relationship are influenced by time or evolve over time. I also argue that if individuals within interdependent situations consider future interactions, this may lead to more integrative negotiations and healthier relationship behaviors. I then discuss concepts within interdependence theory that have not been fully tested within a temporal context, and suggest possible directions for research on the role of time within interdependent situations.

INTERDEPENDENCE THEORY

Interdependence theory provides a framework for understanding social interactions (please see Kelley et al., 2003, and Rusbult & Van Lange, 1996, 2003, for complete reviews of interdependence theory). It explicitly states that any interpersonal interaction can be modeled with a consideration of the environment within which the interaction occurs. It also takes into account how each individual's behavior shapes the outcomes present within that environment. Additionally, the motives and cognitive processes of each individual affect the outcome of each interaction. The motives that shape the interaction are influenced by the nature of the relationship between the interacting members. Kelley (1997b) discusses this interaction between the individual and the environment as a "stimulus field." Finally, an interdependence analysis models the effect of time on the situation or relationship, so that a series of interactions are linked in a meaningful manner (Kelley, 1984, 1997a).

Outcome Matrices

One of the fundamental tools of interdependence theory is the outcome matrix (Kelley et al., 2003; Kelley & Thibaut, 1978; Thibaut & Kelley, 1959; von Neumann & Morgenstern, 1948). Outcome matrices were designed to model situations and display the outcomes possible for a particular interaction. A typical outcome matrix is shown in Table 11.1, in which the columns and rows list the behavioral options of the interaction partners. For example, within Table 11.1, if both countries choose to compete for a particular resource they will receive substantially less than if both had chosen to cooperate. By varying their behaviors each country helps determine the outcomes possible for a particular interaction. The most basic such table is a two-person, two-option situation, but any number of options and/or group members can be represented.

One of the major limitations of the outcome matrix approach to describing interdependent situations is that whereas one can easily describe a static interaction, it is difficult to capture the more dynamic aspects of the relationship that evolve over time using only a single-outcome matrix (Kelley, 1997a, 1997b). Most interdependent relationships and interactions involve a series of decisions or interactions regarding ongoing issues; additionally, participants frequently reevaluate the relationship so as to decide whether to remain in the situation, or select a new one. As such, theorists often rely upon *transition lists* to help incorporate temporal factors into interdependence-based models of behavior.

TABLE 11.1

Outcome Matrix for Israeli/Palestinian Negotiation

		Israel	
		Cooperate	*Compete*
Palestine	*Cooperate*	10/10	0/20
	Compete	20/0	3/3

Note. Numbers to the left of the diagonal lines in each cell are outcomes for the Palestinians; numbers to the right of the diagonal lines are outcomes for Israel.

Transition Lists

Transition lists (see Table 11.2 for an example) were designed to extend outcome matrices to model movement between and within interactions (see Kelley, 1984, 1997a for a complete description of transition lists). Specifically, they add to an outcome matrix a consideration of what will occur after individuals have made their behavioral choices. This is accomplished in the following manner. First, for each set of lists there is a list (list L) that represents the beginning of an interdependent interaction (see Table 11.2 for illustration). Lists within the set specify the behavioral choices for each member, and the immediate outcomes for that choice, as well as the consequences for each combination of behavioral choices. For example, within a particular round of the Israeli/Palestinian negotiation either side may choose to act on self-interest, and if they do they will influence not only their immediate outcomes but also the subsequent interactions. Similar to an outcome matrix, each list represents only a single interaction between the interacting members but these interactions are linked to consequences and other interactions. The consideration of consequences and the implicit consideration of future interactions within transition lists can help to guide researchers who want to examine temporal changes and dynamic relationships beyond the limits of the outcome matrix (Kelley, 1984, 1997a).

Transition lists, then, add to the interdependence model a consideration of time and the future. They also have the ability to incorporate behavioral options that cannot be represented within a simple outcome matrix, but that may impact interdependent situations in which time is a consideration. For example, inaction can result either as a consequence of one or both parties choosing not to act, or because a time limit has been reached. Thus, real time can become a factor within transition lists, in that it can force action or inaction.

The ability to exit an interaction is also possible within a transition list, as opposed to the outcome matrix, which when usually employed limits individuals to behavioral choices that produce a specific outcome. The ability for either interaction partner to exit any given interaction also has important implications for present and future interactions (Drigotas & Rusbult, 1992; Giebels, De Dreu, & Van de Vliert, 2000; Kelley, 1984, 1997). The decision to exit a particular relationship will depend on a number of factors including the person's evaluation of the current relationship and the degree to which he or she depends on the current relationship for positive outcomes.

Comparison level (CL) is a mechanism for evaluating a relationship or the outcome of an interaction (Kelley & Thibaut, 1978; Rusbult & Van Lange, 1996; Thibaut & Kelley, 1959). The comparison level is defined as the expected set of outcomes in a given relationship or interaction that has been influenced

TABLE 11.2

Transition List Based on Outcome Matrix for Israeli/Palestinian Negotiation

List	Israel's and Palenstine's Choice Options for Outcome Matrix in Figure 1.	Possible Choice Combinations	Outcomes for Each Choice Combination: Israel	Outcomes for Each Choice Combination: Palestine	Transition (indicates next list in the interaction)
L	Israel1, Israel2, IsraelN	I1 P1	No payoff for List L		go to List M
	Palestine1, Palestine2,	I1 P2			List N
		I1 PN			List L
		I2 P1			List O
		I2 P2			List Q
		I2 PN			List L
		IN P1			List L
		IN P2			List L
		IN PN			List L
M	I1, I2, IN	I1 P1	10	10	EXIT
	P1, P2, PN	(other pairs)			List L
		(i.e., I1 P2; I1 PN;			
N	I1, I2, IN	I1 P2	0	20	EXIT
	P1, P2, PN	(other pairs)			List L
O	I1, I2, IN	I2 P1	20	0	EXIT
	P1, P2, PN	(other pairs)			List L
Q	I1, I2, IN	I2 P2	3	3	EXIT
	P1, P2, PN	**(other pairs)**			List L

Note. I = Israel. *P* = Palestine. *1* = Cooperate (top row or left column of matrix in Table 11.1). *2* = Compete (bottom row or right column of matrix in Table 11.1). *N* = Inaction. EXIT = indicates actors leave the interaction. Other pairs refers to all other choice pairs, besides the one listed in the box immediately above. From Kelley (1997a). Copyright 1997 by John Wiley & Sons Limited. Reproduced by permission.

by past experiences and relationships. Expected outcomes are the outcomes a person either assumes they will receive based on previous experiences outcomes or has observed others receiving within similar relationships. A complementary piece of information is provided by the *comparison level for alternatives* (CL_{alt}), which can be defined as the weakest set of outcomes a person will accept, given alternative opportunities. So, an individual whose present and anticipated outcomes are better than CL should find the interaction satisfying, but as outcomes move below CL that interaction becomes less satisfying. CL_{alt} and CL can thus be influential in determining both how individuals choose new situations, as well as how behavior might be changed in an ongoing situation (Drigotas & Rusbult, 1992; Giebels et al., 2000; Kelley, 1984, 1997). If a person experiences a lower set of outcomes than CL, that individual will be motivated to leave. Similarly, if an individual is aware of alternative outcomes that are greater than $CL_{alt,}$ the individual will also be more likely to leave the current relationship for the alternative relationship.

The degree to which each partner is dependent on the other for positive outcomes will also influence exit decisions (Kelley et al., 2003; Kelley & Thibaut, 1978; Rusbult & Van Lange, 1996; Thibaut & Kelley, 1959). Within interdependent relationships, each partner is in some way dependent on the actions of the other for positive outcomes. The more control the other has over outcomes, the more dependent a partner is on the other for positive outcomes. In a mutually dependent relationship each partner is motivated to maintain the situation because of the other's ability to provide positive outcomes. This situation becomes more stable over time with each partner exhibiting equal amounts of power in the relationship (Rusbult, 1983). But a unilateral dependent situation is less stable over time because one partner needs the other more than the other needs him or her.

We see, then, that the behavioral choices embedded in transition lists differentiate them from simple outcome matrices, but that the basic determinants of interdependent behavior (i.e., comparison level and dependence) remain constant. In addition, the choice points that interaction members confront are also unique to transition lists (Kelley, 1984, 1997a). A choice point is a moment within an interaction in which a decision regarding the future of the interaction must be made by one or both partners. There are a number of different choice points that interacting partners can face when time is accounted for within an interdependent interaction. Individuals at the beginning of an interaction will often face the choice of how to, or who will, choose the initial situation between the partners, in addition to the type of situation to enter for that particular exchange. The individuals will also have to make decision regarding the possible outcomes and who shall act first within the interaction. Only after all of these determinations are made can the actual behavioral choices (i.e., the actions that produce the specific outcomes) occur. Successful navigation of these many choice points, and reception of outcomes associated with the behavioral choices, can then lead to a new transition list set. But navigation of the choice points depends on how the interacting individuals evaluate and transform any given situation into an effective situation.

Outcome Transformations

Outcome matrices and transition lists are used to describe a particular situation called the *given situation* (Kelley et al., 2003; Kelley & Thibaut, 1978; Thibaut & Kelley, 1959). The given situation does not, however, take into account individual differences in motivation or cognitions. Such factors often do operate on the given situation, such that it is transformed into an *effective situation* that reflects these differences. The effective situation represents outcomes that have been transformed through a number of interpersonal processes (Joireman, Kuhlman, Van Lange, Doi, & Shelley, 2003; Kelley & Thibaut, 1978; Rusbult & Van Lange, 1996; Thibaut & Kelley, 1959). For example, emotions, the perceived nature of the current relationship, and the evaluation of past interactions are all known to influence transformations (Kelley et al., 2003; Kelley & Thibaut, 1978; Thibaut & Kelley, 1959; Yovetich & Rusbult, 1994). Of particular interest for this discussion, there are a number of interpersonal motivational transformations that are directly influenced by both the history and anticipated future interactions of a given interdependent relationship. Aggression and revenge, forgiveness, and generosity are motivations to transform the given situation, which probably depend on past interactions and the anticipation of future interactions for activation (Van Lange, 2004).

Aggression and Revenge. Aggression is a motivation that is usually activated in response to another's behavior. If an individual has encountered an individual who, or situation that, in the past has caused harm, the person may act aggressively, which in interdependence terms is operationalized as assignment of a negative weight to other's outcomes (Van Lange, 2004). One school of thought contends that individuals who act in an aggressive manner are responding to past interactions that were characterized by noncooperation or destructive behavior (Van Lange, 2004). Individuals who act in an aggressive manner appear to "overassimilate" to noncooperative or destructive behavior. Thus, this motive is dependent on past interactions and anticipation of future interactions. But is all aggressive behavior destructive, or can it be adaptive? Tripp, Bies, and Aquino (2002) argue that revenge, or a reaction to destructive behavior, can be viewed as either "destructive and antisocial" or "constructive and prosocial." The extent to which a specific instance of revenge is constructive seems to be a function of the degree of altruism, and of justice inherent in the act of revenge. Altruism in this context is considered to have occurred if the person seeking revenge does not directly benefit from the act of revenge. Justice refers to the equivalence between the severity of the initial act and the act of revenge. So if an individual incurs costs (and gains no direct benefit) and the initial act and the act of revenge are equal, the act of revenge will be perceived as constructive and prosocial (Tripp et al., 2002). But revenge may not be a rational choice (particularly if the person seeking revenge incurs costs) if there was not a consideration of future interactions and was not in reaction to the other's past behavior; in other words, revenge is an irrational choice if the interaction is not considered within the context of time.

Forgiveness. At the other end of the behavioral continuum is forgiveness. Forgiveness has been defined as the motivation to control or suppress revenge behaviors. Forgiveness is also time dependent, in that it is a reaction to past behaviors and also anticipates future interactions with the transgressor. McCullough, Fincham, and Tsang (2003) were interested in expanding traditional forgiveness research, which is typically very static, in order to better represent the temporal nature of the construct. They defined forgiveness as a process of temporal change and a resistance to revenge behaviors, and as being a function of two different processes: *forbearance*, which refers to the motivation to resist the initial revenge urge, and *trend forgiveness*, which is defined as the extent to which there is longitudinal prosocial change (i.e., an increase or maintenance of prosocial behavior over time), and accounts for resistance to revenge over time. According to McCullough and colleagues it is the combination of initial forbearance and consequential trend forgiveness that leads an individual away from a continuation of destructive behavior. We see later that the breaking of a cycle of relationship-destructive behavior is very important within close relationships.

Generosity. Generosity is an additional motivation for which a consideration of future and past interactions guides activation of prosocial behavior. Individuals with a generous motivation place the greatest weight on other's outcomes within an interdependent relationship (Van Lange, 2004). Generosity can be motivated by empathy in situations in which we have information concerning the partner's hardships or we are in a close relationship with the other (Batson, 1991; Batson & Ahmad, 2001; Batson et al., 1995, 1996). So generosity is dependent on the history of the interaction for activation. But unlike forgiveness, which models increases in prosocial motivation over time, generosity can decrease over time, particularly within noncooperative environments (Rumble, 2003). Research on empathy and generous or altruistic behavior has suggested that other concerns (i.e., personal gain or welfare) can distract individuals from their generous or altruistic behaviors and concerns about the self take prominence over other concerns (Batson, 1991; Mansbridge, 1990). Thus, the passage of time can have a mixed effect on the activation of a generous motivation.

APPLICATIONS OF INTERDEPENDENCE THEORY

Interdependence theory has been applied to a vast array of situations (see Kelley et al., 2003, for some examples). Three specific areas of research in which the theory is most prominent are close relationships and negotiation; for the third area, interested readers can consult chapter 12, this volume, for a discussion of interdependence and cooperative behavior.

Close Relationships

Research on close relationships has addressed a number of time-related factors. Of particular interest for this chapter are commitment, dependence, and accommodation (Drigotas & Rusbult, 1992; Le & Agnew, 2003; Rusbult, Bissonnette, Arriaga, & Cox, 1998). These variables are crucial in understanding close relationships, in part, because they consider the influence of time and the consideration of the future on relationship behaviors.

Commitment. Commitment refers to a partner's long-term orientation toward the relationship—to what extent does a partner envision and invest in a positive relationship with an extended time horizon (Drigotas & Rusbult, 1992; Le & Agnew, 2003; Rusbult et al., 1998; Rusbult & Buunk, 1993; Rusbult, Martz, & Agnew, 1998; Rusbult & Van Lange, 1996)? Commitment thus explicitly references cognitions about time horizons, and research has shown it is affected by, and affects, a number of variables that also either explicitly or implicitly imply a consideration of past and future interactions. For example, researchers have studied to what extent degree of commitment to the relationship influences the willingness of romantic partners to temper negative reactions to potentially relationship damaging behavior, a process otherwise known as accommodation (Rusbult et al.,1998; Rusbult, Verette, Whitney, Slovik, & Lipkus, 1991). The research suggests that such temperance (i.e., accommodation) is important, because without a concern for future interactions, partners will be less motivated to react constructively to the destructive behavior, and instead tend to respond in a self-interested fashion, which can lead to a continuation instead of a cessation of the conflict (Rusbult et al.,1991, 1998).

Dependence. Dependence within interdependent relationships refers to a reliance on others for positive outcomes (Kelley et al., 2003; Kelley & Thibaut, 1978; Thibaut & Kelley, 1959). If these outcomes are lower than expected, or lower than desired levels, the result can be a desire to end the relationship. Thus, in order to model close relationships with interdependence theory, dependence must be considered in order to understand why some relationships endure, whereas others end (Drigotas & Rusbult, 1992; Kelley et al., 2003; Kelley & Thibaut, 1978; Thibaut & Kelley, 1959).

Research on close relationships has demonstrated that individuals often face stay/go decisions within their relationships (Drigotas & Rusbult, 1992; Rusbult & Buunk, 1993). Stay/go decision are choices between maintaining the current relationship and leaving the current relationship, either to enter an alternative relationship or to be alone. Although these decisions are shaped by a number of interdependent factors, there are two, commitment and dependence, for which the consideration of time is important (Drigotas & Rusbult, 1992). In our opening example, Robert's choice between working on his marriage and starting a new relationship with someone else is an example of situation selection, and his choice will be influenced by a number of interdependent variables, including dependence.

Drigotas and Rusbult (1992) propose that reduced satisfaction with a current relationship is the result of specific needs (intimacy, emotional involvement, etc.) not being satisfactorily met by the current relationship, and/or a realization that alternative relationships have higher benefit to cost ratios. Either partner's reduced dependence upon the current relationship for positive outcomes may thus lead individuals to exit the relationship and either enter a new, more rewarding relationship, or simply be alone. In two studies with dating couples, Drigotas and Rusbult found that individuals who were more dependent on their partner for need satisfaction had relationships that persisted longer than individuals with less satisfaction. In other words, people who depended on their relationship to meet particular needs (intimacy, companionship, and emotional involvement) were less likely to end that relationship than were people who were less dependent on the relationship for need satisfaction.

Accommodation. Within the context of a close relationship, the consideration of consequences and future interactions is particularly important because these relationships, by definition, exist over an extended time period. Accommodation is an interdependent process that considers relationship behaviors over time, particularly responses to relationship-destructive behavior. For example, Rusbult and colleagues (1991, 1998) have investigated the consequences of destructive behavior within close relationships. A pattern of destructive behavior, whether at the beginning of an interaction or in response to the other's behavior, is detrimental to the long-term health of a relationship. This is similar in nature to the negative-reciprocity cycle (or conflict spiral) that can occur within mixed-motive interactions—an initially negative action is responded to with a negative action, which in turn provokes yet another negative behavior by the original actor, and so on (Axelrod, 1984). A cycle of destructive behavior within close relationships can be reduced or stopped if the individuals respond to destructive behavior with accommodation, rather than another destructive behavior (Rusbult et al., 1991, 1998). In order to accommodate destructive behavior, an individual must be invested in and committed to the relationship, which requires a consideration of future interactions and consequences. By considering consequences and future interactions an individual may be able to respond with either active (voice) or passive (loyalty) relationship maintenance behaviors (or accommodative behavior), which are effective in reducing cycles of destructive behavior (Rusbult et al., 1991, 1998). It is interesting that within the negotiation literature there is a similar strategy for dealing with a negative-reciprocity cycle called the *gradual reduction in tension model* (or GRIT) (Osgood, 1962). GRIT is a process through which parties make small concession to the other in order to reduce tensions and the cycle of noncooperation with one another, and has been proven useful in de-escalating conflicts (Osgood, 1962). This is discussed further again within the section on interdependence and negotiation research.

Inaction is not only a viable choice within interdependent situations; within close relationships it is sometimes the most adaptive choice. Research on the accommodation model has found that if close relationship partners are able to control their desire to even the score when their partner acts in a destructive manner, relationship quality will often be enhanced (Drigotas & Rusbult, 1992; Rusbult et al., 1998). Such restraint may operationalize itself as inaction or tolerance of the negative behavior. Within the accommodation model such inaction is referred to as *loyalty*. More formally, loyalty can be defined as a passive behavior motivated by the hope and expectation that the relationship will improve without direct intervention by a third party (Drigotas & Rusbult, 1992; Rusbult et al., 1998). Inaction in response to destructive behaviors can thus improve future interactions within close relationships, but it does appear to require taking a long-term perspective on the relationship. If one or both partners is unable to take such a perspective, or is less motivated to sustain the relationship long term, accommodative behaviors may be less likely to occur (Drigotas & Rusbult, 1992; Rusbult et al., 1998). Individuals who are focused on current outcomes may act in a manner that serves short-term goals, and by doing so they may well sacrifice the long-term health of their relationship.

Conclusion. Research on close relationships has demonstrated and benefited from a consideration of time and its related behaviors and cognitions. Because close relationships exist within an extended time horizon, it is particularly important to understand how the consideration of past and future interactions shapes behavior.

Negotiation

Though research on negotiation rarely employs interdependence theory to model interaction, negotiation phenomena can nonetheless also be analyzed, and benefit from, an interdependence perspective (see Giebels et al., 2000, and Bazerman, Curhan, Moore, & Valley, 2000, for examples and discussions of the application of interdependence theory to negotiation research). For example, integrative negotiation can be argued to be a type of mixed motive interaction, which in turn can be modeled using interdependence theory (Thompson, 1998). In addition, interdependence concepts such as power and dependence play an important role in the study of negotiation processes (Giebels et al., 2000).

Integrative negotiations are interactions in which partners are able to take full advantage of all the resources at their disposal during the negotiation (Carnevale & Pruitt, 1992; Thompson, 1998). In integrative negotiations, negotiators are able to see the interaction as a non-zero-sum game, as opposed to a zero-sum game or distributive negotiation. Though most negotiations are integrative in nature, negotiators tend to bargain as if the situation is distributive in nature, a misperception that is called the "fixed-pie perception" (Bazerman & Neale, 1992; Carnevale & Pruitt, 1992; Harinck, De Dreu, & Van Vianen, 2000; Thompson, 1998; Thompson & Hastie, 1990). A particular strategy for achieving integrative agreements is logrolling. Logrolling involves negotiators making concessions on a particular issue that is important to the opponent but not to themselves (Carnevale & Pruitt, 1992; Lindskold & Han, 1988; Moran & Ritov, 2002; Thompson, 1990a, 1990b, 1998). It is possible that, in order to understand the integrative nature of a negotiation and the potential benefits of techniques such as logrolling, negotiators may need to able to envision future interactions. Negotiations that make use of logrolling could be modeled using transition lists. In a simple case, the transition list could help highlight that if one negotiation partner (the initial logroller) made an initial concession on an issue they deemed unimportant, this person could (in the long run) receive more on an issue they considered important, and this pattern could repeat until the parties reach an agreement or exit. Being able to see the integrative potential of a negotiation and use techniques such as logrolling increases the likelihood they will reach the best possible (i.e., pareto-optimal) agreement (Carnevale & Pruitt, 1992; Thompson, 1998). Achieving the pareto-optimal outcome may involve repeated interactions to successfully negotiate the relevant issues. The negotiation between Israel and Palestine is an example of a repeated situation that has the potential to be an integrative negotiation. The two parties negotiating on a peace accord may interact and make concessions to the other on numerous occasions over the same issues, and these concessions could move the parties closer to achieving a superior outcome.

Negotiations are affected by the ability to exit. Giebels et al. (2000) studied how the presence of exit options would affect negotiations. In their study, it was sometimes the case that neither partner had the option to exit; sometimes only one partner had such an option; and sometimes both partners had the ability to exit the relationship. When only one partner had the ability to exit the negotiation, that person focused less on the integrative possibilities within the negotiation, and instead acted as if it were a distributive negotiation (Giebels et al., 2000). This implies that such negotiators are possibly less focused on future interactions. In terms of time issues, it may be that focusing more attention on the power differential (or one party's ability to leave) acts to reduce the perceived time horizon. It is interesting that the researchers did not find any significant difference between both parties and neither party having the ability to exit. This may be because power was equal in both situations, which would allow negotiators to focus on the integrative possibilities instead of on the power differential. Also, within a situation where both negotiators can exit, the choice to exit has the potential to be a riskier choice than to stay. If that is the case, then negotiators may focus more on the potential for future interactions with their partners, which could lead to more integrative agreements.

Real time and time pressures are two additional factors of interest to negotiation researchers (Mosterd & Rutte, 2000; Stuhlmacher & Champagne, 2000; Stuhlmacher, Gillespie, & Champagne, 1998). Time pressure can be caused either by a true time limit placed on a negotiation or through an increase in the costs associated with any negotiation. Time pressure has been shown to increase a negotiator's tendency to be more risk averse, influence the nature of concession making (i.e., quicker to make concessions), and lead to less than pareto-optimal agreements (Mosterd & Rutte, 2000; Stuhlmacher & Champange, 2000; Stuhlmacher et al., 1998). De Dreu (2003) also found that time pressure increases the heuristic processing of negotiators. In particular, he found that increases in time pressure were linked to fixed-pie perceptions, which lead to more distributive as opposed to integrative negotiating. This limits the ability of negotiator to see the potential for mutually satisfying agreements. In other words, the role of real time (in particular time pressure) can lead negotiators to focus on the present instead of focusing on the long-term consequences of a particular negotiation.

Conclusion. Even though negotiation research has not always directly benefited from an interdependence theory perspective, any number of issues important to negotiation researchers can benefit from such an analysis. The influence of exit options on power within negotiations and the influence of time pressure are just two examples of how temporal concerns influence negotiation.

FUTURE DIRECTIONS AND CONCLUDING REMARKS

Interdependence theory has added to our understanding of interpersonal interactions, including research on close relationships and negotiation. Interdependence theory has potential to continue to guide research on interpersonal interactions particularly when temporal variables and constraints are considered. In fact, a more realistic model of human behavior is possible when time is considered within interpersonal interactions.

Within the field of close relationships, researchers have used interdependence theory to model relationship maintaining and destructive behaviors. In particular, the role of dependence, commitment, and accommodation has been explored in both marriages and dating couples. In future research it might be fruitful to investigate how close relationships evolve and change over time; in other words how do accommodation, commitment and dependence change over the course of a relationship? Through the use of transition lists it may be possible to model the evolution of close relationships through time. For example, the conflicts and decision making partners engage in could be modeled using transition lists, which could help to predict how the couple will manage conflict in the future or to suggest alternatives to ending conflict.

In addition to modeling using transition lists, Kelley and colleagues (2003) have identified a number of interpersonal structures that may apply to close relationships when time is considered. Research on willingness to sacrifice begins to answer the question of how couples coordinate their efforts, but how this process evolves over time within the relationship or the course of an interaction is unclear (Van Lange, Agnew, Harinck, & Steemers, 1997; Van Lange, Rusbult et al., 1997). In other words, it is unclear whether partners take turns sacrificing their own self-interest for the relationship or if over time one partner becomes more accommodating.

Negotiation researchers, though they have not always directly modeled negotiations using interdependence theory, could nonetheless benefit from such a perspective. The concept of choice points within interdependent interactions is one area where negotiation researchers could expand their understanding of the evolution of a negotiation. Negotiation research is particularly suited to test the assertions of interdependent theorists concerning choice points because of the dynamic nature of negotiations. Negotiators faced many of these choices over the course of an interactions; for example, they must choose who will make the opening offer or over what resources to negotiate. Modeling these choices using interdependence theory as a guide not only would add to the negotiation literature but would expand our basic understanding of interdependent processes over time.

One of the potential benefits in modeling interactions with interdependence theory is the ability to model behaviors that generalize over a large number of situations and relationships (Kelley et al., 2003). Finding similarities in the interactions between close relationship partners and negotiation partners provides a better understanding of basic human social behavior. For example, the similarities between accommodation in close relationships and GRIT in negotiations, indicates that if individuals are able to extend the time horizon of their relationship and act generously with the other, the interaction has the potential to lead to greater cooperation instead of noncooperation (Osgood, 1962; Rusbult et al., 1998).

In conclusion, although there are different methods of modeling interactions, the main focus of the preceding chapter was on interdependence theory. Interdependence theory guides research in a number of areas including close relationships and to a lesser extent negotiation processes. In particular, interdependence theory can help guide research where time is a consideration. The work by interdependent theorists also has the ability to expand our knowledge of basic interaction processes across a variety of situations.

ACKNOWLEDGMENTS

I would like to thank Craig D. Parks and Paul Van Lange for their comments on an earlier draft of this chapter.

Work on this chapter was supported by Grant T32-MH19728 from the National Institute of Mental Health.

REFERENCES

Axelrod, R. (1984). *The evolution of cooperation.* New York: Basic Books.

Batson, C. D. (1991). *The altruism question: Toward a social-psychological answer.* Hillsdale, NJ: Lawrence Erlbaum Associates.

Batson, C. D., & Ahmad, N. (2001). Empathy-induced altruism in a prisoner's dilemma II: What if the target of empathy has defected? *European Journal of Social Psychology, 31,* 25–36.

Batson, C. D., Batson, J. G., Todd, R. M., Brummett, B. H., Shaw, L. L., & Aldeguer, C. M. (1995). Empathy and the collective good: Caring for one of the others in a social dilemma. *Journal of Personality & Social Psychology, 68,* 619–631.

Batson, C. D., Sympson, S. C., Hindman, J. L., Decruz, P., Todd, R. M., Weeks, J. L., Jenning, G., & Burris, C. T. (1996). "I've been there, too": Effect on empathy of prior experience with a need. *Personality & Social Psychology Bulletin, 22,* 474–482.

Bazerman, M. H., Curhan, J. R., Moore, D. A., & Valley, K. L. (2000). Negotiation. *Annual Review of Psychology, 51,* 279–314.

Bazerman, M. H., & Neale, M. A. (1992). *Negotiating rationally.* New York: The Free Press.

Carnevale, P. J., & Pruitt, D. G. (1992). Negotiation and mediation. *Annual Review of Psychology, 43,* 531–582.

De Dreu, C. K. W. (2003). Time pressure and the closing of the mind in negotiation. *Organizational Behavior and Human Decision Processes, 91,* 280–295.

Drigotas, S. M., & Rusbult, C. E. (1992). Should I stay or should I go? A dependence model of breakups. *Journal of Personality and Social Psychology, 62,* 62–87.

Giebels, E., De Dreu, C. K. W., & Van de Vliert, E. (2000). Interdependence in negotiation: Effects of exit options and social motive on distributive and integrative negotiations. *European Journal of Social Psychology, 30,* 255–272.

Harinck, F., De Dreu, C. K. W, & Van Vianen, A. E. (2000). The impact of conflict issues on fixed-pie perceptions, problem solving, and integrative outcomes in negotiation. *Organizational Behavior and Human Decision Processes, 81,* 329–358.

Holmes, J. (2002). Interpersonal expectations as the building blocks of social cognition: An interdependence theory perspective. *Personal Relationships, 9,* 1–26.

Joireman, J. A., Kuhlman, D. M., Van Lange, P. A. M., Doi, T., & Shelley, G. P. (2003). Perceived rationality, morality, and power of social choice as a function of interdependence structure and social value orientation. *European Journal of Social Psychology, 33,* 413–437.

Kelley, H. H. (1984). The theoretical description of interdependence by means of transition lists. *Journal of Personality and Social Psychology, 47,* 956–982.

Kelley, H. H. (1997a). Expanding the analysis of social orientations by reference to the sequential-temporal structure of situations. *European Journal of Social Psychology, 27,* 373–404.

Kelley, H. H. (1997b). The "stimulus field" for interpersonal phenomena: The source of language and thought about interpersonal events. *Personality and Social Psychology Review, 1,* 140–169.

Kelley, H. H., Holmes, J. G., Kerr, N. L., Reis, H. T., Rusbult, C. E., & Van Lange, P. A. M. (2003). *An atlas of interpersonal situations.* Cambridge, England: Cambridge University Press.

Kelley, H. H., & Thibaut, J. (1978). *Interpersonal relations: A theory of interdependence.* New York: Wiley.

Le, B., & Agnew, C. R. (2003). Commitment and its theorized determinants: A meta-analysis of the investment model. *Personal Relationships, 10,* 37–57.

Lewin, K. (1946). Behavior and development as a function of the total situation. In L. Carmicheal (Ed.), *Manual of child psychology* (pp. 791–844). New York: Wiley.

Lindskold, S., & Han, G. (1988). GRIT as a foundation for intergrative bargaining. *Personality and Social Psychology Bulletin, 14,* 335–345.

Mansbridge, J. J. (1990). On the relation of altruism and self-interest. In J. J. Mansbridge (Ed.), *Beyond self-interest* (pp. 133–143). Chicago: University of Chicago Press.

McCullough, M. E., Fincham, F. D., & Tsang, J. (2003). Forgiveness, forbearance, and time: The temporal unfolding of transgression-related interpersonal motivations. *Journal of Personality and Social Psychology, 84*, 540–557.

Moran, S., & Ritov, I. (2002). Initial perceptions in negotiations: Evaluation and response to "logrolling" offers. *Journal of Behavioral Decision Making, 15*, 101–124.

Mosterd, I., & Rutte, C. G. (2000). Effects of time pressure and accountability to constituents on negotiation. *The International Journal of Conflict Management, 11*, 227–247.

Osgood, C. E. (1962). An alternative to war or surrender. Urbana: University of Illinois Press.

Pruitt, D. (1998). Social conflict. In D. Gilbert, S. Fiske, & G. Lindzey (Eds.), *The handbook of social psychology* (4th ed., Vol. 2, pp. 470–503). Boston: McGraw-Hill.

Rumble, A. C. (2003). *The role of empathetic behavior in an extended social dilemma.* Unpublished doctoral dissertation, Washington State University.

Rusbult, C. E. (1983). A longitudinal test of the investment model: The development (and deterioration) of satisfaction and commitment in heterosexual involvements. *Journal of Personality & Social Psychology, 45*, 101–117.

Rusbult, C. E., Bissonnette, V. L., Arriaga, X. B., & Cox, C. L. (1998). Accommodation processes during the early years of marriage. In T. N. Bradbury (Ed.), *The developmental course of marital dysfunction* (pp. 74–113). New York: Cambridge University Press.

Rusbult, C. E., & Buunk, B. P. (1993). Commitment processes in close relationships: An interdependent analysis. *Journal of Social and Personal Relationships, 10*, 175–204.

Rusbult, C. E., Martz, J. M., & Agnew, C. R. (1998). The investment model scale: Measuring commitment level, satisfaction level, quality of alternatives, and investment size. *Personal Relationships, 5*, 357–391.

Rusbult, C. E., & Van Lange, P. A. M. (1996). Interdependence processes. In E. T. Higgins & A. W. Kruglanski (Eds.), *Social psychology: Handbook of basic principles* (pp. 564–596). New York: Guilford Press.

Rusbult, C. E., & Van Lange, P. A. M. (2003). Interdependence, interaction, and relationships. *Annual Review of Psychology, 54*, 351–375.

Rusbult, C. E., Verette, J., Whitney, G. A., Slovik, L. F., & Lipkus, I. (1991). Accommodation processes in close relationships: Theory and preliminary empirical evidence. *Journal of Personality and Social Psychology, 60*, 53–78.

Stuhlmacher, A. F., & Champagne, M. V. (2000). The impact of time pressure and information on negotiation process and decisions. *Group Decision and Negotiation, 9*, 471–491.

Stuhlmacher, A. F., Gillespie, T. L., & Champagne, M. V. (1998). The impact of time pressure in negotiation: A meta-analysis. *International Journal of Conflict Management, 9*, 97–116.

Thibaut, J. W., & Kelley, H. H. (1959). *The social psychology of groups.* New Brunswick, NJ: Transaction.

Thompson, L. (1990a). An examination of naive and experienced negotiators. *Journal of Personality and Social Psychology, 59*, 82–90.

Thompson, L. (1990b). The influence of experience on negotiation performance. *Journal of Experimental Social Psychology, 26*, 528–544.

Thompson, L. (1998). *The mind and heart of the negotiator.* Upper Saddle River, NJ: Prentice Hall.

Thompson, L., & Hastie, R. (1990). Social perception in negotiation. *Organizational Behavior and Human Decision Processes, 47*, 98–123.

Tripp, T. M., Bies, R. J., & Aquino, K. (2002). Poetic justice or petty jealously? The aesthetics of revenge. *Organization Behavior and Human Decision Processes, 89*, 966–984.

Van Lange, P. A. M. (2004). From generosity to aggression: Five interpersonal orientations relevant to social dilemmas. In R. Suleiman, D. Budescu, I. Fisher, & D. M. Messick (Eds.), *Advances in social dilemma research.* Cambridge University Press: Cambridge, UK.

Van Lange, P. A. M., Agnew, C., Harinck, F., & Steemers, G. (1997). From game theory to real life: How social value orientation affects willingness to sacrifice in ongoing close relationships. *Journal of Personality and Social Psychology, 73,* 1330–1344.

Van Lange, P. A. M., Rusbult, C. E., Drigotas, S. M., Arriaga, X., Watcher, B. S., & Cox, C. C. (1997). Willingness to sacrifice in close relationships. *Journal of Personality and Social Psychology, 72,* 1373–1395.

von Neumann, J., & Morgenstern, O. (1948). *Theory of games and economic behavior.* Princeton, NJ: Princeton University Press.

Yovetich, N. A., & Rusbult, C. E. (1994). Accommodative behavior in close relationships: Exploring transformation of motivation. *Journal of Experimental Social Psychology, 30,* 138–164.

12

Temporal Factors in Social Dilemma Choice Behavior: Integrating Interdependence and Evolutionary Perspectives

Craig D. Parks
Donelle C. Posey
Washington State University

Mixed-motive conflict has been a popular topic of study within psychology for decades. An enduring focus within this research domain has been on social dilemmas, which can be broadly defined as a situation in which there is a conflict between individual gain and collective gain—the best personal outcome is earned by behaving selfishly, but if all persons act in this way, everyone receives a poorer outcome than if all had acted in a collectively beneficial manner (Dawes, 1980; Komorita & Parks, 1995). For example, it is personally more lucrative to keep one's money and watch public television broadcasts than to give money to the station and watch its shows, but if all viewers act in this way, then the station will receive no money, will cease broadcasting, and all persons will be in a worse state than if all had made donations.

The study of (at least human) cooperation in social dilemma–type situations is usually approached from one of two complementary perspectives. One approach emphasizes the *how* of cooperation, and the other the *why* of cooperation. The first is typically approached from an interpersonal/social standpoint. Here, researchers inquire about how cooperation develops (or fails to develop), how features of the situation affect cooperative choice, and what role cognitive and individual difference variables play in determining cooperative behavior. These researchers typically rely on Thibaut and Kelley's (1959) seminal work on interdependent behavior as the theoretical basis, and so this work is often referred to as the *interdependence* approach to cooperation. Other researchers are concerned with the more basic question of why cooperation exists as a behavior and usually attack the issue within an evolutionary framework. Of what survival value is cooperation? Is it a behavior that evolutionary pressures should favor, or select out? Do the dynamics of human cooperation mimic those of animal cooperation? How neurally simple can an organism be and still exhibit basic cooperative behavior? What areas of the human brain are activated during cooperative choice? Such questions, not surprisingly, are generally referred to as the *evolutionary* approach to the study of cooperation.

A central issue in the study of cooperation is the impact of time on cooperative behavior. What factors affect whether and how people alter their behavior in repeated-trial interactions? This issue is attractive because many real-world social dilemmas involve multiple choice points. Many charities and nonprofit entities conduct yearly, or even more frequent, pledge drives to solicit donations; during warm weather homeowners must continually decide whether the water supply can support their using water for nonessential endeavors (e.g., watering the lawn); and every time we leave the house, we must decide whether to "contribute" to air pollution by driving a car or use an alternate means of transportation. Time as a factor in cooperative choice has long been a part of interdependence-oriented research, and it can be argued that evolutionary theorists, through their historical concern with the stability of cooperation in repeated interactions between organisms (e.g., Axelrod, 1984), also have an interest in time and cooperation. This work, however, is largely descriptive or based on simulations; investigation of how evolutionary principles might help explain the impact of time within an experimental, or more short-term temporal, setting is much less common. As such, examining the specific contribution of time to cooperation in both research realms seems reasonable, and in particular asking how we might integrate across the two lines of research could be revealing.

The purpose of this chapter, then, is threefold. First, we want to outline the general issues pertaining to temporal phenomena and cooperation. Next, we look more closely at some of the time-specific issues within both interdependence and evolutionary research. Finally, we want to discuss how and why integrating across the two approaches would be fruitful for mixed-motive theorists.

The reader should note that, though our discussion will focus specifically on social dilemmas, time issues are important in other mixed-motive venues as well, most notably negotiation and conflict resolution. There exist a number of excellent resources for learning about time issues in these areas (most recently, Bazerman, Curhan, Moore, & Valley, 2000).

COOPERATION AND THE PASSAGE OF TIME

The impact of time on cooperative choice can be examined from two vantage points. The *intrasubject* perspective addresses how an individual's tendency to cooperate changes over time and across settings, and the *intrasetting* perspective focuses on how cooperation alters within a single, repeated-trial social situation. The first has been examined primarily from a developmental standpoint, and has focused on both the trajectory of frequency of cooperative behavior from prekindergarten through early adulthood (e.g., Eisenberg, Carlo, Murphy, & Van Court, 1995) as well as the development of individual differences that predispose a young person toward cooperation (e.g., Eisenberg et al., 1999). This work has been thoroughly reviewed elsewhere (Eisenberg, 2003), and is additionally beyond the scope of this chapter, so we do not attend to this line of inquiry here.

The intrasetting perspective is the one that has received the most attention from social psychologists and game theorists, and this attention itself separates into two distinct foci: a behavioral focus, which addresses how people alter their behavior in a repeated-trial situation, and a cognitive focus, which addresses to what extent people take a short-run versus long-run perspective when planning out their behavior. Let us consider each in turn.

Repeated-Trial Behavior. Many real social dilemmas involve repeatedly acting within the same situation, with the same coactors. Residents of a neighborhood that draw from the same aquifer make daily, independent decisions regarding how much water to take from the aquifer; city dwellers must decide each day whether to use their cars or ride on public transportation; viewers of a public television station are asked multiple times during a year to contribute to the station. One's initial choice is generally assumed to be influenced by individual difference factors alone, because at that point one has no information about the preferences or tendencies of others. It is what happens on subsequent trials, when one has experienced success or failure from cooperation, and has had the opportunity to collect some information about the actions of others, that interests many researchers. A complicating factor that further adds intrigue is that, in most real dilemmas, anything more than limited communication among group members is practically impossible. The viewing audience for a public television station simply could not assemble and plan out a course of action for supporting the station, for example. This lack of interaction introduces a variety of social psychological phenomena into the dilemma—attributional processes, anonymity, framing biases, and cultural differences, to name just a few. How, then, do individual differences, a history of group success or failure, and social psychological variables interact to determine cooperative choice on the current trial? This question has occupied mixed-motive researchers since at least the 1960s, and remains a popular topic of study (see, e.g., Komorita & Parks, 1996).

Perceptions of Time. The other research focus emphasizes cognitive variables, and especially mental representations and perceptions of time. In particular, researchers in this area are interested in the deceptively simple question, "When do people think the dilemma will end?" It has long been known that, if the uncertainty is removed and people are simply told when they will have to make their last choice, endgame behavior becomes predictable: People rarely cooperate on the last trial, and are only slightly more cooperative on very late trials (Kreps, Milgrom, Roberts, & Wilson, 1982).[1] But of course, in real social dilemmas the end is almost never known with certainty, and for certain types of dilemmas, like resource dilemmas, if the end seems near group members can sometimes exercise temporary restraint, and the resource will partially replenish itself. When the objective end is unknown, "the end" becomes a subjective assessment on the part of each group member, and researchers pursue how one makes this assessment, and how it influences cooperative choice.

It is worth noting that the intrasubject and intrasetting perspectives might also interact, in the guise of a person–situation interaction. Indeed, in recent years no small amount of attention has been paid to how such an interaction might affect cooperation over time. For example, Laursen and Hartup (2002) have demonstrated that the nature of reciprocity within a friendship can alter over time, and the degree of alteration can be predicted from both the specific exchange behaviors favored by the friends and the cognitive representations of reciprocity that the friends maintain.

These, then, are the major variables that theorists who are interested in temporal factors in cooperation confront. Next, we look at how at least some of

[1]Some economists have criticized this "backward induction" approach on the grounds that its predictions cannot be supported in situations in which the endpoint is not known (e.g., Dow & Werlang, 1994).

these questions can be answered, from both an interdependence standpoint and an evolutionary standpoint.

THE INTERDEPENDENCE APPROACH

Behavior

Thibaut and Kelley (1959) explicitly took up the question of how frequency of cooperative choice should change over time. They argued that an initial behavior will be repeated until the outcome associated with it has lost its subjective reward value (i.e., the reward had satiated), and/or repetition of the behavior becomes too fatiguing. At that point, the person will change his or her behavior: In a discrete-choice situation, a cooperator will become competitive, and a competitor cooperative.[2] Thus, the cooperator/competitor will continue to cooperate/compete until the reward from the behavior is no longer reinforcing, and/or performance of the behavior is too effortful or boring, at which point the person will adopt a new behavior. Note that this same logic will apply to the new choice, and eventually it too will lose its reward value and become tiresome. As such, Thibaut and Kelley predicted that, given a long enough duration of interaction, a person will "shift, over time, from one item to another in their repertoires with very little immediate repetition of the same items" (p. 18). In other words, the person will eventually revert to choice alternation on consecutive trials.

One of the difficulties in empirically evaluating this predicted pattern is that the typical laboratory study has an endgame confound built into it. Participants know that they will be in the lab for the specific amount of time indicated on the experimental sign-up sheet, or, if they are students for whom study participation is an expected part of a class, they can infer the time limit from the number of participation credit hours the study is worth. So, although there are many studies in which decline in rate of cooperation occurs (e.g., Komorita, Parks, & Hulbert, 1992), it is not clear whether such declines are products of the satiation/boredom predicted by Thibaut and Kelley; or are induced by the expected end-of-game. Specific tests of choice frequency over time have involved prisoner's dilemma games (PDGs) with multiple hundreds of trials, and these studies have somewhat supported Thibaut and Kelley's assertions. Specifically, Anatol Rapoport and colleagues (Rapoport & Chammah, 1965; Rapoport, Guyer, & Gordon, 1976) observed a decline and then increase in the initial behavior when players had complete information about outcomes, which Thibaut

[2]The question of how choice will change over time in a more continuous-choice situation, under which there are degrees of cooperation, has not yet been taken up by researchers.

and Kelley predicted should happen early in the sequence. Rapoport's players did not, however, eventually resort to choice switching, which is in contrast to Thibaut and Kelley's model. Furthermore, Rapoport observed that frequency of cooperation deteriorated to near zero in all players if the outcome matrix was not available, which is clearly at odds with Thibaut and Kelley. As Rapoport had people play for upwards of 700 trials, it seems unlikely that he did not allow enough time for the entire process to play itself out. Also problematic, Terhune (1974), conducting a 150-trial PDG, showed that patterns of cooperative choice differed across subgroups as classified by the Thematic Apperception Test. What is important is not the classification per se, but Terhune's demonstration that individual differences play a part in determining repeated-trial cooperation, which is difficult to reconcile with Thibaut and Kelley's emphasis on satiation and boredom.[3]

As might be guessed from the dates of the studies referenced earlier, further attempts to tease out the exact course of cooperative choice have not been made. Conducting multiple hundred trials of a PDG is resource-intensive and draining on participants. Interestingly, some emerging work in the animal behavior literature may provide some insight into the dynamics of long-term cooperation. McSweeney and colleagues have argued that many instances of seeming satiation to a stimulus during a multitrial situation are actually instances of habituation (McSweeney & Murphy, 2000; McSweeney & Roll, 1998). Important for our purposes, this line of inquiry has shown that interrupting the trials can result in a behavioral reset, in that when the trials restart, rather than the organism picking up where it left off, choices instead follow the pattern established at the start of the session (Cannon & McSweeney, 1998). This suggests that a competitor who was gravitating toward cooperation when the session was interrupted would revert back to competition when trials recommenced, and vice versa for a cooperator.

All of this suggests that the interdependence perspective of choice in a long-term dilemma being a function of satiation and fatigue is likely not accurate. Unfortunately, we have little more than a few fragments of evidence to suggest how the model should be modified.

Time Outlook

Interest in how time perspective affects interpersonal relations can be traced back at least to World War II–era studies of the psychological impact of uncer-

[3]In fairness to Thibaut and Kelley, it should be noted that the study of individual differences in cooperation did not begin in earnest until after publication of their text.

tain time horizons on the behavior of prisoners (Bondy, 1943; Cohen, 1953; Farber, 1944; Lewin, 1948), with the consistent observation being that uncertainty about the duration of internment provoked conflict among the prisoners. Thibaut and Kelley (1959) seem to have been the first to consider the impact of time perspective on general interpersonal behavior. They assumed that behavioral repertoires eventually become automatic in stable social situations. Because automatic behaviors require little thought, it follows that a short-term perspective will dominate. However, should the situation change (e.g., the group acquires a new member), or the person learns new behaviors, choice will no longer be automatic, and as such the person will begin to consider the long-term implications of various actions. Thibaut and Kelley suggested that, unless there are "strong instigations" to return to a short-term outlook, the long-term perspective will become the guiding principle for future choice. The primary value of a long-term outlook, according to Thibaut and Kelley, is that it reduces one's dependence on the current interaction partner—if immediate outcomes are unimportant, then how the other behaves is of little consequence—and as such reduces the other's power over the person. Thus, a long-term thinker is freer to sample across the behavioral repertoire than a short-term thinker.

Thibaut and Kelley; then, saw long-term thinking as a means of becoming unencumbered by the whims and behaviors of others. Being exploited right now is tolerable, because one is planning for the long run. Perhaps the clearest manifestation of this argument is in theory on the efficacy of the tit-for-tat (TFT) strategy in mixed-motive encounters. TFT prescribes that one should begin a multitrial encounter with a cooperative choice, and thereafter behave as one's opponent did on the previous trial. Axelrod (1984) demonstrated in his now-classic computer simulations that, though TFT can never conclude an encounter with a greater total reward than the opponent, across a wide variety of situations and partners it will emerge with a larger average payoff than any other (tested) strategy. The TFT user can thus do quite well over time, but has to accept that, in any given encounter, he or she will finish with less, or possibly the same as, the other person. One can thus predict that short-term thinkers should be less enamored of TFT than long-term thinkers, but, although there is much empirical data on Axelrod's predictions (see Komorita & Parks, 1996, for a summary), only recently have researchers begun to examine the role of time outlook in TFT adoption (e.g., Van Lange, Ouwerkerk, & Tazelaar, 2002).

Like Thibaut and Kelley, Pruitt and Kimmel (1977) also considered a long-term outlook to be important, but not because it offered protection from powerful others (though, like Thibaut and Kelley, they believed that short-term thinking engenders a defensive mind-set). Rather, Pruitt and Kimmel saw a long-term perspective as essential for developing mutual cooperation among group members, because only under such a perspective could participants develop an expectation that others will be cooperative. In considering Rapoport and Chammah's (1965) U-shaped cooperation functions, Pruitt and Kimmel argued that the nadir of the function is the point at which outlooks change from short-term to long-term. At this point, outcomes have been poor for a while, and according to Pruitt and Kimmel this leads participants to question their approach to the situation. Such a questioning produces the twin realizations that participants are dependent on one another, and that consistent exploitation of the other (i.e., being uncooperative when the other is cooperative) is hopeless. The situation has thus devolved into a choice between mutual cooperation and mutual noncooperation, and of the two the former is more lucrative. Hence, participants reorient their priorities, and behave so as to achieve mutual cooperation. The key here is the discovered futility of striving for consistent exploitation, a discovery that can occur only when one projects into the future. In revising Pruitt and Kimmel's theory to make it more applicable to large groups, Yamagishi (1986) retained the emphasis on long-term thinking, and reiterated the idea that U-shaped cooperation functions are produced by short-term thinking followed by long-term thinking. It must be pointed out, though, that there exists almost no empirical evidence demonstrating that group members do in fact experience such an epiphany at the low point of the function. Pruitt and Kimmel relied solely on anecdotal evidence reported by Terhune (1974), Yamagishi provided no additional support, and current researchers have not taken up the issue.

An Individual Difference? At present, scholars who follow the interdependence approach have attempted to predict and understand the time perspective that the person brings to the situation. Unlike Thibaut and Kelley, Pruitt and Kimmel, and Yamagishi, these scholars do not automatically assume that participants in an interdependent situation begin with a short-term focus. There is evidence in this literature that time perspective sometimes manifests itself as an individual difference trait (Joireman, Lasane, Bennett, Richards, & Solaimani, 2001; Strathman, Gleicher, Boninger, & Edwards, 1994) and is sometimes a function of the situation (Axelrod, 1984; Joireman et al., 2001; Mannix, 1991; Strathman et al., 1994), which thus suggests that time perspective may follow the classic person–situation interaction seen in so many other research venues.

The "person" side of this equation has concentrated on a particular individual difference referred to as "consideration of future consequences" (Strathman et al., 1994). The trait has been shown to be predictive of rate of cooperation in interdependent situations (Insko et al., 1998). Little is known about the development of the trait, though researchers have documented gender and age differences, with women and older individuals tending to adopt long-term outlooks than men and younger individuals (Lindsay & Strathman, 1997). Situational factors largely orient around the "shadow of the future," or the expectation that the interaction partners will be encountered again in the future. In particular, Axelrod (1984) argued that the shadow is an important consideration in the decision to cooperate, with cooperation more likely as the shadow elongates (i.e., future interaction increases in likelihood). Recall that Axelrod was concerned with the efficacy of TFT at inducing cooperation. As such, he argued that a short shadow eliminates opportunities for reciprocation, and undermines TFT's properties. A longer but finite shadow will produce initially high rates of cooperation that consistently decline as the end of the interaction nears. As the shadow continues to grow, participants will take an increasingly long-term perspective on the interaction; if the end cannot be "seen," then one should observe the highest rates of cooperation. Axelrod thus suggested that the shadow of the future should be elongated whenever possible. Insko et al. provided some indirect support for this notion by demonstrating that an extended shadow can make group-level interactions more cooperative. Whether it will also work at the individual level remains uncertain.

What should be clear from this most recent body of work is that time outlook seems to be more complicated than originally thought. In light of this evidence it seems unlikely that people automatically adopt a short-term outlook at the commencement of the interaction. The person–situation work, however, does not negate the possibility that outlook changes at some point during the interaction.

Summary of the Interdependence Approach

The interdependence approach to cooperation offers models and predictions of time-related behavior as well as the perception of time within a mixed-motive task. The behavioral model has emphasized satiation and fatigue as factors in determining how and when people change their behavior over time. Empirical research has only partially supported the predictions of this model. Though further work on this issue has been dormant, there is some emerging research in the animal behavior literature that has the potential to inform our thinking about cooperative behavior over time.

Theory on the importance of time perspective has addressed short-term versus long-term thinking. The consistent assumption is that long-term thinking is a stimulant of cooperation, and short-term thinking is not. Empirical tests of the dynamics of short- and long-term thinking in the mixed-motive context are virtually nonexistent. Current thought is that time perspective may be a type of person–situation interaction.

THE EVOLUTIONARY APPROACH

Behavior

Evolutionary perspectives on cooperation have historically emphasized the role of genetic relatedness in cooperation. Theories of kin selection predict a direct relationship between degree of relatedness (i.e., proportion of shared genetic material) and extent of cooperation, in that we should cooperate most frequently with those to whom we are most closely related, less regularly with those who are distant relations, and rarely with those who are unrelated. Because the proportion of shared genetic material remains constant over the life span (i.e., we do not become less related to people as we age), it follows that rate of cooperation among genetically related individuals should also be relatively constant across time. Segal and Hershberger (1999) specifically addressed this by having monozygotic and dizygotic twins play 100 trials of a PDG.[4] They found that, overall, monozygotic twins earned a higher per-person payoff than dizygotic twins, and more important for our purposes, monozygotic twins actually *increased* their cooperation over time, relative to dizygotic twins. This is consistent with the notion that degree of relatedness can predict cooperative choice over time.

The challenge for evolutionary approaches to cooperation is to explain short-run prosocial behavior among small groups of unrelated others. That it happens, and with some frequency, is beyond question. There exist a number of attractive models that explain how, over the long run, stable cooperation might be achieved among large groups of such actors (e.g., Axelrod, 1984; Macy & Skvoretz, 1998; Parks & Komorita, 1997; Simon, 1990), and recent theorizing has postulated that certain group-regarding traits (e.g., intelligence, group-regardingness, communicability) are selected for (Brewer & Caporael, 1990; Caporael, Dawes, Orbell, & van de Kragt, 1989), but to what extent do these models account for more immediate, small-group cooperation? Can we predict with whom someone will cooperate, and how frequently, and how the pattern of

[4]Research has consistently shown that monozygotic twins are more cooperative and empathic with each other than are dizygotic twins (e.g., Constantino & Todd, 2000; Matthews, Batson, Horn, & Rosenman, 1981; Rushton, Fulker, Neale, Nias, & Eysenck, 1986; Segal, 2002), making these two groups attractive for studying genetic factors in cooperation.

behavior might alter over time? Efficient empirical evaluation of the aforementioned models typically requires some form of a single-trial social dilemma game so as to eliminate opportunities for revenge and retribution, crucial for supporting an evolutionary line of reasoning but, of course, incapable of providing insight into time effects. One exception is a study by Kameda and Nakanishi (2002), which was designed to determine if, over time, cooperators and selfish individuals could come to exist in equilibrium, or if one group would dominate, and eventually eliminate, the other. They ran 60 trials of a learning game in which people could either search for task-relevant information (cooperation) or take advantage of the search efforts of others (selfishness). They found that cooperators and selfish individuals did indeed come to coexist, in a proportion predictable from the nature of the task: As the task became more challenging, selfish individuals increasingly dominated cooperators; as the task became easier, cooperators dominated selfish individuals; and the two existed in equal numbers when the task was of moderate difficulty. Kameda and Nakanishi's findings thus suggest that time allows the members of a group to find a tolerable balance of group-regarding and self-regarding behaviors. The data also suggest that some individuals adjust their behavior to meet the needs of the situation, whereas others adopt a uniform strategy that is applied in all situations.

The reader should note that we do not offer this up as a criticism of the evolutionary approach to cooperation. Far from it. Instead, we see it as an exciting opportunity to meld two complementary perspectives on cooperation (an opportunity that, as we document later, we are not the first to identify). In fact, it is likely that more in-depth analysis of evolutionary phenomena, time, and cooperative behavior will soon develop. Kenrick, Li, and Butner (2003) have recently offered a theoretical framework for integrating evolutionary-models and dynamic-systems approaches to social behavior, and the authors explicitly show how cooperative processes can be analyzed from this perspective. As this model gains recognition, one expects that we will come to learn more about how cooperative behavior over time is affected by evolution-based factors.

Time Outlook

Evolutionary psychologists have not taken up the issue of time perspective per se, though there are a number of topics that have been addressed that can be brought to bear on the issue of time outlook. For example, a number of researchers have suggested that people have the ability to identify and recall selfish individuals, so that future encounters with those individuals can be avoided (e.g., Brown & Moore, 2000; de Vos & Zeggelink, 1997). Oda (1997) has provided some empirical support for this notion by showing that people can more accurately recall details of photographs of "competitors" than of "cooperators" up to 7 days after exposure to the photographs. This suggests that there may be an evolutionary tendency to think about, and plan for, future mixed-motive exchanges. Research on mortality salience, which has been argued to have an evolutionary basis (Pyszczynski, Greenberg, & Solomon, 1997),[5] also raises the possibility that short- and long-term thinking may be influenced by our biological history. In brief, this research has demonstrated that cognitions about one's own death can produce behavior change. Relevant to this chapter, it has been shown that people who are cognizant of their ultimate demise engage in more frequent prosocial behaviors than do people for whom death is not salient (Jonas, Schimel, Greenberg, & Pyszczynski, 2002). If we consider death awareness to be a type of long-term thinking, then it can be argued that these data provide a partial test of the interdependence theory proposition that long-term thinking produces more cooperation than short-term thinking. Jonas et al.'s results additionally suggest that the effect may be at least partially grounded in evolutionary principles. It should be noted that Jonas et al. ascribe their findings to the death thoughts acting to make a particular cultural worldview salient, and make no argument that such thoughts induce long-term thinking. However, it is possible (and testable) that cognitions about one's death—which presumably one would expect to occur far in the future—leads to a more general long-term mind-set.

Summary of the Evolutionary Approach

To date evolutionary psychologists have not been especially concerned with time-based influences on cooperation, though elements of this research lend themselves to speculation about how evolved tendencies might affect dynamic cooperation. The strongest evidence exists for a connection between degree of genetic relatedness and consistency of cooperation over time. Theory on time and cooperation among unrelated others has to date focused on the stability of cooperative and selfish subsets within the group. Even less is known about how cognitions about time relate to evolved processes, though there exist some databases that seem possibly linked to research on time perception. The recent development of theoretical approaches that connect evolutionary models and dynamic processes seem likely to spur research on this issue.

[5]Pelham (1997) has criticized the assumption that terror management is a basic human motive.

AN INTEGRATIVE PERSPECTIVE

We have shown how interdependence theorists attempt to explain temporal influences on cooperation, and we have identified some key evolutionary psychological studies that suggest a role for temporal variables in evolutionary theories of cooperative choice. We hope to have conveyed two key points: First, the predictions of classic interdependence models of both cooperative behavior over time and cognitions about such behavior are somewhat at odds with the (relatively scant) empirical studies of such behavior; and second, temporal factors have not been a component of evolutionary approaches to cooperation, though there is reason to believe that they could be. As time is such an important aspect of theorizing about cooperative processes (Messick & Liebrand, 1997), it is reasonable to argue that any evolution-based theory of human cooperation is incomplete without a consideration of the role of time progression.

Each approach to the study of cooperation, then, seems to struggle with the dynamic aspect of the behavior. We suggest that much of this struggle could be minimized by integrating the two perspectives. In fact, others have made this argument before us. In particular, Sheese and Graziano (2002; see also Graziano, Hair, & Finch, 1997) have suggested that an evolutionary perspective can be an important link between theories of social exchange and models of social development, in that evolutionary principles and longitudinal situations may interact to influence exchange behaviors. Thus, our call for integration is really nothing more than an elaboration on their argument. What follows is an elaboration of the benefits of integration as we see it.

The two sets of data that we have summarized in this chapter complement each other in many ways; as such, they can be combined to form a more complete picture of how time influences cooperative choice. For example, we mentioned that actual choice from lengthy PDGs does not completely follow the pattern that would be expected if satiation and/or fatigue were affecting behavior. Recall that Kameda and Nakanishi's (2002) data suggested that behavior change may be driven in part by situational factors. It may thus be that repeated-trial behavior is affected by satiation *if the situation is held constant,* and by situational factors otherwise. Such a rule is reasonable from a behavior-analytic standpoint—a well-established principle is that satiety diminishes when behavioral outcomes are altered—and also makes sense from an interdependence point of view, as it is known that people transform the value of outcomes as a mixed-motive game progresses (Kelley & Thibaut, 1978), which produces a psychologically "new" situation for which some people will alter their behavior, and some will not. Note that this logic is also consistent with both Pruitt and Kimmel's (1977) argument that the low point of the U-shaped cooperation function is the point at which people reframe the situation as having two possible outcomes rather than four, and with Thibaut and Kelley's (1959) suggestion that situational changes move people away from automatic responding and toward a conscious evaluation of the long-term consequences of their actions.

A second issue for which the two perspectives provide an interesting complement is that of time horizons. Interdependence theory has shown us that, all else being equal, the length of the "shadow of the future" affects for how long cooperative behavior will occur. We are more frequently cooperative as the endpoint of the interaction seems ever-farther away. Evolutionary psychologists have demonstrated that people maintain identifying information about selfish others in memory, presumably to facilitate identification of those individuals in the future so as to avoid interactions with them. Given the limitations on for how long information can be accurately maintained in memory (recall that Oda, 1997, waited just 1 week before testing memory for the competitor and cooperator photographs), it is reasonable to ask what other strategies people have developed for spotting and avoiding selfish others if they anticipate a lengthy shadow. Research on motivated-person perception suggests that such strategies do exist. For example, Devine, Sedikides, and Fuhrman (1989) showed that people most accurately recall a target person's attributes, and have that information most efficiently organized in memory, when they anticipate future interaction with that person, though this accuracy seems to be limited to situations in which the target and person are of the same gender (Sedikides, Devine, & Fuhrman, 1991). These same researchers have also shown that anticipated interaction has no special effect on recall of multiple targets (e.g., all members of a group considered simultaneously) (Sedikides et al., 1991). This seems to be a function of efficiency: If individuals cannot determine who among the multiple targets will be the dominant interaction partner (and Sedikides et al.'s participants could not make such a determination), they do not attempt to maintain individuating information about each person. Though Devine and colleagues did not attempt to link their findings to evolutionary processes, their data nonetheless suggest that people exert considerable mental effort to accurately store information about (same-gender) future interaction partners, and this work could be fruitfully connected to the evolutionary research on detection of noncooperators to allow for development of models of the evolutionary value of time outlook.

CONCLUSION

Our goal has been to introduce the reader to the importance of time and time perspective to the study of human cooperation in mixed-motive situations. Our review has been necessarily brief, and was primarily intended to demonstrate that the two dominant perspectives on cooperation—namely, interdependence and evolutionary theories—offer complementary findings on how time and time outlook impact cooperative choice. We hope to both stimulate more, and more systematic, research on time factors in cooperation, and encourage some more unified theory on the influence of time on cooperation, theory that incorporates both interdependence and evolutionary principles.

REFERENCES

Axelrod, R. (1984). *The evolution of cooperation.* New York: Basic Books.

Bazerman, M. H., Curhan, J. R., Moore, D. A., & Valley, K. L. (2000). Negotiation. *Annual Review of Psychology, 51,* 279–314.

Bondy, C. (1943). Problems of internment camps. *Journal of Abnormal and Social Psychology, 38,* 453–475.

Brewer, M. B., & Caporael, L. R. (1990). Selfish genes vs. selfish people: Sociobiology as origin myth. *Motivation and Emotion, 14,* 237–243.

Brown, W. M., & Moore, C. (2000). Is prospective altruist-detection an evolved solution to the adaptive problem of subtle cheating in cooperative ventures? Supportive evidence using the Wason selection task. *Evolution and Human Behavior, 21,* 25–37.

Cannon, C. B., & McSweeney, F. K. (1998). The effects of stopping and restarting a session on within-session patterns of responding. *Behavioral Processes, 43,* 153–162.

Caporael, L. R., Dawes, R. M., Orbell, J. M., & van de Kragt, A. J. C. (1989). Selfishness examined: Cooperation in the absence of egoistic incentives. *Behavioral and Brain Sciences, 12,* 683–699.

Cohen, E. A. (1953). *Human behavior in the concentration camp.* New York: Norton.

Dawes, R. M. (1980). Social dilemmas. *Annual Review of Psychology, 31,* 169–193.

de Vos, H., & Zeggelink, E. (1997). Reciprocal altruism in human social evolution: The viability of reciprocal altruism with a preference for "old-helping-partners." *Evolution and Human Behavior, 18,* 261–278.

Devine, P. G., Sedikides, C., & Fuhrman, R. W. (1989). Goals in social information processing: The case of anticipated interaction. *Journal of Personality and Social Psychology, 56,* 680–690.

Eisenberg, N. (2003). Prosocial behavior, empathy, and sympathy. In M. H. Bornstein, L. Davidson, C. L. M. Keyes, & K. A. Moore (Eds.), *Well-being: Positive development across the life course* (pp. 253–265). Mahwah, NJ: Lawrence Erlbaum Associates.

Eisenberg, N., Carlo, G., Murphy, B. C., & Van Court, P. (1995). Prosocial development in late adolescence: A longitudinal study. *Child Development, 66,* 1179–1197.

Eisenberg, N., Guthrie, I. K., Murphy, B. C., Shepard, S. A., Cumberland, A., & Carlo, G. (1999). Consistency and development of prosocial dispositions: A longitudinal study. *Child Development, 70,* 1360–1372.

Farber, M. L. (1944). Suffering and time perspective of the prisoner. *University of Iowa Studies of Child Welfare, 20*(409), 153–227.

Graziano, W. G., Hair, E. C., & Finch, J. F. (1997). Competitiveness mediates the link between personality and group performance. *Journal of Personality and Social Psychology, 73,* 1394–1408.

Insko, C. A., Schopler, J., Pemberton, M. B., Wieselquist, J., McIlrath, S. A., Currey, D. P., & Gaertner, L. (1998). Long-term outcome maximization and the reduction of interindividual-intergroup discontinuity. *Journal of Personality and Social Psychology, 75,* 695–711.

Joireman, J. A., Lansane, T. P., Bennett, J., Richards, D., & Solaimani, S. (2001). Integrating social value orientation and the consideration of future consequences within the extended norm activation model of proenvironmental behavior. *British Journal of Social Psychology, 40,* 133–155.

Jonas, E., Schimel, J., Greenberg, J., & Pyszczynski, T. (2002). The Scrooge effect: Evidence that mortality salience increases prosocial attitudes and behavior. *Personality and Social Psychology Bulletin, 28,* 1342–1353.

Kameda, T., & Nakanishi, D. (2002). Cost-benefit analysis of social/cultural learning in a nonstationary uncertain environment: An evolutionary simulation and an experiment with human subjects. *Evolution and Human Behavior, 23,* 373–393.

Kelley, H. H., & Thibaut, J. W. (1978). *Interpersonal relations.* New York: Wiley.

Kenrick, D. T., Li, N. P., & Butner, J. (2003). Dynamical evolutionary psychology: Individual decision rules and emergent social norms. *Psychological Review, 110,* 3–28.

Komorita, S. S., & Parks, C. D. (1995). Interpersonal relations: Mixed-motive interaction. *Annual Review of Psychology, 45,* 178–204.

Komorita, S. S., & Parks, C. D. (1996). *Social dilemmas.* Boulder, CO: Westview Press.

Komorita, S. S., Parks, C. D., & Hulbert, L. G. (1992). Reciprocity and the induction of cooperation in social dilemmas. *Journal of Personality and Social Psychology, 62,* 607–617.

Kreps, D. M., Milgrom, P., Roberts, J., & Wilson, R. (1982). Rational cooperation in the finitely repeated prisoner's dilemma. *Journal of Economic Theory, 27,* 245–252.

Laursen, B., & Hartup, W. W. (2002). The origins of reciprocity and social exchange in friendships. In B. Laursen & W. G. Graziano (Eds.), *Social exchange in development* (pp. 27–40). San Francisco: Jossey-Bass.

Lewin, K. (1948). Time perspective and morale. In G. Lewin (Ed.), *Resolving social conflicts* (pp. 103–124). New York: Harper.

Lindsay, J. J., & Strathman, A. (1997). Predictors of recycling behavior: An application of a modified health behavior model. *Journal of Applied Social Psychology, 27,* 1799–1823.

Macy, M. W., & Skvoretz, J. (1998). The evolution of trust and cooperation between strangers: A computational model. *American Sociological Review, 63,* 638–660.

Mannix, E. A. (1991). Resource dilemmas and discount rates in decision-making groups. *Journal of Experimental Social Psychology, 27,* 379–391.

McSweeney, F. K., & Murphy, E. S. (2000). Criticisms of the satiety hypothesis as an explanation for within-session decreases in responding. *Journal of the Experimental Analysis of Behavior, 74,* 347–361.

McSweeney, F. K., & Roll, J. M. (1998). Do animals satiate or habituate to repeatedly presented reinforcers? *Psychonomic Bulletin and Review, 5,* 428–442.

Messick, D. M., & Liebrand, W. B. G. (1997). Levels of analysis and the explanation of the costs and benefits of cooperation. *Personality and Social Psychology Review, 1,* 129–139.

Oda, R. (1997). Biased face recognition in the prisoner's dilemma game. *Evolution and Human Behavior, 18,* 309–315.

Parks, C. D., & Komorita, S. S. (1997). Reciprocal strategies for large groups. *Personality and Social Psychology Review, 1,* 314–322.

Pruitt, D. G., & Kimmel, M. J. (1977). Twenty years of experimental gaming: Critique, synthesis, and suggestions for the future. *Annual Review of Psychology, 28,* 363–392.

Pyszczynski, T., Greenberg, J., & Solomon, S. (1997). Why do we need what we need? A terror management perspective on the roots of human social motivation. *Psychological Inquiry, 8,* 1–20.

Rapoport, A., & Chammah, A. M. (1965). *Prisoner's dilemma.* Ann Arbor: University of Michigan Press.

Rapoport, A., Guyer, M. J., & Gordon, D. G. (1976). *The 2 x 2 game.* Ann Arbor: University of Michigan Press.

Sedikides, C., Devine, P. G., & Fuhrman, R. W. (1991). Social perception in multitarget settings: Effects of motivated encoding strategies. *Personality and Social Psychology Bulletin, 17,* 625–632.

Segal, N. L. (2002). Co-conspirators and double-dealers: A twin film analysis. *Personality and Individual Differences, 33,* 621–631.

Segal, N. L., & Hershberger, S. L. (1999). Cooperation and competition between twins: Findings from a prisoner's dilemma game. *Evolution and Human Behavior, 20,* 29–51.

Sheese, B. E., & Graziano, W. G. (2002). Evolutionary perspectives on the development of social exchanges. In B. Laursen & W. G. Graziano (Eds.), *Social exchange in development* (pp. 73–85). San Francisco: Jossey-Bass.

Simon, H. A. (1990). A mechanism for social selection and successful altruism. *Science, 250,* 1665–1668.

Strathman, A., Gleicher, F., Boninger, D. S., & Edwards, C. S. (1994). The consideration of future consequences: Weighing immediate and distant outcomes of behavior. *Journal of Personality and Social Psychology, 66,* 742–752.

Terhune, K. W. (1974). "Wash-in," "wash-out," and systemic effects in extended prisoner's dilemma. *Journal of Conflict Resolution, 18,* 656–685.

Thibaut, J. W., & Kelley, H. H. (1959). *The social psychology of groups.* New York: Wiley.

Van Lange, P. A. M., Ouwerkerk, J. W., & Tazelaar, M. J. A. (2002). How to overcome the detrimental effects of noise in social interaction: The benefits of generosity. *Journal of Personality and Social Psychology, 82,* 768–780.

Yamagishi, T. (1986). The structural goal/expectation theory of cooperation in social dilemmas. *Advances in Group Processes, 3,* 51–87.

13

Considering the Future Consequences of Aggressive Acts: Established and Potential Effects in the Context of the General Aggression Model

Kathryn B. Anderson
Our Lady of the Lake University

Mark D. Wood
University of Rhode Island

> Anybody who wants to harm American troops will be found and brought to justice. There are some that feel like if they attack us that we may decide to leave [Iraq] prematurely.... There are some who feel like the conditions are such that they can attack us there. My answer is, bring 'em on.
>
> —George W. Bush (Fireman, 2003)

One of the main concerns of individuals who criticized the U.S. decision to engage in the second Gulf War in Iraq is that the Bush administration may not have been able to foresee important potential future consequences of this action, such as heightened Muslim anger at the United States, protracted civil unrest in Iraq, and lack of credibility with the United Nations (U.N.) for initiating war contrary to U.N. consensus. Certainly some types of aggressive behavior in some situations (such as engaging in a war) are the result of careful consideration of likely consequences to the aggressor and to others. But a lot of the time, aggression occurs more spontaneously, when the aggressor lacks the time, ability, or motivation to think about what may happen as a result of her or his action. For example, a study of car crashes on the Washington, D.C. Capital Beltway in 1993 and 1994 found that aggressive driving caused about 50% of the crashes. Of the aggressive-driving crashes, 18% were caused by a "sideswipe cut-off," which occurs when one vehicle hits another while changing lanes, typically related to frequent lane changes and traffic congestion (National Highway Traffic Safety Administration, 1996). Aggressive driving often occurs when one driver perceives that another driver dominated or attacked her or him and the driver spontaneously reacts by moving their car in a way that endangers the other car. In doing so, the driver is often so focused on "getting him back" or "not letting her win" that the driver fails to think about the damage that could come to their car, to themselves, and to others on the road. In focus groups conducted with Capital Beltway drivers in 1997, 75% of aggressive drivers said that they *always* or *often* compete with other drivers; one driver described driving on the Beltway as a "competitive sport" (National Highway Traffic Safety Administration, 1998). This type of aggression is termed *impulsive*, as it occurs relatively quickly, with very little, if any, thought.

Impulsive and Deliberative Aggression

Impulsive aggression is contrasted with *premeditated* or *deliberative* aggression (C. A. Anderson & Huesmann, 2003), which is characterized by effortful thought about why and how the individual aggresses and, presumably, the likely outcomes of the aggressive act. Instrumental aggression is typically deliberative, in that the aggressor acts to obtain a goal separate from inflicting harm on the target. The goal-striving motivation forces future thought in the form of "if–then thinking" (i.e., "if I belittle her, then she will leave me alone"). Impulsive aggression is hypothesized to be more likely among individuals for whom current situational cues dominate their phenomenological experience, at the expense of consideration of the possible future consequences of their actions. In contrast, individuals who tend to think about the future consequences of their actions may be more likely to think of possible positive and negative outcomes associated with various behavioral choices (e.g., aggression, nonaggressive negotiation, flight) before acting. If the likely negative outcomes are perceived to outweigh the likely positive outcomes (which should be the case for nonhostile individuals), a nonaggressive behavioral choice may be made to avoid negative consequences. If perceived positive outcomes outweigh the negative ones, instrumental aggression may result in order to receive the positive outcomes.

Consideration of Future Consequences

Many studies have demonstrated the significant relation between impulsivity and aggression (e.g., Buss & Perry, 1992; Joireman, J. Anderson, & Strathman, 2003). However, few studies have assessed if impulsive individuals consider the consequences of their aggression less than do less impulsive individuals, and if consequence consideration affects the likelihood of aggressing. As detailed later, Joireman et al. conducted four studies that tested the relations between impulsivity, consideration of future consequences (using the Consideration of Future Consequences Scale by Strathman, Gleicher, Boninger, & Edwards, 1994), and aggression likelihood and found support for the notion that consideration of the consequences of one's actions mediates the link between impulsivity and aggression. The Joireman et al. findings suggest that the extent to which individuals consider the consequences of their actions is an aspect of future orientation that is important in the prediction of aggression. Impulsive people are less likely to consider the consequences of their actions, and are also more likely to engage in impulsive aggression.

The integrative General Aggression Model, by Craig Anderson and colleagues, seeks to integrate the roles of situational and personological factors in the genesis of human aggression, and thereby provides a useful framework for the discussion of how future time orientation can inhibit or facilitate aggression.

THE GENERAL AGGRESSION MODEL

Over the past decade, C. A. Anderson and colleagues have developed a framework that integrates aspects of the major theories of the causes of aggression called the General Aggression Model (GAM; C. A. Anderson, K. A. Anderson, & Deuser, 1996; C. A. Anderson & Bushman, 2002b; C. A. Anderson, Deuser, & DeNeve, 1995). The model illustrates how factors about the person and the situation can independently or interactively can cause increases in aggressive feelings, aggressive thoughts and/or perceived or physiological arousal (see Fig. 13.1). The automatic activation of aggressive feelings and thoughts (including aggressive behavioral scripts) as a function of a hostile stimulus is similar to priming mechanisms described by Berkowitz's theory of Cognitive Neoassociationism (CN; e.g., Berkowitz, 1994), which states that negative affect primes

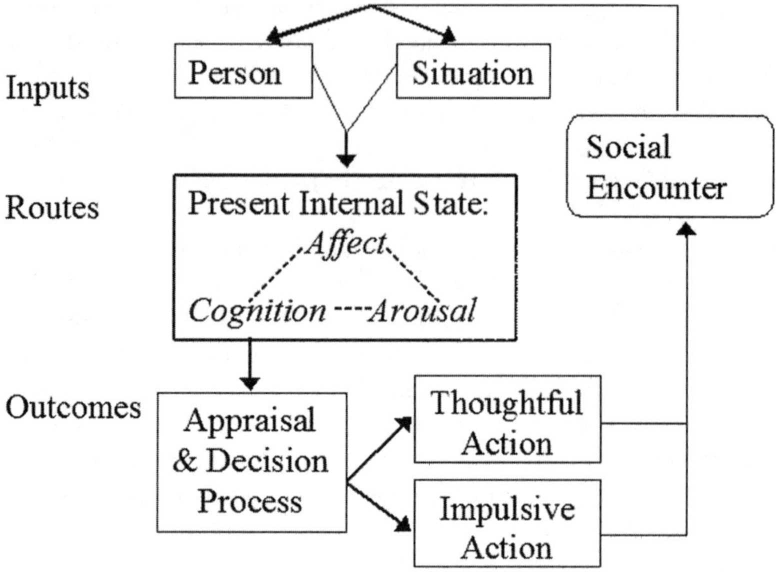

FIG. 13.1. The general aggression model episodic processes. From Anderson and Bushman, 2002b. Reprinted, with permission, from the *Annual Review of Psychology, Volume 53.* Copyright 2002 by Annual Reviews (www.annualreviews.org).

hostile thoughts, physiological responses, and hostile expressive-motor responses or escape tendencies. For example, a person with aggressive tendencies (e.g., high in trait irritability or hostility) who is insulted may have thoughts about hurting the person who insulted her. She may also feel angry and her heart rate may elevate. The main difference between the priming processes described by the GAM and the CN is that CN posits that generalized negative affect is elicited in response to a hostile situational stimulus and the negative affect activates hostile cognitive, physiological, and expressive-motor responses. The GAM states that hostile cognitions, affect, and/or physiological arousal can be automatically activated by hostile situational cues (or simply by aggressive dispositional tendencies), in the absence of negative affect.

Next, the GAM posits that the individual undergoes an appraisal process akin to that articulated in Gilbert's three-stage model of attribution (e.g., Gilbert, Pelham, & Krull, 1988). If the individual does not have sufficient time or cognitive resources, she may rapidly characterize the situation as a hostile one (attribute blame to the other person in the situation) and may act impulsively and aggressively toward the target. Alternatively, if the person does have sufficient time and ability (e.g., is not cognitively busy), she may allocate those resources to determining the guilt or innocence of the target and generating alternative explanations for the incident before acting (i.e., a thoughtful action). Finally, whether a thoughtful or an impulsive action is taken, that action is perceived by the "provocateur" and influences the provocateur's future actions toward the individual. This pattern of recursive escalation of aggressive interactions can strengthen the individuals' aggressive scripts (see Crick and Dodge's Script Theory, 1994), model aggressive conflict resolution (see Bandura's Social Learning Theory, 1973), and reduce their overall social competence (see Huesmann, 1988; 1998).

Future-oriented individuals who tend to think about the potential consequences of their actions may be more likely to engage thoughtful, effortful reappraisal of situational events prior to acting than are present-oriented individuals. This makes sense because pondering future consequences to the self is part of the cost–benefit analysis that results in the type of action chosen. In addition, some situational factors may encourage a person's focus on the current social context and inhibit reappraisal processes. Therefore, some situation and personality factors may facilitate or inhibit the person's ability and motivation to think about the future consequences of an aggressive act.

In the following sections, we summarize research on the prediction of aggression by the two main input variables of the GAM: situation and personality variables. Subsequently, we discuss how social information processing (i.e., the priming of aggressive scripts and the appraisal of the provacateur's intent) influences the aggressive or nonaggressive behavioral choice. Across all of these areas, we attempt to place particular emphasis on the construct of future orientation as it relates to each of these domains. Finally, we provide suggestions for future research as well as implications for preventive interventions to reduce aggression and its deleterious consequences.

SITUATIONAL FACTORS ASSOCIATED WITH AGGRESSION

The GAM posits that the situational context can independently and/or in interaction with person factors, affect hostile thoughts, hostile feelings, and arousal, and may affect the choice of whether or not to engage in aggression. Five situational factors that acutely affect the individual and have been clearly linked to aggression are addressed here: provocation, frustration, discomfort, observation of media violence, and alcohol and other drugs. We also consider the relevance of future orientation in relation to these situational factors and how one's ability to think about the potential consequences of aggression may be affected by these situational conditions.

Provocation

Verbal or physical attack from another person can range from a subtle sarcastic insult or a jovial shove to threatening profanity and inflicting bodily pain. An important determinant of whether or not attacks are reciprocated is if the target believes that the provocateur intended harm (e.g., Dodge, Murphy, & Buchsbaum, 1984; Epstein & Taylor, 1967). Also, when individuals think that attacks on them are maliciously perpetrated, they become more physiologically aroused than when they believe the act was not malicious (Zillmann & Cantor, 1976).

These findings can easily be explained by the GAM. First, a potentially provoking stimulus is experienced by the individual and immediate appraisal occurs, which often assumes a dispositional cause of the other person's behavior (e.g., the person is rude). If the individual has the time, cognitive resources, and motivation to think about the cause of the provocation, the person may take into account the situation that the provocateur was in when she or he acted and his or her intent. If the perceived intent is nonmalicious (e.g., an accident), the behavior choice would more likely be to respond nonaggressively. However, if the intent is construed to be malicious, the individual may engage in retaliatory aggression. Rumination about the causes of the attack may foster desire for retribution and may enhance the likelihood of an intense aggressive response. For example, Caprara, Renzi, Alcini, D'Imperio, and Travaglia (1983) found that individuals who tend to feel continued or enhanced anger following a provocation (i.e., ruminate) show greater retaliatory aggression than do individuals who are less likely to ruminate. Also, when targets view that a future threat is likely from a provocateur, they are more likely to assume that the act was malicious (Dodge & Somberg, 1987). That is, the *salience of the future threat* may affect the interpretation of the provacateur's act, as well as the resulting behavioral choice. If the likelihood of a future threat from the provocateur is high, the person may be highly motivated to aggress as a result of the perception of malicious intent, but may stifle an aggressive reaction for fear of retaliation from the provocateur. Predictions about the relation between perceived immediate and delayed consequences of aggression following provocation (and other situational factors) and aggression are provided in a later section.

Frustration

The perceived blocking of goal attainment has been considered a main cause of aggressive behavior since the Frustration–Aggression Hypothesis was developed in the 1930s (Dollard, Doob, Miller, Mowrer, & Sears, 1939). Multiple experiments that induced frustration, such as that produced by an insolvable jigsaw puzzle (Geen, 1968), have demonstrated the causal impact of frustration. When people feel like their frustrations are illegitimately caused and unjustified, they are more likely to lash out at an available target, particularly if that target performed a goal-blocking behavior (J. Dill & C. A. Anderson, 1995; Pastore, 1952).

The relations between frustration, future orientation, and aggression have not yet been explored, but have interesting possibilities. The very experience of frustration involves some degree of future orientation, for the individual seeks to perform an action to obtain a goal in the immediate or distant future. Future orientation may moderate the relationship between frustration and aggression, such that those with higher levels of future orientation may become particularly frustrated (and experience hostility and aggression) when they experience goal blocking because the future goals they seek are salient to them. Instrumental aggression is likely to occur when a person is frustrated if the individual anticipates that aggression will help her or him to attain the goal, especially if the aggression will eradicate the goal-blocking object (e.g., another person who is impeding goal attainment). Alternatively, future-oriented individuals may be better at generating solutions to obtain their goals and therefore experience less hostility and aggression when frustrated.

Discomfort/Pain

When people become physically uncomfortable for any reason (e.g., heat, loud noise, bodily pain), they usually experience increased negative affect and hostile feelings (e.g., C. A. Anderson et al., 1995), aggressive thoughts are primed (K. B. Anderson, C. A. Anderson, K. E. Dill, & Deuser, 1998), and aggressive behavior can result (Berkowitz, Cochran, & Embree, 1981). Stress associated with discomfort is especially acute when the cause of the discomfort is beyond the person's control (Glass & Singer, 1972), as in the case of a neighbor's car alarm going off in the night.

Thinking about the future may decrease the impact of discomfort on stress and hostile feelings, in that a future-oriented person may think about the typically short-term nature of discomfort. Also, individuals who are future oriented may be more likely to think of strategies that they can engage in to reduce the discomfort and anticipate being more comfortable in similar situations in the future. It is likely that future-oriented people plan more effectively (in order to anticipate positive future outcomes) and may be more able to cope with discomfort. Joireman, Werner, and Kwon (2004) observed a relation between the consideration of future consequences (CFC) and stress that is consistent with this prediction. Their longitudinal data revealed that people low in CFC at the first measurement (Time 1) reported higher depression levels at a later measurement (Time 2), possibly due to stress and negative affect resulting from their failure to anticipate long-term consequences of their actions. Investigations of the relations between future orientation, coping style, stress, and aggression would greatly add to this literature.

Viewing Media Violence

Observing violent media and playing violent video games have been shown to enhance the experience of aggressive thoughts (including aggressive behavioral scripts), feelings, and behavior (C. A. Anderson & Bushman, 2002a; C. A. Anderson & K. E. Dill, 2000). Bandura (1973) showed that violent media can be highly influential on aggressive behavior when the aggressor is rewarded for the aggression. High-CFC people may be particularly susceptible to imitation of rewarded aggression because they ostensibly think about the potential future consequences of their behavior and viewing rewarded violence may make these positive consequences more accessible. Low-CFC individuals may impulsively imitate aggressive characters' actions and/or act in accordance with accessible aggressive behavioral scripts without thinking about potential future consequences.

Alcohol and Other Drugs

Results from a large body of experimental studies conducted over the past four decades indicate that alcohol consumption has a strong but not invariant facilitative effect on aggressive behavior (Bushman, 1997; Bushman & Cooper, 1990; Taylor & Chermack, 1993). In a meta-analysis of laboratory studies of alcohol and aggression, average effect size estimates (*d*) were .61 for alcohol versus placebo (i.e., expect alcohol, receive placebo) condition comparisons and .25 for alcohol versus control (i.e., neither expect nor receive alcohol) group comparisons (Bushman & Cooper, 1990). Although much less research has been done with other drugs, a number of experimental studies demonstrate that central nervous system (CNS) depressants (e.g., diazepam) increase aggression, whereas nicotine, amphetamines, and delta-9-tetrohydrocannabinol (THC; the primary active ingredient in marijuana) were found to be unrelated to aggression (Bushman, 1993; Taylor & Chermack, 1993).

Consistent with the GAM, the impact of alcohol (and presumably other CNS depressants) is moderated by an array of situational (e.g., provocation) and individual difference (e.g., trait aggressiveness) factors (Sher & Wood, in press). Although explicit tests of potential pharmacological and information processing mechanisms are exceedingly rare, consistent with "alcohol myopia" theory (Steele & Josephs, 1990), Wood (1996) found that both aspects of social information processing and participants' motives mediated the alcohol–aggression relation. Specifically, intoxicated participants exhibited less concern about controlling the "opponent's" responses, and this lack of concern was significantly related to higher levels of aggressive behavior. Moreover, hostile motives (i.e., desiring to hurt the opponent or impair his performance) also mediated the alcohol–aggression relation such that intoxicated participants expressed these motives more frequently, and the motives were positively associated with aggressive behavior.

Future orientation has not been examined with respect to the alcohol—aggression relation. Extension of the work of Joireman et al. (2003) involving manipulation of both alcohol consumption and the expectation of future interaction within the context of a laboratory study of aggression could provide interesting results. Analyses could determine whether hypothesized reductions in aggression among high future-oriented individuals expecting to interact would still occur under conditions of intoxication.

Situational Variables and Ability to Think About Aggression Consequences

As stated earlier, provocation, frustration, discomfort, media violence, and alcohol and other drugs were highlighted in this review because they have been clearly linked to aggressive behavior. These factors also dominate the individual's current experience and may therefore reduce the person's ability and motivation to think about the future consequences of their actions. These situational factors may cause individuals to act similarly to those low in CFC in that they focus the attention of the individual on the present situation. Table 13.1 presents the main causes for a person's inability to think about the consequences of aggression and the focus of attention while experiencing the five situational factors.

Consistent with Berkowitz's CN and the GAM, general negative and/or hostile affect typically arises following the experience of provocation, frustration, and discomfort. This may cause the individual to focus only on current phenomenological experience (e.g., "I'm hot") and allocate effort only to think about actions that pertain to the reduction of the negative affective or physiological experience. The individual may mainly focus on salient situational cues linked to their negative experience (i.e., the provacateur, the goal-blocking object, or the cause of the discomfort) and may impulsively "lash out" aggressively,

TABLE 13.1

Main Causes of Inability to Think of Future Aggression Consequences and Attention Foci of Individuals Currently Experiencing Five Situational Factors

Situational Factor	*Cause of Inability to Think of Aggression Consequences*	*Attention Focus*
Provocation	Negative/hostile affect	Anger at provocateur
Frustration	Negative/hostile affect	Goal-blocking object
Discomfort/Pain	Negative/hostile affect	Reducing discomfort if aware of discomfort
Viewing Media Violence	Priming of aggressive scripts	Behavior and images of characters
Alcohol and Drugs	Impaired information-processing ability and motivation	Immediate situational cues

without undergoing reappraisal, because of distraction caused by the often intense negative affective experience. Visual images and aggressive memories and scripts primed by violent TV, movies, and video games may similarly dominate the viewer's cognitions, thereby acting as a "cognitive load," rendering the viewer unable and unmotivated to consider the future consequences of their aggression. Direct imitation of television characters' aggressive behavior is potentially increased. Finally, cognitive impairment caused by alcohol and other drugs can also reduce the user's ability and motivation to think about aggression consequences and can cause them to focus on the immediate social context and aggress impulsively.

In sum, these five situational factors may increase the tendency to focus on the immediate situation and accessible thoughts and feelings and, in turn, decrease consideration of future consequences. Cognitive mechanisms underlying the effects of these situational factors on aggression therefore may approximate those of low-CFC individuals.

PERSONALITY FACTORS ASSOCIATED WITH AGGRESSION

Overview

Throughout the past century, consideration of the potential antecedent role of person factors in aggressive behavior has waxed and waned mirroring the dominant theoretical zeitgeist across various historical eras. For example, near the turn of the 20th century, when instinct theories proliferated, McDougall (1908) conceptualized aggression as a by-product of the "instinct of pugnacity" (p. 240), which he conceptualized as an *innate predisposition* with cognitive, affective, and behavioral components. Subsequent influential theories of aggression (e.g., Dollard et al., 1939) heavily emphasized situational antecedents (e.g., frustration) with little consideration of how the enduring traits individuals brought to these situations might affect behavior. When personality was considered, it was more often examined as a "marker" of psychopathology (e.g., violent crime) and not in the moderational role in which it is typically cast in more contemporary models of aggression (e.g., C. A. Anderson & Bushman, 2002b; Geen, 1990).

Consistent with the GAM (C. A. Anderson & Bushman, 2002b; see also Fig. 13.1) we view person factors as interacting with situational "inputs" to affect the expression of aggressive behavior through cognitive, affective, and arousal states which, in turn, influence information appraisal and reappraisal processes. Thus, from this perspective, personality is best conceptualized as a moderator of aggressive behavior or as affecting the *processes* by which aggression unfolds. Here we focus predominantly on personality factors and social cognitive processes (e.g., Crick & Dodge, 1994) with known or potential connections with future orientation. A host of other person factors, such as gender (Bettencourt & Miller, 1996), beliefs (Huesmann & Guerra, 1997), attitudes (K. B. Anderson, 1996; Malamuth, Linz, Heavey, Barnes, & Acker, 1995), values (Nisbett & Cohen, 1996), goals (Horowitz & Schwartz, 1974), and scripts (Huesmann, 1998), that have been discussed in connection with aggression are not considered here (although see C. A. Anderson & Bushman, 2002b).

Definitions and Organization

Research examining the role of personality in aggressive behavior has largely focused on "narrow-band" personality traits (e.g., impulsivity, irritability) with much less empirical attention aimed at understanding how more comprehensive personality taxonomies (e.g., Costa & McCrae, 1992a; S. B. G. Eysenck & H. J. Eysenck, 1975; Zuckerman, Kuhlman, Thornquist, & Kiers, 1991) may influence interpersonal aggression or how narrow-band traits can be integrated into these models. Previous factor analyses of a broad range of personality scales by Zuckerman and colleagues (Zuckerman, Kuhlman, & Camac, 1988; Zuckerman, Kuhlman, Joireman, Teta, & Kraft, 1993) provide a useful means for integration of narrow-band traits with more comprehensive personality models. These findings, coupled with research by Joireman et al. (2003), provide the framework for our discussion of traits, future orientation, and aggression.

Accordingly, this section provides a brief overview of research examining relations between a three-factor conceptualization of personality and aggression, and then summarizes the larger body of research focusing on narrow-band personality traits with reference to their known or hypothesized representation within a three-factor model of personality. We recognize that the "appropriate" number of traits represented in comprehensive models of personality has been the subject of both wide debate (Costa & McCrae, 1992b; H. J. Eysenck, 1992) and empirical investigation (Zuckerman et al., 1988, 1993), and that other conceptualizations may be equally defensible. Nevertheless, we believe that this organization is useful for several reasons. First, most of the personality traits considered here, including future orientation, can be subsumed by a three-factor model of personality, which provides a relatively parsimonious means of examining personality–aggression relations. Second, the three-factor structure allows for integration and examination of previous theorizing suggesting that dispositional characteristics may moderate situational antecedents of aggression in multiple ways. Specifically, Geen (1990) proposed that traits associated with hostility or emotionality may make certain individuals more susceptible to situational factors influencing aggression. Alternatively, traits associated with "impulsive-unsocialized-sensation seeking" (e.g., impulsivity, sensation seeking, psychoticism) (Zuckerman et al., 1988), may make other individuals less likely to inhibit aggressive tendencies. In the former case, individual difference factors may make an individual more sensitive to situational instigators, whereas in the latter, disposition may be more related to self-regulatory processes that inhibit aggression. The extent to which future orientation is relevant to either (or both) of these hypothesized processes is considered here, we believe, for the first time.

Third, increasingly, researchers have begun to examine the neuropsychological underpinnings of aggressive behavior and the potential links between "executive control function," impulsivity, and other personality traits subsumed under the rubric of impulsive-unsocialized-sensation seeking (Giancola, 1995; Hoaken, Shaughnessy, & Pihl, 2003). Specifically, the prefrontal cortex is thought to be important for a range of higher order cognitive abilities, including temporal organization, behavioral inhibition, and flexibility in response to environmental contingencies (Kolb & Whishaw, 1990). Individual differences in any of these aspects of frontal lobe functioning can easily be incorporated into contemporary models of aggression. For example, it has been suggested that frontal lobe functioning is an index of behavioral inhibition, similar to Berkowitz's self-regulation construct (Lau, Pihl, & Peterson, 1995); and that poor temporal organization may lead to behavior that is "governed more by impulse, current focus of attention, or salient stimulus cues than by rules or plans"

(Lau et al., 1995, p. 150). This notion is also consistent with the GAM (C. A. Anderson & Bushman, 2002b) in that individual differences in frontal lobe functioning might be associated with a tendency to act more on the basis of the initial appraisal of aggression-evoking stimuli and less as a result of reappraisal processes. It is also reasonable to speculate that deficiencies in executive control functioning are likely to be associated with personality factors, including future orientation.

C. A. Anderson and Huesmann (2003) noted that biological predispositions may influence the development of knowledge structures such as schemas and behavioral scripts, as well as affective reaction patterns to certain stimuli. Specifically, certain individuals may inherit an enhanced biological "preparedness" to rapidly learn connections between certain stimuli (e.g., frustrating situations) and anger and aggression. Similarly, other individuals may inherit a preparedness to readily learn negative consequences of aggression.

Relations Between Broad-Band Personality Traits and Aggression

The three-factor conceptualization of Zuckerman et al. (1988) adopted here closely approximates S. B. G. Eysenck and H. J. Eysenck's (1975) model of personality, with the first factor labeled "Sociability" (Sy.) as characterized by tendencies toward extraversion and activity. Factor 2 is labeled "Impulsive-Unsocialized-Sensation Seeking" (IMP-U-SS) and, in addition to impulsivity and sensation seeking, subsumes trait factors such as Eysenck's psychoticism, aggression (Jackson, 1984), and quick decision time (Buss & Plomin, 1975). Factor 3 is labeled "Neuroticism/Emotionality" (N-Emot.) and indexes emotional lability, hostility, and anger.

Most studies examining relations between broad-band personality traits and aggression are consistent with the "marker" investigations of psychopathology noted earlier and have not examined personality as a moderator of situational factors. Moreover, overwhelmingly, these studies have relied on self-report measures of aggressive tendencies and have not directly observed or assessed aggressive behavior. Several of these studies have observed positive, significant associations between N-Emot. and self-reported aggressive tendencies (Buss & Perry, 1992; Caprara, Cinanni, D'Imperio, Passerini, Renzi, & Travaglia, 1985; Harkness, McNulty, & Ben-Porath, 1995; Hernandez & Mauger, 1980; Lynn, Hampson, & Agahi, 1989). Likewise, other studies have observed significant positive relations between aggressive tendencies and IMP-U-SS (Hernandez & Mauger, 1980; Lynn et al., 1989; Rushton, Fulker, Neale, Nias, & H. J. Eysenck, 1989). Two studies of schoolchildren (Mynard & Joseph, 1997; Slee & Rigby,

1993), using the Junior Eysenck Personality Inventory, observed that aggressive behavior (bullying) was associated with higher levels of psychoticism, although only one of the two found evidence for a hypothesized association with neuroticism (Slee & Rigby, 1993).

In a notable exception to the trend toward assessing aggressive tendencies in lieu of actual behavior, Asendorpf and Van Aken (2003), in a 9-year longitudinal study, examined concurrent relations between a five-factor model of personality and a range of behaviors, including aggressiveness toward peers, among children assessed at ages 4–6, 10, and 12. Across multiple raters (parents, teachers, friends) and time points, regression analyses controlling for the other four personality dimensions revealed significant negative associations between agreeableness (subsumed under IMP-U-SS) and aggressive behavior. Less consistent associations were observed for the neuroticism—aggression relation, although parent ratings of neuroticism and aggression were significantly and positively related at age 12.

Taken together, these studies provide evidence consistent with the notion of heightened aggressiveness as a function of individual differences in traits related to both negative emotionality and impulsive-unsocialized sensation seeking. These associations are also consistent with Geen's (1990) hypothesis regarding differing interaction patterns with situational inputs. This more complex pattern of association and the potential role for future orientation have not been examined using broad-band measures of personality. Accordingly, we turn now to the larger literature examining narrow-band traits and aggressive behavior.

Relations Between Narrow-Band Personality Traits and Aggression

Consistent with Joireman et al. (2003), one method of examining links between aggression, narrow-band personality traits (e.g., impulsivity, sensation seeking) and future orientation is with reference to the broader construct of IMP-U-SS. As noted earlier, in addition to impulsivity and sensation seeking, this trait also encompasses Eysenck's psychoticism factor, as well as tendencies toward disinhibited aggression (Jackson, 1984) and quick decision time (Buss & Plomin, 1975). Next, we review studies examining links between aggression, future orientation, and traits indexed by IMP-U-SS, and then consider traits subsumed by N-Emot. (irritability, hostility, anger), as well as other narrow-band traits (e.g., self-esteem, narcissism) that have been examined in connection with aggressive behavior.

IMP-U-SS and aggression. Impulsivity has been linked with violence and aggressive tendencies or behavior in several correlational and a few experimental studies (Buss & Perry, 1992; Hynan & Grush, 1986; Joireman et al., 2003; Stanford, Greve, & Dickens, 1995). Additionally, links between impulsivity and conduct-disorder/antisociality have been observed (Luengo, Carillo-de-la-Pena, Otero, & Romero, 1994); and impulsivity/behavioral dysregulation is an important component of influential interaction theories of antisociality (e.g., Moffitt, 1993).

Joireman et al. (2003) recently examined associations between the four facet scales of the Sensation Seeking Scale (Zuckerman, 1979) and the Aggression Questionnaire (Buss & Perry, 1992). In this study, verbal aggression showed the strongest associations with boredom susceptibility, whereas physical aggression was most highly correlated with disinhibition. Path model analyses yielded direct associations between boredom susceptibility and verbal aggression for both men and women, with direct associations from boredom susceptibility and physical aggression observed only among men. Of particular relevance to our focus here, aggression was found to be higher among those lower on future orientation, and the future orientation–aggression link was moderated by anticipated future interaction. Namely, when future interaction with a person who had insulted them was *expected*, high-future-oriented individuals were less verbally aggressive, but when interaction was either not expected or ambiguous, high- and low-future-oriented participants did not differ in their responses. Joireman and Becker (2004) have replicated this effect in a lab experiment measuring aggressive behavior. These results (and others discussed later with respect to N-Emot.) extend our understanding of the GAM by demonstrating that future orientation interacts with situational factors (e.g., anticipated interaction) to influence aggressive responding, possibly due to differential appraisal/reappraisal processes. However, no studies, to our knowledge, have demonstrated that dispositional impulsivity moderates the effects of situational factors such as provocation or frustration on aggression, or that relations between either impulsivity or sensation seeking and aggression may be moderated by individual differences in future orientation. Research explicating these links, particularly experimental studies measuring aggressive behavior, would constitute additional systematic advances in understanding the role of traits related to IMP-U-SS in the expression of aggressive behavior.

N-Emot. and aggression. Most of the dispositional factors that are typically referenced with respect to "aggressive personality" (e.g., irritability, hostility, trait aggression) can be subsumed by the broad trait of N-Emot. Understandably then, this personality domain has been investigated more than any other as a predictor of aggression. Moreover, unlike most of the personality traits reviewed here, several studies have tested and observed Person × Situation interactions involving various aspects of N-Emot. and have done so with a greater preponderance of behavioral measures of aggression than has been seen elsewhere.

Caprara and colleagues have investigated relations between individual differences related to N-Emot. and aggressive behavior in a series of studies. The three most consistent relations with aggressive behavior have been irritability (e.g., readiness to explode), emotional susceptibility (e.g., helplessness, inadequacy, vulnerability), and dissipation-rumination (e.g., tendency to hold a grudge following provocation). Main effects have been observed in several laboratory studies such that higher levels of these facets are associated with increased aggression (Caprara, 1982; Caprara, Barbaranelli, Pastorelli, & Perugini, 1994; Caprara et al., 1983). Moreover, both irritability and dissipation-rumination have demonstrated moderating effects with situational factors such as negative feedback (Caprara, 1986; Caprara et al., 1983). Bushman and Geen (1990) also found that irritability moderated the relationship between violent media viewing and aggressive thought listing among male participants.

Self-Esteem, narcissism, and aggression. Intriguing recent research has called to question a well-entrenched popular notion; that low self-esteem is predictive of a wide range of maladaptive behaviors, including aggression. Specifically, Baumeister, Smart, and Boden (1996) have found support for the hypothesis that individuals with *high and unstable* levels of trait self-esteem are prone to aggressive behavior, particularly when their self-image is threatened (see also Kernis, Grannemann, & Barclay, 1989). Additional recent research has refined the focus from self-esteem to trait narcissism, a tendency to have an inflated sense of self, with a strong emotional investment in ego protection. Bushman and Baumeister (1998) observed a significant Narcissism × Ego Threat interaction, such that individuals higher on narcissism who received an ego threat demonstrated the highest level of aggression observed in the study. Finally, Bushman, Bonacci, van Dijk, and Baumeister (2003) recently extended examination of narcissism to sexual aggression, and observed that high-trait-narcissism individuals awarded less money to female confederates who had refused to read a sexually arousing passage to them.

SITUATION X PERSON INTERACTIONS ON AGGRESSION

As posited by the GAM, the effects of situation and person variables on hostile thoughts, feelings, and behavior, can be exacerbated (or in some cases ameliorated) when both types of predictors are present. C. Anderson and colleagues have examined direct and indirect situation and person influences on aggressive thoughts, feelings, and behavior within the context of the GAM. For example, K. B. Anderson et al. (1998), using a slightly modified version of Caprara et al.'s (1983) irritability scale labeled as "trait hostility," observed a significant interaction effect of trait hostility and pain on perceptions of aggressive content in a word similarity task. They found that individuals high on trait hostility who were in pain exhibited the highest level of aggressive thoughts. C. A. Anderson and K. E. Dill (2000), using Buss and Perry's (1992) aggression questionnaire, observed that high aggressive tendencies were significantly positively associated with self-reports of aggressive behavior and interacted with exposure to violent video games such that the aggressive behavior–violent video relation was much higher among those with high aggressive tendencies, particularly for men. Interestingly, C. A. Anderson (1997) found that viewing a violent media clip increased aggressive thoughts only for those who were *low* on dispositional hostility. Consistent with the GAM, these findings, in combination with observed relations between trait and state hostility, suggest that both cognitive and affective processes may be implicated in the links between situational factors (e.g., pain, media violence), trait hostility, and aggressive behavior.

As described earlier, Joireman and colleagues have found an interesting interaction between CFC and the perceived immediacy of a negative consequence on self-reported (Joireman et al., 2003) and actual (Joireman & Becker, 2004) measures of aggression: Low-CFC individuals are less aggressive than high-CFC people when an immediate negative consequence is anticipated, and high-CFC individuals are less aggressive than low-CFC people when a delayed negative consequence is anticipated. Predictions for the likelihood of an aggressive response to provocation (or while experiencing other situational predictors of aggression) based on this effect are summarized in Table 13.2. Also included in Table 13.2 is aggression likelihood when the consequence of aggression is anticipated to be positive or negative. When an immediate anticipated consequence is negative, low-CFC individuals are less likely to be aggressive than high-CFCs because low-CFCs are more present oriented and sensitive to immediate situational cues. Alternatively, high-CFCs are less likely to be aggressive than low-CFCs when a more distal negative consequence is anticipated because high-CFCs are more future oriented and think more about future outcomes. The same basic pattern is predicted when the consequence valence is positive, except that the pattern is inverted to reflect more, rather than less, aggression likelihood associated with consequence consideration because of the desire to attain positive consequences. That is, low-CFCs are more aggressive when anticipating immediate positive consequences and high-CFCs are more aggressive when anticipating delayed positive consequences.

TABLE 13.2

Predicted Likelihood of Aggressive Response to Provocation (or Other Situational Predictors) as a Function of CFC and Consequence Immediacy and Valence

	Immediate Consequence	
Valence of Consequence of Aggression	*Consideration of Future Consequences of Actions*	
	Low	High
Negative	Highly Unlikely	Likely
Positive	Highly Likely	Likely
	Delayed Consequence	
Valence of Consequence of Aggression	*Consideration of Future Consequences of Actions*	
	Low	High
Negative	Likely	Highly Unlikely
Positive	Likely	Highly Likely

Not included in Table 13.2 are aggression predictions that account for *both* the immediate and delayed consequences of aggression. Namely, when positive immediate consequences are followed by negative delayed consequences, then high-CFCs would not act aggressively and low-CFCs would. When negative immediate consequences are following by delayed positive ones, low-CFCs would not act aggressively and high-CFCs would.

SOCIAL INFORMATION PROCESSING AND APPRAISAL

According to the GAM, after situation factors prime aggressive thoughts, aggressive feelings, and/or arousal, the individual immediately engages in an initial, automatic effortless appraisal of the situation. Then, if the individual if motivated and able, she will engage in a more effortful process of reappraising the situation. The determination of a perceived provacateur's intent (e.g., accident or malicious attack) is typically the focus of the appraisal, along with potential perceived rewards and costs associated with an aggressive response. The GAM posits that the appraisal process can either be limited to the initial appraisal, such as when cognitive resources are limited (e.g., under conditions of distraction, cognitive load, or intoxication), or it can include reappraisal, which is slower and more deliberative (e.g., thinking about potential multiple causes for the provocation, conducting a cost–benefit analysis on the aggressiveness of the response).

Crick and Dodge (1994) proposed a six-stage model of social information processing that includes: (a) encoding of social cues in an interaction, (b) interpretation of those cues, (c) identification of the person's goals of the interaction, (d) searching and retrieving accessible potential responses, (e) deciding on an available response, and (f) acting on the selected response. Huesmann (1988, 1998) proposed a similar model that describes reciprocal effects of the person's emotional state and activated schemas on encoding, interpretation, retrieval of accessible scripts, response choice, and the individual's interpretation of the environmental response to her or his behavior.

All three models (the GAM; Crick & Dodge, 1994; Huesmann, 1988, 1998) agree that situations and knowledge structures interact in affecting the person's construal of the social situation (C. A. Anderson & Huesmann, 2003). Several studies have shown that trait-aggressive individuals experience hostile attributional biases in that they perceive hostility in nonhostile or ambiguously hostile situations (Dodge & Coie, 1987; Nasby, Hayden, & DePaulo, 1979; Steinberg & Dodge, 1983). For example, K. E. Dill, C. A. Anderson, K. B. Anderson, and Deuser (1997) found that participants with aggressive tendencies were more likely to complete ambiguously aggressive story stems (e.g., two people who get in a car accident) aggressively and were more likely to perceive aggression in dyadic interactions that varied in aggressiveness, than participants with less aggressive tendencies. An interesting extension of this work would be to test whether individuals with hostile tendencies who are high in future orientation show a stronger hostile perception bias for anticipated social interactions than do those who are low in future orientation.

FUTURE ORIENTATION AND AGGRESSIVE TENDENCIES

The Joireman et al. (2003; Joireman, Werner, & Kwon, 2004) findings indicate that individuals with aggressive tendencies tend not to consider the future consequences of their actions. Others have reported conceptually similar findings. For example, Slaby and Guerra (1988) found that high-aggressive students presented with interpersonal conflict scenarios generated less effective solutions to the conflicts, showed more of a hostile attribution bias, and thought of fewer future consequences of an aggressive response in the interactions than did less aggressive students. Crick and Dodge (1994) speculate that the inability to generate a wide range of possible solutions to conflict is a symptom of cognitive deficits and precipitates the social maladjustment experienced by highly aggressive children. However, Lochman, Lampron, and Rabiner (1989) found that when children with aggressive tendencies were forced to deliberate about a variety of responses to interpersonal conflict, they subsequently did not differ from children with less aggressive tendencies in the number of aggressive responses to conflict that they generated. Geen (1998) suggested that it may therefore not be the case that aggressive children are unable to think of nonaggressive solutions to conflict; rather they act impulsively on accessible aggressive responses that are part of well-learned scripts without processing all relevant situational information (including negative potential future consequences).

An interesting direction for future research would be to assess aggressive responses of individuals who fit the uncommon profile of high aggressive tendencies and high future orientation. These individuals may have a greater tendency to think about the future consequences of their actions, especially when negative or positive outcomes are anticipated. They may also tend to consider a wider range of possible responses and not act on the most immediately accessible scripts, as compared to those lower on this trait. The aggressive behavior of aggressive, future-oriented individuals may be moderated by consideration of more distal outcomes. For example, they may be more likely to think about both nonaggressive and aggressive future goals and motives related to others (e.g., wanting to both ingratiate themselves to and hurt their boss) and retrieve scripts that blend these motives and goals (i.e., ridiculing the boss behind her back). Future-oriented aggressive individuals may therefore be more likely to engage in a more deliberative appraisal process than low-future-oriented aggressive individuals, especially when negative or positive future consequences are anticipated. Alternatively, low future-oriented aggressive individuals are more likely to be affected by positive and negative salient immediate consequences of aggression and not by potential future outcomes.

Dodge and Somberg (1987) found that aggressive boys' hostile attributions were greater if the situation included threat of a future encounter with threatening peer. The GAM would predict that a salient threat of future aggression from a peer is a hostile stimulus that primes aggressive scripts, which may result in a hostile attribution bias. This development of a hostile attribution bias increases the likelihood that when individuals with aggressive tendencies engage in effortful processing prior to acting, the cognitions generated (i.e., positive consequences and hostile scripts) support an aggressive behavioral choice.

FUTURE RESEARCH DIRECTIONS

The relation between future orientation and aggression, and situational and personality factors that mediate and moderate the relation, is a topic that is ripe for future research. The GAM provides a useful framework for this research in that it explicates how situation and person factors impact the behavioral choice via the activation of cognitive, affective, and arousal processes. The inclusion of experimental manipulations of aggression-eliciting situational factors (e.g., provocation, alcohol) and the measurement of future orientation (and perhaps other personality factors such as IMP-U-SS) as independent variables, followed by the measurement of aggressive behavior, would significantly advance understanding of the boundary conditions of these effects.

Most of the discussion of future orientation in this chapter has focused on the conceptualization utilized by Joireman et al. (2003, 2004), namely the extent to which people think about the future consequences of the current actions, operationalized as the Strathman et al. (1994) Consideration of Future Consequences Scale (CFC). CFC is a useful operationalization of future orientation in studying its prediction of aggression because aggressive acts almost always elicit a response from the social environment that impacts the individual. However, the broader propensity to think about one's self and others in the future (i.e., the future-orientation scale of the Stanford Time Perspective Inventory; Zimbardo & Boyd, 1999) also may yield interesting results when paired with aggressive tendencies and behavioral opportunities. For example, aggressive individuals who are future oriented may be more likely to think about, and perhaps strive for personal goals than those who are less future oriented. The salience of these goals and even "possible selves," as described by Markus (e.g., Cross & Markus, 1991), may moderate the relation between future orientation and aggressive tendencies. Similarly, as stated earlier, future-oriented people with hostile tendencies may be more likely to engage in instrumental aggression in order to attain an outcome other than harm of the target, such as power.

Most of the research on future orientation conceptualizes it as a disposition. Future research that provides situational manipulations of future orientation by encouraging participants to think about the future and about the consequences of their actions (perhaps by making positive or negative action consequences salient to them) would be interesting in that such research may suggest further interventions to reduce aggression. It will also be useful to determine if invoking future orientation is sufficient in reducing aggression, or if the provision of future outcomes of aggression that would negatively impact the participant is necessary to reduce aggression. As indicated earlier, future research is also needed in which several factors about the appraisal processes are manipulated, including the time, cognitive resources (e.g., distraction, cognitive load), and motivation that participants have during the initial perception of a provoking event and after they have formed an initial impression of the intent of the provocateur.

Finally, there is also good reason to anticipate that the tendency to maintain a negative past orientation predicts aggressive tendencies and perhaps behavior. Zimbardo and Boyd (1999) found a significant positive correlation ($r = .57$) between negative past orientation and self-reported aggression. Aggressive individuals who think about the past may be likely to ruminate on perceived wrongdoings of others that have negatively impacted them. Obsessive rumination of this sort may lead to increased anger and impulsive or instrumental aggression. Future investigations of aggressive behavior that include both Zimbardo and Boyd's time perspective scale, including the future and negative past orientation scales, and the dissipation-rumination scale created by Caprara et al. (1983) would further illuminate the relationship between time orientation and aggression.

APPLICATIONS

One main goal of enhanced understanding of the determinants of aggression is the application of this knowledge to preventive interventions that reduce aggressive behavior. As detailed in this chapter and summarized in the preceding section, many basic questions about the role of future orientation in the expression of aggression remain. Nonetheless, some preliminary suggestions that focus on future orientation within the context of the GAM are made in the following paragraphs that may be helpful for those interested in developing aggression interventions. These suggestions are purely speculative, as research linking future orientation, impulsivity, aggressive tendencies, and behavior is sparse.

The only published work to date on future orientation and aggression (Joireman et al., 2003) revealed that: (a) low-CFC individuals are higher in impulsivity and trait aggression than high-CFCs, (b) low-CFCs reported that they would be more aggressive in a hypothetical situation than did high-CFCs when facing an *immediate* negative consequence of acting aggressively, and (c) high-CFCs reported that they would be more aggressive than did low-CFCs when facing a *delayed* negative consequence of acting aggressively. These findings suggest that aggressive responses may be reduced when people are encouraged to think about possible immediate and long-term negative consequences that may occur to them as a result of aggressing *before* they respond to a hostile provocation. This is likely to be more difficult for individuals high in traits indexed under IMP-U-SS, for impulsive people tend not to think before they act, and for those low in CFC. In the context of the GAM, IMP-U-SS, and low CFC are dispositions that may be linked to the automatic activation of aggressive scripts, attributions, hostile feelings, and possibly physiological arousal following a hostile stimulus. Impulsive and low-CFC individuals may also be unlikely to reevaluate (reappraise) initial hostile attributions for the intentions of others' perceived provoking behaviors, and fail to think about how their aggression may end up harming them in the future.

Some cognitive and cognitive-behavioral approaches to psychological treatment of aggressive individuals (e.g., Aaron Beck's cognitive therapy; Beck & Emory, 1979) emphasize the process of how stimuli are construed or distorted according to the client's depressed or hostile mind-set (hostile attribution bias) and encourage the client to effortfully replace initial hostile attributions with less hostile ones. Practice of this reappraisal process may even result in the development of a nonhostile attribution bias. Furthermore, less aggressive behavioral reactions can presumably be adopted by learning to associate nonhostile attributions of potential provacateur's intent (e.g., he made a mistake) with nonhostile behavioral reactions (e.g., walk away). In conjunction with this revaluation/reappraisal process, aggression interventions may also benefit from teaching clients to practice thinking about how their aggressive responses to situations can have immediate and long-term negative consequences for them and how nonaggressive responses result in more positive outcomes. It may also be useful for them to reflect on previous situations in which they have responded aggressively and think about the negative outcomes that resulted for them and for others. Generally speaking, this type of approach is consistent with an emerging literature on school-based violence prevention efforts that suggests that the development of social competence is one important means by which childhood aggressive behavior may be reduced (Conduct Problems Prevention Research Group, 1999; Flannery et al., 2003; Grossman et al., 1997; Kellam, Ling, Merisca, Hendricks, & Ialongo, 1998; Reid, Eddy, Fetrow, & Stoolmiller, 1999).

In sum, research and theory on aggression suggests that interventions designed to reduce aggression should encourage individuals (particularly impulsive individuals) to: (a) slow down and create time before responding following a perceived provocation or other situational precipitating factors, (b) develop automatic links/associations between perceived provocation and nonaggressive intent so that the provocateur is given "the benefit of the doubt," (c) develop automatic associations with the newly primed nonaggressive intent of the provocateur and a nonaggressive response of the individual, (d) think effortfully about positive outcomes of nonaggressive acts and negative consequences of aggressive acts before responding, and (e) practice these steps in interactions to revise existing and create new behavioral scripts until nonaggressive responses are incorporated into the person's daily behavioral repertoire. At that ideal point, the individual would then no longer experience a hostile attribution bias and would display less aggressive tendencies.

REFERENCES

Anderson, C. A. (1997). Effects of violent movies and trait irritability on hostile feelings and aggressive thoughts. *Aggressive Behavior, 23,* 161–178.

Anderson, C. A., Anderson, K. A., & Deuser, W. E. (1996). Examining an affective aggression framework: Weapon and temperature effects on aggressive thoughts, affect, and attitudes. *Personality and Social Psychology Bulletin, 22,* 366–376.

Anderson, C.A., & Bushman, B. J. (2002a). The effects of media violence on society. *Science, 295,* 2377–2378.

Anderson, C. A., & Bushman, B. J. (2002b). Human aggression. *Annual Review of Psychology, 53,* 27–51.

Anderson, C. A., Deuser, W. E., & DeNeve, K. (1995). Hot temperatures, hostile affect, hostile cognition, and arousal: Tests of a general model of affective aggression. *Personality and Social Psychology Bulletin, 21,* 434–448.

Anderson, C. A., & Dill, K. E. (2000). Video games and aggressive thoughts, feelings, and behavior in the laboratory and life. *Journal of Personality and Social Psychology, 78,* 772–790.

Anderson, C. A., & Huesmann, L. R. (2003). Human aggression: A social-cognitive view. In M. A. Hogg & J. Cooper (Eds.), *Handbook of social psychology* (pp. 296–323). London: Sage.

Anderson, K. B. (1996). *Cognitive and personality predictors of male-on-female aggression: An integration of theoretical perspectives.* Unpublished doctoral dissertation, University of Missouri–Columbia.

Anderson, K. B., Anderson, C. A., Dill, K. E., & Deuser, W. E. (1998). The interactive relations between trait hostility, pain, and aggressive thoughts. *Aggressive Behavior, 24,* 161–171.

Asendorpf, J. B., & Van Aken, M. A. G. (2003). Validity of Big Five personality judgments in childhood: A 9 year longitudinal study. *European Journal of Personality, 17,* 1–17.

Bandura, A. (1973). *Aggression: A social learning analysis.* Englewood Cliffs, NJ: Prentice-Hall.

Baumeister, R. F., Smart, L., & Boden, J. M. (1996). Relation of threatened egotism to violence and aggression: The dark side of self-esteem. *Psychological Bulletin, 117,* 497–529.

Beck, A. T., & Emory, G. (1979). *Cognitive therapy of anxiety and phobic disorders.* Philadelphia: Center for Cognitive Therapy.

Berkowitz, L. (1994). Is something missing? Some observations prompted by the cognitive-neoassociationist view of anger and emotional aggression. In L. R. Huesmann (Ed.), *Aggressive behavior: Current perspectives* (pp. 35–57). New York: Plenum Press.

Berkowitz, L., Cochran, S. T., & Embree, M. C. (1981). Physical pain and the goal of aversively stimulated aggression. *Journal of Personality and Social Psychology, 40,* 687–700.

Bettencourt, B. A., & Miller, N. (1996). Gender differences in aggression as a function of provocation: A meta-analysis. *Psychological Bulletin, 119,* 422–447.

Bushman, B. J. (1993). Human aggression while under the influence of alcohol and other drugs. An integrative research review. *Current Directions in Psychological Science, 2,* 148–152.

Bushman, B. J. (1997). Effects of alcohol on human aggression: Validity of proposed explanations. In M. C. Galanter (Ed.), *Recent developments in alcoholism: Vol. 13. Alcoholism and violence* (pp. 227–243). New York: Plenum Press.

Bushman, B. J., & Baumeister, R. F. (1998). Threatened egotism, narcissism, self-esteem, and direct and displaced aggression: Does self-love or self-hate lead to violence. *Journal of Personality and Social Psychology, 75,* 219–229.

Bushman, B. J., Bonacci, A. M., van Dijk, M., & Baumeister, R. F. (2003). Narcissism, sexual refusal, and aggression: Testing a narcissistic reactance model of sexual coercion. *Journal of Personality and Social Psychology, 84,* 1027–1040.

Bushman, B. J., & Cooper, H. M. (1990). Effects of alcohol on human aggression: An integrative research review. *Psychological Bulletin, 107*(3), 341–354.

Bushman, B. J., & Geen, R. G. (1990). Role of cognitive-emotional mediators and individual differences in the effects of media violence on aggression. *Journal of Personality and Social Psychology, 58,* 156–163.

Buss, A. H., & Perry, M. (1992). The aggression questionnaire. *Journal of Personality and Social Psychology, 63*(3), 452–459.

Buss, A. H., & Plomin, R. (1975). *A temperament theory of personality development.* New York: Wiley.

Caprara, G. V. (1982). A comparison of the frustration-aggression and emotional susceptibility hypotheses. *Aggressive Behavior, 8,* 234–236.

Caprara, G. V. (1986). Indicators of aggression: The dissipation-rumination scale. *Personality and Individual Differences, 7,* 763–769.

Caprara, G. V., Barbaranelli, C., Pastorelli, C., & Perugini, M. (1994). Individual differences in the study of aggression. *Aggressive Behavior, 20,* 291–303.

Caprara, G. V., Cinanni, V., D'Imperio, G., Passerini, S., Renzi, P., & Travaglia, G. (1985). Indicators of impulsive aggression: Present status of research on irritability and emotional susceptibility scales. *Personality and Individual Differences, 6,* 665–674.

Caprara, G. V., Renzi, P., Alcini, P., D'Imperio, G., & Travaglia, G. (1983). Instigation to aggress and escalation of aggression examined from a personological perspective: The role of irritability and emotional susceptibility. *Aggressive Behavior, 9,* 345–351.

Conduct Problems Prevention Research Group. (1997). Initial impact of the Fast Track prevention trial for conduct problems: I. The high-risk sample. *Journal of Consulting and Clinical Psychology, 67,* 631–647.

Costa, P. T., Jr., & McCrae, R. R. (1992a). Four ways five factors are basic. *Personality and Individual Differences, 13,* 653–665.

Costa, P. T., Jr., & McCrae, R. R. (1992b). Reply to Eysenck. *Personality and Individual Differences, 13,* 861–865.

Crick, N. R., & Dodge, K. A. (1994). A review and reformulation of social information-processing mechanisms in children's social adjustment. *Psychological Bulletin, 115,* 74–101.

Cross, S., & Markus, H. (1991). Possible selves across the life span. *Human Development, 34,* 230–255.

Dill, J., & Anderson, C. A. (1995). Effects of justified and unjustified frustration on aggression. *Aggressive Behavior, 21,* 359–369.

Dill, K. E., Anderson, C. A., Anderson, K. B., & Deuser, W. E. (1997). Effects of aggressive personality on social expectations and social perceptions. *Journal of Research in Personality, 31,* 272–292.

Dodge, K. A., & Coie, J. D. (1987). Social information-processing factors in reactive and proactive aggression in children's peer groups. *Journal of Personality and Social Psychology, 53,* 1146–1158.

Dodge, K. A., Murphy, R. R., & Buchsbaum, K. (1984). The assessment of intention-cue detection skills in children: Implications for developmental psychopathology. *Child Development, 55,* 163–173.

Dodge, K. A., & Somberg, D. R. (1987). Hostile attribution biases among aggressive boys are exacerbated under conditions of threat to self. *Child Development, 58,* 213–224.

Dollard, J., Doob, L. W., Miller, N. E., Mowrer, O. H., & Sears, R. R. (1939). *Frustration and aggression.* New Haven, CT: Yale University Press.

Epstein, S., & Taylor, S. P. (1967). Instigation to aggression as a function of degree of defeat and perceived aggressive intent of the opponent. *Journal of Personality, 35,* 265–289.

Eysenck, H. J. (1992). Four ways five factors are not basic. *Personality and Individual Differences, 13,* 667–673.

Eysenck, S. B. G., & Eysenck, H. J. (1975). *Manual of the EPQ (Personality Questionnaire)*. San Diego: Educational and Industrial Testing Service.

Fireman, K. F. (2003, July 3). Words of defiance: Bush challenges Iraqi militants, defends invasion of country. *Newsday,* p. A05.

Flannery, D. J., Vazsonyi, A. T., Liau, A. K., Guo, S., Powell, K. E., Atha, H., Vesterdal, W., & Embry, D. (2003). Initial behavior outcomes for the PeaceBuilders universal school-based violence prevention program. *Developmental Psychology, 39,* 292–308.

Geen, R. G. (1968). Effects of frustration, attack, and prior training in aggressiveness upon aggressive behavior. *Journal of Personality and Social Psychology, 9,* 316–321.

Geen, R. G. (1990). *Human aggression*. Pacific Grove, CA: Brooks/Cole.

Geen, R. G. (1998). Aggression and antisocial behavior. In D. T. Gilbert, S. T. Fiske, & G. Lindzey (Eds.), *The handbook of social psychology* (Vol. 2, 4th ed., pp. 317–356). New York: McGraw-Hill.

Giancola, P. R. (1995). Evidence for dorsolateral and orbital prefrontal cortical involvement in the expression of aggressive behavior. *Aggressive Behavior, 21,* 431–450.

Gilbert, D. T., Pelham, B. W., & Krull, D. S. (1988). Of thoughts unspoken: Social inference and the self-regulation of behavior. *Journal of Personality and Social Psychology, 55,* 685–694.

Glass, D. C., & Singer, J. E. (1972). *Urban Stress*. New York: Academic Press.

Grossman, D., Neckerman, H., Koepsell, T., Liu, P. Y., Asher, K. N., Beland, K., Frey, K., & Rivara, F. P. (1997). Effectiveness of a violence prevention curriculum among children in an elementary school: A randomized controlled trial. *Journal of the American Medical Association, 277,* 1605–1611.

Harkness, A. R., McNulty, J. L., & Ben-Porath, Y. S. (1995). The Personality Psychopathology Five (PSY-5) constructs and MMPI-2 scales. *Psychological Assessment, 7,* 104–114.

Hernandez, S. K., & Mauger, P. A. (1980). Assertiveness, aggressiveness, and Eysenck's personality variables. *Personality and Individual Differences, 1,* 143–149.

Hoaken, P. N. S., Shaughnessy, V. K., & Pihl, R. O. (2003). Executive cognitive functioning and aggression: Is it an issue of impulsivity? *Aggressive Behavior, 29,* 15–30.

Horowitz, R., & Schwartz, G. (1974). Honor, normative ambiguity, and gang violence. *American Sociological Review, 39,* 238–251.

Huesmann, L. R. (1988). An information-processing model for the development of aggression. *Aggressive Behavior, 14,* 13–24.

Huesmann, L. R. (1998). The role of social information processing and cognitive schema in the acquisition and maintenance of habitual aggressive behavior. In R. G. Geen & E. Donnerstein (Eds.), *Human aggression: Theories, research, and implications for social policy* (pp. 73–109). San Diego: Academic Press.

Huesmann, L. R., & Guerra, N. G. (1997). Children's normative beliefs about aggression and aggressive behavior. *Journal of Personality and Social Psychology, 72,* 408–419.

Hynan, D. J., & Grush, J. E. (1986). Effects of impulsivity, depression, provocation, and time on aggressive behavior. *Journal of Research in Personality, 20,* 158–171.

Jackson, D. N. (1984). *Personality research form manual*. Port Huron, MI: Research Psychologists Press.

Joireman, J., Anderson, J., & Strathman, A. (2003). The aggression paradox: Understanding links among aggression, sensation seeking, and the consideration of future consequences. *Journal of Personality and Social Psychology, 84,* 1287–1302.

Joireman, J., & Becker, C. (2004). *Aggression and the consideration of future consequences*. Unpublished data.

Joireman, J., Werner, N., & Kwon, P. (2004). *A longitudinal study on the cycle of aggression*. Unpublished data.

Kellam, S., Ling, X., Merisca, R., Hendricks, B. C., & Ialongo, N. (1998). The effects of the level of aggression in first grade classrooms on the course and malleability of aggressive behavior in middle school. *Developmental Psychopathology, 10,* 165–185.
Kernis, M. H., Grannemann, B. D., & Barclay, L. C. (1989). Stability and level of self-esteem as predictors of anger arousal and hostility. *Journal of Personality and Social Psychology, 56,* 1013–1022.
Kolb, B., & Whishaw, I. Q. (1990). *Fundamentals of human neuropsychology* (3rd ed.). New York: Freeman.
Lau, M. A., Pihl, R. O., & Peterson, J. B. (1995). Provocation, acute alcohol intoxication, cognitive performance, and aggression. *Journal of Abnormal Psychology, 104,* 150–155.
Lochman, J. E., Lampron, L. B., & Rabiner, D. L. (1989). Format differences and salience effects in the social problem-solving assessment of aggressive and nonaggressive boys. *Journal of Clinical Child Psychology, 18,* 230–236.
Luengo, M. A., Carillo-de-la-Pena, M. T., Otero, J. M., & Romero, E. (1994). A short-term longitudinal study of impulsivity and antisocial behavior. *Journal of Personality and Social Psychology, 66,* 542–548.
Lynn, R., Hampson, S., Agahi, E. (1989). Television violence and aggression: A genotype-environment correlation and interaction theory. *Social Behavior and Personality, 17,* 143–164.
Malamuth, N. M., Linz, D., Heavey, C. L., Barnes, G., & Acker, M. (1995). Using the confluence model of sexual aggression to predict men's conflict with women: A 10-year follow-up study. *Journal of Personality and Social Psychology, 69,* 353–369.
McDougall, W. (1908). *An introduction to social psychology.* London: Methuen.
Moffitt, T. E. (1993). Adolescence-limited and life-course-persistent anti-social behavior. *Psychological Review, 100,* 674–701.
Mynard, H., & Joseph, S. (1997). Bully/victim problems and their association with Eysenck's personality dimensions in 8 to 13 year-olds. *British Journal of Educational Psychology, 67,* 51–54.
Nasby, H., Hayden, B., & DePaulo, B. M. (1979). Attributional bias among aggressive boys to interpret unambiguous social stimuli as displays of hostility. *Journal of Abnormal Psychology, 89,* 459–468.
National Highway Traffic Safety Administration. (1996, March). *Analysis of the Capital Beltway crash problem.* Washington, DC: U.S. Department of Transportation.
National Highway Traffic Safety Administration. (1998, April). *Aggressive drivers view traffic differently Capital Beltway focus groups find* (Traffic Tech. 175). Retrieved May 23, 2004, from www.nhtsa.dot.gov/people/outreach/traftech/pub/tt175.html
Nisbett, R. E., & Cohen, D. (1996). *Culture of honor: The psychology of violence in the South.* Boulder, CO: Westview.
Pastore, N. (1952). The role of arbitrariness in the frustration-aggression hypothesis. *Journal of Abnormal and Social Psychology, 47,* 728–731.
Reid, J., Eddy, J. M., Fetrow, R., & Stoolmiller, M. (1999). Description and immediate impacts of a preventive intervention for conduct problems. *American Journal of Community Psychology, 27,* 483–517.
Rushton, J. P., Fulker, D. W., Neale, M. C., Nias, D. K., & Eysenck, H. J. (1989). Aging and the relation of aggression, altruism, and assertiveness scales to the Eysenck Personality Questionnaire. *Personality and Individual Differences, 10,* 261–263.
Sher, K. J., & Wood, M. D. (in press). Subjective effects of alcohol II: Individual differences. In M. Earlywine (Ed.), *Mind altering drugs.* New York: Oxford University Press.
Slaby, R. G., & Guerra, N. G. (1988). Cognitive mediators of aggression in adolescent offenders: 1. Assessment. *Developmental Psychology, 24,* 580–588.
Slee, P. T., & Rigby, K. (1993). The relationship of Eysenck's personality factors and self-esteem to bully/victim behaviour in Australian schoolboys. *Personality and Individual Differences, 14,* 371–373.

Stanford, M. S., Greve, K. W., & Dickens, T. J. (1995). Irritability and impulsiveness: relationship to self-reported impulsive aggression. *Personality and Individual Differences, 19,* 757–760.

Steele, C. M., & Josephs, R. A. (1990). Alcohol myopia: Its prized and dangerous effects. *American Psychologist, 45,* 921–933.

Steinberg, M. D., & Dodge, K. A. (1983). Attributional bias in aggressive boys and girls. *Journal of Social and Clinical Psychology, 1,* 312–321.

Strathman, A., Gleicher, F., Boninger, D. S., & Edwards, C. S. (1994). The consideration of future consequences: Weighing immediate and distant outcomes of behavior. *Journal of Personality and Social Psychology, 66,* 742–752.

Taylor, S. P., & Chermack, S. T. (1993). Alcohol, drugs, and human physical aggression. *Journal of Studies on Alcohol* (Suppl. 11), 78–88.

Wood, M. D. (1996). *Alcohol and aggression: An examination of the roles of cognitive functioning and personality.* Unpublished doctoral dissertation, University of Missouri–Columbia.

Zillmann, D., & Cantor, J. R. (1976). Effect of timing of information about mitigating circumstances on emotional responses to provocation and retaliatory behavior. *Journal of Experimental Social Psychology, 12,* 38–55.

Zimbardo, P. G., & Boyd, J. N. (1999). Putting time in perspective: A valid, reliable individual-difference metric. *Journal of Personality and Social Psychology, 77,* 1271–1288.

Zuckerman, M. (1979). *Sensation seeking: Beyond the optimal level of arousal.* Hillsdale, NJ: Lawrence Erlbaum Associates.

Zuckerman, M., Kuhlman, D. M., & Camac, C. (1988). What lies beyond E and N? Factor analyses of scales believed to measure basic dimensions of personality. *Journal of Personality and Social Psychology, 54,* 96–107.

Zuckerman, M., Kuhlman, M., Joireman, J., Teta, P., & Kraft, M. (1993). A comparison of three structural models for personality: The Big Three, the Big Five, and the Alternative Five. *Journal of Personality and Social Psychology, 65,* 757–768.

Zuckerman, M., Kuhlman, D. M., Thornquist, M., & Kiers, H. (1991). Five (or three): Robust questionnaire scale factors of personality without culture. *Personality and Individual Differences, 12,* 929–941.

14

Future Focus and Depth in Organizations

Allen C. Bluedorn
University of Missouri–Columbia

Depending on how one looks at it, this chapter deals with what either is a phenomenon closely related to future orientation or is a vital, yet often overlooked dimension of the future orientation construct itself. This phenomenon or dimension is temporal depth, a concept I introduced a few years ago (Bluedorn, 2000) and proposed as a variable orthogonal to a person or group's (i.e., culture-carrying collectivity's) general tendency to emphasize the past, present, or future. Making such a distinction, temporal depth is about the questions: How far behind? How far ahead? It is not about the question of direction (i.e., past or future). In this chapter I consider temporal depth in an organizational context and thereby examine the relationships temporal depth has and may have with the behavior of people in organizations. Furthermore, temporal depth is also considered as a culture property of organizations, which leads to a discussion of its relationships with other attributes of the organization as a whole. This examination begins with a discussion of the temporal depth concept, closely related concepts, and such nascent theory that has been developed. Following that discussion I examine existing research on temporal depth and build upon that discussion and current theory to address prospects for application and future research.

CONCEPTS AND THEORY

Before presenting the temporal depth concept, it is necessary to describe the temporal concepts that led to its development. The original conceptual progenitor, at least in one scholarly tradition, is *time orientation*, a very general concept developed by Kluckhohn and Strodtbeck (1961) who described past, present, and future as "the possible cultural interpretations of the temporal focus of human life" (p. 13). They then built upon Spengler's conceptualizations of present and future in *The Decline of the West* to describe the past as emphasizing traditions and their maintenance or restoration, the present as more or less timeless, lacking tradition, and ignoring the future, and the future itself as realizable (Kluckhohn & Strodtbeck, 1961, p. 14). Although the trichotomy of past, present, and future obviously did not originate in Kluckhohn and Strodtbeck's thinking, their Spengler-inspired descriptions do provide both precision compared to lay usage and insight into the likely relationships between each orientation and other cultural and individual attributes. Similar labels, such as temporal perspective and temporal orientation, have been applied to this same basic trichotomy (Bluedorn, Kaufman, & Lane, 1992).

Before moving on to the next major point in the development of the temporal depth concept, I should mention that I have objected elsewhere to the use of phrases such as time orientation or temporal perspective as labels for this concept (Bluedorn, 2000). The reasons for my objections were that I felt such labels would impede the development of temporal scholarship in the social sciences. My reasoning was that labels such as time orientation or temporal perspective *imply* that an individual's or a culture's emphasis on past, present, or future subsumes everything about time that there is to know about time concerning that individual or culture. As I and many others have demonstrated (e.g., Bluedorn, 2002; Lauer, 1981; Levine, 1997; McGrath & Kelly, 1986), there is more about time to study than whether an individual or a society focuses on the past, present, or future. I mention this to alert readers to the point that there are many other important temporal phenomena of relevance to individuals and groups than the tripartite distinction described by Kluckhohn and Strodbeck and others (see Bluedorn, 2002, for a discussion of other temporal phenomena).

That said, it seemed to me that simply because someone might, for example, be oriented to the future, that orientation did not necessarily mean the person was oriented to the *distant* future. It seemed that someone who was very focused on things in the next 20 years is just as oriented to things in the future as someone who is focused on things in the next 200 years. Both are focused on things in the future, and it did not seem reasonable to describe one as more future oriented than the other because, as indicated, both are focused on the future, just different regions of it. (A similar logic seemed to apply to the past as well.) So it seemed to me that two things were involved here, and these two things needed to be distinguished.

One matter was simply the general direction where one's attention and concern were directed, and to simplify things, I left the present out of this, in part because one could argue that what is described as a present focus might actually be a very short-term future or recent past orientation. Indeed, Hofstede and Bond (1988) used a dichotomy that contrasted a future orientation with a past-and-present orientation. This point could certainly be debated, but we need not do so here because, after all, this volume is about future orientation, so we are devoting most of our attention to the future's temporal domain.

In part because of my reservations about labels like time orientation, I elected to use the temporal focus label to describe the issue of direction and defined it as the degree of emphasis on past, present, and future (Bluedorn, 2000, p. 124.) Certainly, authors such Kluckhohn and Strodbeck (1961) used the phrase temporal depth in passing (see the quote from them earlier), but they chose to use the time orientation label for this concept. To me temporal focus is more descriptive, but our principal concern here is not focus, for a focus on the future is a given. Instead, our interest is in the issue of how far into the future is the individual or culture primarily concerned? And that question brings us to the concept of temporal depth.

Originally described in Bluedorn (2000), I subsequently expanded the definition of temporal depth to include both individuals and groups, making the definition of *temporal depth:* the temporal distances into the past and future that individuals and collectivities typically consider when contemplating events that have happened, may have happened, or may happen (Bluedorn, 2002, p. 114). As a dimension orthogonal to temporal focus, temporal depth can be applied to the past, the future, or both. Indeed, as research discussed later in this chapter demonstrates, not only can past and future temporal depths be combined to describe a person or group's total temporal depth, but such a combination is a part of several important findings about organizations.

So there are two concepts, temporal focus and temporal depth, and my contention has been that the two are independent, that they are orthogonal. Indeed, research from several studies of individuals reveals that measures of these concepts are either uncorrelated (i.e., the correlations are not statistically significant), or when the correlations are statistically significant, the correlations are very small (magnitudes smaller than .19; Bluedorn, 2000, p. 124; 2002, pp. 271–272). This empirical evidence supports the conceptual claim of the two concepts' independence.

Thus there are two orthogonal variables: temporal focus, which is a dichotomy—future and past-and-present (a single category) if one follows Hofstede and Bond (1988)—and temporal depth, which is a continuum. But continua are sometimes dichotomized to promote the development of typologies, and that is what I did with temporal depth (Bluedorn, 2000). I did so to develop a typology by cross-classifying temporal focus and a dichotomized temporal depth to yield four distinctive types: the shallow and deep pasts, which are the short- and long-term pasts, respectively; and the shallow and deep futures, which are the short- and long-term futures, respectively (see Bluedorn, 2000, p. 125, especially Fig. 7.1). Such a set of categories suggests major differences among the categories, and one would suspect that those differences ought to be related to differences in other variables. The theoretical work specifying which variables these types would be related to and how they would be related has yet to be developed systematically. Nevertheless, the important differences suggested by an emphasis on the shallow past contrasted with an emphasis on the deep future ought to provide motivation for theoretical development. And the distinction between the shallow and deep futures has certainly been assumed to be important for some time now, with the shallow future generally regarded as morally and pragmatically inferior to deeper futures (see Bluedorn, 2002, chap. 5, for an overview of the general attitudes and assumptions about the differences between the shallow and deeper futures and the presumed virtues of the latter).

To illustrate the theoretical potential of this typology, I build upon March's (1991) insightful work on the organizational strategies of exploration and exploitation. Exploration is about examining and developing new possibilities whereas exploitation is about profiting from well-established certainties (technologies, markets, etc.). Both strategies are about the future, but each focuses on a different temporal depth. Exploration takes time, more time, so it requires a deeper temporal depth than exploitation, which is much more oriented to the short term. So here we can see that two very different strategic orientations—exploration and exploitation—seem to be related systematically to different future orientations in the typology.

These theoretical linkages can be developed even further by combining some of March's ideas with those from the classic work of Lawrence and Lorsch (1967). March saw exploitation and exploration strategies differing in characteristics such as the certainty, speed, and clarity of feedback about the results of the strategy, with such feedback being more certain, faster, and clearer for exploitation than for exploration strategies (please note that a strategy is a pattern in a stream of decisions; Mintzberg, 1978, p. 935.) Lawrence and Lorsch's work, which was a major foundation of the environmental contingency theory of organizations, included important findings about the time span of feedback (how long it takes to learn the consequences of decisions). What they found was that departments such as research and development, which had to wait relatively long times to learn the consequences of their decisions, developed longer time horizons (future temporal depths) than departments such as production, which had to wait only fairly short times to learn the outcomes of their decisions.

Feedback is part of both theoretical works. Lawrence and Lorsch found that the length of time required to receive feedback about decisions was positively correlated with the length of the temporal depths typically employed in different organizational departments. Thus feedback about decisions seems to play a central role in the development of time horizons (i.e., future temporal depths). And how long it takes to receive feedback is basically what March meant in references to the speed of feedback. So if slower feedback (having to wait longer) seems to generate longer temporal depths whereas faster feedback (waiting shorter periods) produces shorter temporal depths (Lawrence & Lorsch, 1967), and if faster feedback is associated with exploitation strategies (March, 1991), it would be expected on logical grounds that exploitation strategies rather than exploration strategies would be associated with shorter temporal depths. This is, of course, how March described the two strategies—exploitation with short-term depths and exploration with deeper depths—but we now have a more thorough theoretical explanation for why.

The small nomological net just presented could easily generate several hypotheses and questions for empirical examination, hypotheses with both theoretical and applied importance. Examples of such questions include the following issues:

1. Do organizations with shallower future temporal depths engage in a greater proportion of exploitation strategies than organizations with deeper future temporal depths?
2. Do organizations with shallower future temporal depths implement exploitation strategies more successfully than organizations with deeper temporal depths?

These two questions are intended to illustrate the kinds of issues that can be generated from the modest theoretical nexus just presented. Certainly, comparable questions could be developed about organizations with deep future temporal depths and exploration strategies, and these and the two questions just presented could be stated as formal hypotheses, but the point is not to just develop hypotheses. Rather, it is to illustrate some of the important questions that can be asked that involve temporal depth and whose answers promise important information for both theory and application. However, to test these and questions with similar importance, a measure of temporal depth is required, and upon that foundation a body of research can begin to be built. Indeed, a measure is required to even ask simple descriptive questions about temporal depth (e.g., How far ahead do people tend to think about things?). Thus in the next section, such a measure is described along with a variety of findings based on it.

RESEARCH

Until recently, little research on organizations has investigated anything about time, and this is the case regardless of whether the unit of analysis is the individual in an organization or a culture-carrying collectivity such as the organization as a whole. This generalization applies to the focus of this chapter as well (i.e, temporal depth). A small amount of work has been done on what would be labeled future temporal depth using the terminology discussed in this chapter, but almost none was conducted on the future as a general orientation (temporal focus). The past has been almost completely ignored empirically, either as a general orientation (temporal focus) or in terms of how far into the past individuals or cultures consider things (past temporal depth). Thus almost all that has been done empirically addresses future temporal depth, and I have reviewed and discussed this very small body of work elsewhere (Bluedorn, 2002, chap. 5). In that review, I noted that future temporal depth tended to be measured on questionnaires with single-item indicators, and these seemed to have an ad hoc format (i.e., they are unstandardized). So from the standpoint of psychometric scale development, this state of measurement suggested the need for a multi-item measure.

And so did some insightful research by Judge and Spitzfaden (1995), who proposed a very intriguing idea. Their idea was that executives might not have a single time horizon (in this chapter's terms, a single future temporal depth) that they employed while making decisions for their companies. Instead, they proposed that executives had multiple time horizons they employed regularly—which Judge and Spitzfaden conceptualized as a *portfolio* of time horizons—and that they then proceeded to investigate empirically and relate to organizational performance (more on this shortly). Conceptually, Judge and Spitzfaden's portfolio-of-time-horizons concept moves one from the singular to the plural—a huge change in conceptual space—and this change certainly suggests the need for multi-item scales if one is going to measure temporal depth psychometrically.

There is, thus, both methodological and conceptual justification for multi-item psychometric measures of temporal depth, and I developed one—the Temporal Depth Index (TDI; see Bluedorn, 2002, pp. 265–272, for the complete index and a thorough account of its psychometric development). The TDI is a six-item measure, three of which are about the past and three of which are about the future. The item about the short-term future illustrates the format of all six items: "When I think about the *short-term future*, I usually think about things this far ahead." The "this far ahead" refers to a set of 15 standard choices ranging from "One day" to "More than twenty-five years." (The two other items about the future substitute the phrases "mid-term future" and "long-term future" in place of "short-term future," respectively.)

These three items load on a strong future temporal depth factor in principle components analysis, and a scale formed by summing the three items has produced alpha coefficients (Cronbach, 1951) as high as .90 (Bluedorn, 2002, pp. 268–270). I was also able to support the claim that future temporal focus (orientation) was orthogonal to future temporal depth by reporting nonsignificant correlations between the three individual future items in the TDI and three questions that asked how important the short-, mid-, and long-term futures were to the respondent (indicators of orientation/focus rather than depth). Similarly, the correlation between the summed future TDI items and Usunier and Valette-Florence's (1994) Orientation Towards the Future scale, though statistically significant, was a small $r = .15$ (Bluedorn, 2002, pp. 271–272), which, being a total of 2.25% shared variance, is so little shared variance that it too indicates that the measures, and hence the concepts, are basically independent (orthogonal). Similar results were obtained for the TDI's items about the past and comparable measures of past orientation, including Usunier and Valette-Florence's Orientation Towards the Past scale.

Thus the TDI appears to be a multi-item scale with sound psychometric properties. But such measurement work is not an end in its own right; the reason for developing such measures is to learn about the concepts they measure and especially about their relationships with other variables.

Benchmarks

Despite the widespread use of terms such as time horizon and planning horizon, and many years of searching the time literature, I never came across empirically based descriptive data that provided answers to such simple questions as, when people say they are dealing with the near-term future, how far into the future do they mean? Or when they speak of the long term, how far ahead are they thinking? Although it would be at least a logical possibility that such answers would not vary by culture, that possibility seems exceptionally unlikely. Yet even so, I never encountered empirical data that addressed such issues for specific cultures either.

So these were some of the first questions to which I applied the TDI. Although I had nothing like a random sample of the American population, I did have a large and what should have been a reasonably representative sample of several hundred American college students complete the TDI (students from a large introductory management course at the University of Missouri–Columbia). Based on the frequency distributions I published from those data (Bluedorn, 2002, p. 113), the mean response for the short-term future was 218.11 days (.60 years); for the midterm future, 892.56 days (2.45 years); and for the long-term future, 3,100.89 days (8.50 years). As already acknowledged, this was not a random sample of the American population, but it is at least a start and provides a base of comparison for results from more representative samples. And a second study does provide results from a sample that is representative of an American population.

In May and June of 1999, Steve Ferris and I sent questionnaires to the CEOs and one other top executive at each firm in a large, randomly selected sample of all publicly traded companies in the United States. We received responses from 15.1% of the companies, and when the responding firms were compared to the nonresponding on several important organizational attributes (e.g., size, profitability), the results of the comparisons revealed only one difference if means were compared, and none if medians were (Bluedorn & Ferris, 2004). Hence the sample appears to be quite representative of the population from which it was drawn: all publicly traded American companies.

The questionnaires we sent to the top executives in this sample contained a modified version of the TDI: to wit, a version that asked the executives to respond about people in their companies overall rather than just about themselves. Analyses of their responses to this adapted form of the TDI closely parallel the psychometric results conducted on individuals discussed earlier in terms of a principal-components analysis, the general absences of significant correlations between future- and past-orientation items and the individual items in the TDI (five of the six correlations were not statistically significant), and an alpha coefficient of .88 for the three future items in the TDI. As for the descriptive statistics, the mean short-term future was 148.78 days (.41 years); the midterm future, 532.82 days (1.46 years); and the long-term future was 1,534.39 days (4.20 years) (all from Bluedorn & Ferris, 2004). And as far as I can tell, these were the first published data about the definitions of these time horizons at the organizational level of analysis—which does provide a solid benchmark against which to judge horizons in specific organizations and results from other samples. Even so, the theoretically more interesting use of the TDI is its use to examine relationships between temporal depth and other variables, relationships that are now considered.

The Past and Organizational Age

I developed the TDI over almost a decade as I tried different formats, different wordings, and so forth. Something I noticed as I examined data from the different versions I administered was a very consistent relationship, one that had apparently been noted in only one earlier study (El Sawy, 1983). And that relationship was a positive correlation between the length of past and future temporal depths (the TDI measures both). From virtually each sample to which I would administer a new version of the scale, a statistically significant positive correlation would emerge between the length of the past and future temporal depths. As I report elsewhere (Bluedorn, 2002), in four samples of college students from the University of Missouri–Columbia totaling nearly 1,200 respondents, the correlations between past and future temporal depths ranged from $r = .31$ to $r = .43$. Thus there appeared to be a consistent relationship between these two temporal depths. At least at the individual level of analysis, the further people tended to look into the future, the further they tended to look into the past as well.

And the same relationship held at the organizational level of analysis. In the data from the Bluedorn and Ferris study (2004), we found a similar statistically significant positive correlation, $r = .29$. Though far from determinant, this relationship appears consistently at both the individual and the organizational level of analysis.

We also found the age of the organizations positively correlated with both temporal depths: $r = .35$ with future depth and $r = .40$ with past depth (Bluedorn & Ferris, 2004). The exact theoretical mechanisms underlying these relationships await development, though. But as important as that development may be, from the perspective of both organization theory and management practice, the most important potential relationships involving temporal depth are relationships with organizational performance.

Organizational Performance

Despite their truism-like status (see Bluedorn, 2002, chap. 5 for a review), one finds surprisingly little empirical research that *directly* examines the relationships—often assumed to be causal, though the causality could be reciprocal or from outcomes to temporal depth rather than the other way around—of variance in future temporal depth and individual and organizational outcomes. Especially noteworthy by their absence are studies that address the relationships between future temporal depth and either individual or organizational performance. Some work at the individual level does address what could be considered performance outcomes if performance is defined broadly. For example, Joireman, Kamdar, Daniels, George-Falvy, and Duell (2004) found relationships between the consideration of future consequences (CFC), a variable dealing with an individual's concern for short-term or long-term consequences of their actions (Strathman, Gleicher, Boninger, & Edwards, 1994), and organizational citizenship behaviors (e.g., those high on CFC reported being more likely to engage in organizational citizenship behaviors, but those high on CFC were less willing to do so than those low on CFC if they believed they would leave the organization in the near future). But even in such work, the distinction is blurred or unclear about whether the temporal variable being investigated is what has been described in this chapter as temporal focus or temporal depth. And as far as research on future temporal depth and individual performance in organizations is concerned, there is almost none unless one attempts to reframe and extract relevant findings from the voluminous research on goals (e.g., Locke, 1994) and the research on deadlines (see Bluedorn & Denhardt, 1988, for a review). But the research on these phenomena seldom deals directly enough with variance in future temporal depth for it to usefully inform a discussion of future temporal depth.

At the level of organizational performance, the situation is slightly better because some research has been conducted that directly addresses relationships between either a future focus or future temporal depth and organizational performance. Lim and Seers (1993) found a statistically significant positive correlation between future orientation (focus) and managers' perceptions of how well their organizations were performing as measured on a scale that tapped 16 different aspects of organizational performance. But this is a relationship with temporal focus rather than depth.

Judge and Spitzfaden (1995) addressed questions of temporal depth more directly. As discussed earlier in the chapter, these investigators developed the idea that managers would use a portfolio of time horizons (future temporal depths) for their organizations rather than just a single time horizon. Judge and Spitzfaden found that managers did use such portfolios, and that the greater the variance of the time horizon portfolios the managers used, the better was their organizations' financial performance.

So there is a bit of research showing that a similar phenomenon (Lim & Seers', 1993, future orientation findings) and future temporal depths themselves (Judge & Spitzfaden, 1995) are related to organizational performance. These findings, as well as the accepted wisdom that not only does future temporal depth matter but longer depths are better than shorter depths, led Steve Ferris and me to investigate the relationship between temporal depth and organizational performance in our study of publicly traded American companies.

As described earlier in the chapter, we obtained data from the top executives at the companies in our sample by having the executives complete the TDI, which they completed based on their perceptions of the temporal depths typically used by the people in their companies. As already discussed, the TDI includes three items about future temporal depth and three items about past temporal depth, allowing us to use all six items together as a measure of total temporal depth in these organizations. After statistically controlling for several variables known to affect organizations' financial performance (i.e., organizational size and industry conditions), one measure of financial performance remained positively related to total temporal depth: earnings per share. Positive relationships with three other measures—return on assets, equity, and sales—though statistically significant at the level of zero-order correlations, decreased to become statistically nonsignificant after the control variables entered the equations. Nevertheless, total temporal depth remained positively related to earnings per share even after adding the control variables. What's more, the temporal depth–earnings per share relationship was moderated by organizational age. The relationship was much stronger in the younger half of the companies in the sample than in the older half.

Overall, these studies suggest there is some validity in the truisms about future temporal depth, but my research with Steve Ferris—which I regard as especially important because its results generalize to the entire population of American publicly traded companies—suggests the relationship is neither as consistent nor as simple as it is often assumed to be. But simple or complex, well established or merely suggested, the issue becomes: How can temporal depth and our knowledge of it be *used* in an organizational context? This is the question we consider next.

APPLICATIONS

Despite a modest empirical base, modest at least compared to the volume of work in an area such as goal setting, suggestions, sometimes strong suggestions, have been made about temporal depth and the management of organizations. Aside from the general perspective that longer time horizons seem to improve organizational performance (e.g., Ouchi, 1981), the strongest recommendations for using temporal depth seem to be found in the work of Jaques (1998a). Without introducing all of Jaques' terminology, his recommendations are based on three fundamental tenets: (a) Formal positions in organizations have characteristic future temporal depth requirements associated with them and these depths vary by position; (b) individuals vary from one another and throughout their lifetimes in their ability to deal with temporal depths of differing lengths;

and (c) individual and organizational performance will be optimized when people who have the capacity to deal with the future temporal depths required by organizational positions are placed into those positions. These points are straightforward, the logic is at least intuitively appealing, and Jaques developed a methodology for measuring both individual temporal depth capacities and positions' temporal depth requirements (Jaques, 1998b). A key point in Jaques' model and methodology is that individuals' capacities for dealing with differing temporal depths varies across individual life cycles in characteristic, predictable ways, and where an individual ends up in this capacity is determined largely by the individual's initial capacity, which varies across individuals.

The purported invariance along these life cycle curves in Jaques' model may depict things a bit too mechanically, because research conducted by El Sawy (1983) indicates people may vary more in their temporal within-person depths than Jaques proposes. As described earlier, El Sawy may have been the first investigator to find and note the positive correlation between past and future temporal depths. Because his method was that of a controlled experiment, El Sawy could show that it was the length of the past depth that led to the length of the future depth and not the other way around. This suggests that lengthening past temporal depths will increase the length of future temporal depths. Interestingly, some research has shown that future orientations (temporal focus) can also be altered, in this case by counseling interventions (Marko & Savickas, 1998). So both future temporal focus and depth may be altered, at least temporarily, and may not be quite as determined as Jaques indicated. And even if the change is temporary, for perhaps a day or even just for a few hours, that may be enough to enhance some organizational tasks such as long-range planning.

A reasonable question might be, how would some tasks benefit from longer temporal depths? I learned at least one answer to this question when I interviewed physicist Gregory Benford, who had been involved with several projects that involved attempts to communicate over thousands, sometimes millions of years (Benford, 1999). Benford proposed that a virtue of having longer temporal depths (the really long ones are often called deep time) is the ability to detect patterns or trends in data, patterns or trends that would be invisible or undetectable if shorter depths were used (Bluedorn, 2002, pp. 138, 192). The idea is very straightforward. If one is looking at a curve for a phenomenon that takes several hundred years to complete its cycle, but data are collected for only 20 years, it may not even be noticeable that a curve is involved at all, let alone the form of the curve, its inflection points, and so forth. Thus the importance of the interval used to collect longitudinal data is gaining increasing attention in methodological discussions (e.g., Mitchell & James, 2001; Zaheer, Albert, &

Zaheer, 1999). So if one is dealing with trends, regardless of whether one is doing basic research or long-range planning, it is best to know those trends are there and what their forms are. And a longer temporal depth into both the past and the future should aid in their detection.

These are only a few potential applications, and I chose to discuss them because they have at least been tried or used. But to make informed decisions about these and other potential applications of temporal depth, not surprisingly, more knowledge is needed, hence more research will be required. So it is to the question of what are some of the important issues that require more research that we now turn.

FUTURE DIRECTIONS

Many questions involving temporal depth within an organizational context, and temporal focus for that matter, will not really be limited to an organizational context and will, instead, likely involve general patterns of human behavior. Thus in the suggestions presented in this section, many may involve temporal focus and depth that transcend the boundaries of the organizational domain at both the individual and group levels. And because so little research has been conducted on these phenomena, especially within organizational contexts, the first suggestion is easy: Almost any research on temporal depth and focus has the potential to significantly advance our knowledge about these two phenomena. For many topics this would be an inaccurate, even irresponsible statement, but for temporal depth and focus in an organizational context, I do not fear contradiction.

That said, what would be some reasonable next questions to address given what we do know about temporal depth and focus in organizations? Referring back to the brief theory development discussion focused on March's material about exploration and exploitation strategies, the issues suggested in that discussion would be good topics for empirical investigation (i.e., Are differing temporal depths related to the two different strategies? Are these strategies more successful if they are accompanied by the temporal depths proposed by March—or by some other temporal depths, if there is any relationship with temporal depth at all?) Given the addition of the time-span-of-feedback material from Lawrence and Lorsch (1967) to March's ideas in the earlier discussion, I believe there is a small but tidy theoretical basis for investigating these possible relationships and stating them as formal hypotheses. Their investigation would be a significant contribution to our understanding of organizational processes and would also hold the potential for providing useful guidance for managers attempting to implement one or the other of these strategies.

A second set of questions involves El Sawy's (1983) findings. El Sawy's findings about the correlation between past and future temporal depths has always fascinated me, and as an earlier discussion indicated, I have replicated this relationship many times in my own research—at both the individual and organizational levels. The question is, why does it occur? There is really no extant theory that explains this association, and an equally intriguing finding was El Sawy's experimental demonstration that it is the length of the individual's past depth that influences the length of the future depth, but length of future depth has no causal impact on the length of past depth. Again, why? We really have no theory that explains this, yet this would seem to be an important empirical generalization in need of a theory. Perhaps Weick's (1979) analysis of future perfect thinking may offer some clues (i.e., the past might be more dominant because it is easier to envision details about it, so it seems more real, which gives it primacy over the future). This is simply a suggestion, but clearly a theoretical explanation is required, both for the relationship and for the causal role of past depth.

Other very application-oriented questions also surround El Sawy's findings. To what extent can future temporal depth be controlled? A research design to investigate this question would involve asking individuals to think about events at differing times in the past and then comparing their responses on tasks involving thoughts about events in the future to learn the extent to which the length of the future horizon can be altered. And altered for how long? One suspects the effect would not be permanent, but does it last for only a few minutes, hours, or days—or even longer? The ability to help people engage events at differing temporal depths could have a major impact on a variety of organizational activities, but planning seems the most obvious. El Sawy's original work provides a research design that could readily be adapted to address these questions. Thus El Sawy provides both a methodological and an empirical foundation upon which to build, two attributes generally absent from considerations of the relationship between organizational performance and temporal depth.

As the discussion of the relationship between temporal depth and organizational performance indicated, there does seem to be a limited relationship between temporal depth and at least one indicator of an organization's financial performance. But similar relationships with other indicators disappeared after statistical controls were applied. This leads to a more sociological question: Why has the belief that the long term is generally better than the short term been so widely accepted? This is especially interesting because there is so little empirical evidence of any kind about this relationship. In some ways this seems similar to the situation with time management: lots of faith and enthusiasm, little data. The most rigorous study of time management practices found that although there were a number of outcomes associated with time management practices, many of them desirable, individual performance was not one of them (Macan, 1994). As indicated earlier, faith in the efficacy of longer temporal depths, especially future temporal depths, seems to have achieved truism status. The level of belief in this proposition deserves an explanation, and I suspect it will be sociological in nature.

For example, differing cognitive and behavioral orientations to the future have been related to social class (Lauer, 1981, pp. 44–45). Hence might the life experiences associated with one or more social classes be related to beliefs that a long-term perspective is both instrumentally and ethically preferable to the short term? Or might a dominant cultural belief that the long term is preferable to the short be related to religious and economic complexes of the sort Max Weber (1958) analyzed in his classic work on the Protestant ethic? I speculate about this potential relationship due to that famous analysis' implications for achievement motivation and the deferral of gratification, both of which might promote a general view that the long term is preferable on both pragmatic and moral grounds. The Protestant ethic and social class may provide, perhaps, in full or in part, answers to the question of why a long-term temporal depth is seen as better than a short-term depth. But even if the answers are to be found elsewhere, the question certainly deserves to be examined.

Finally, even if temporal depth or focus were not related to many other things, which is not the case, these phenomena seem like such fundamental attributes of individuals and collectivities that they would deserve explanation in their own right. Thus what produces a particular temporal depth or focus? Sometimes we might answer that cultures do, but what we really need to know is, what attributes and processes within a culture produce a long-term depth rather than a short one, a future orientation rather than one focused on the past? Although we have some descriptive material about focus or depth in some cultures, we don't have that much of a theoretical explanation for them at the cultural level. Nor at the individual level.

What we do have are a few empirical findings that provide hints. Thus Das (1986) found a positive relationship between a manager's personal future temporal depth and the temporal depth the manager would use on behalf of the organization when making decisions for it. This could partially explain the organization's temporal depth (a cultural phenomenon) if one extends this to a people-make-the place attraction, selection, attrition (ASA) theory explanation (Schneider, 1987). But as indicated, this would be at most a partial explanation, for other empirical work found temporal depth linked to the volatility of the organization's environment: the more the environment changed, the shorter the organization's temporal depth (Bluedorn & Ferris, 2004). The previously discussed Lawrence and Lorsch (1967) work provides yet another factor (time span of feedback) that may be involved. And what about individuals? Certainly the culture into which we are born has an impact on our depths and orientations, but what else does too?

These are just a few of the worthy questions about temporal depth and focus awaiting empirical and theoretical investigation. Many others could be asked right now, and as more research is undertaken, even more questions will be generated. At present, the asking has just begun.

REFERENCES

Benford, G. (1999). *Deep time: How humanity communicates across millennia*. New York: Bard.

Bluedorn, A. C. (2000). Time and organizational culture. In N. M. Ashkanasy, C. P. M. Wilderom, & M. F. Peterson (Eds.), *Handbook of organizational culture and climate* (pp. 117–128). Thousand Oaks, CA: Sage.

Bluedorn, A. C. (2002). *The human organization of time: Temporal realities and experience*. Stanford, CA: Stanford University Press.

Bluedorn, A. C., & Denhardt, R. B. (1988). Time and organizations. *Journal of Management, 14*, 299–320.

Bluedorn, A. C., & Ferris, S. P. (2004). Temporal depth, age, and organizational performance. In C. F. Epstein & A. L. Kalleberg (Eds.), *Fighting for time: Shifting boundaries of work and social life* (pp. 113–149). New York: Russell Sage Foundation.

Bluedorn, A. C., Kaufman, C. F., & Lane, P. M. (1992). How many things do you like to do at once? An introduction to monochronic and polychronic time. *Academy of Management Executive*, 6(4), 17–26.

Cronbach, L. J. (1951). Coefficient alpha and the internal structure of tests. *Psychometrika, 16*, 297–334.

Das, T. K. (1986). *The subjective side of strategy making: Future orientations and perceptions of executives*. New York: Praeger.

El Sawy, O. A. (1983). Temporal perspective and managerial attention: A study of chief executive strategic behavior. *Dissertation Abstracts International, 44*(05A), 1556–1557.

Hofstede, G., & Bond, M. H. (1988). The Confucius connection: From cultural roots to economic growth. *Organizational Dynamics, 16*(4), 4–21.

Jaques, E. (1998a). *Requisite organization: A total system for effective managerial organization and managerial leadership for the 21st century* (Rev. 2nd ed.). Arlington, VA: Cason Hall.

Jaques, E. (1998b). *Time-span handbook*. Arlington, VA: Cason Hall.

Joireman, J., Kamdar, D., Daniels, D., George-Falvy, J., & Duell, B. (2004). *A social dilemma analysis of organizational citizenship behaviors*. Manuscript submitted for publication.

Judge, W. Q., & Spitzfaden, M. (1995). The management of strategic time horizons within biotechnology firms: The impact of cognitive complexity on time horizon diversity. *Journal of Management Inquiry, 4*, 179–196.

Kluckhohn, F. R., & Strodtbeck, F. L. (1961). *Variations in value orientations*. Evanston, IL: Row, Peterson.

Lauer, R. H. (1981). *Temporal man: The meaning and uses of social time*. New York: Praeger.

Lawrence, P. R., & Lorsch, J. W. (1967). *Organization and environment: Managing differentiation and integration*. Boston: Harvard University, Graduate School of Business Administration.

Levine, R. (1997). *A geography of time: The temporal misadventures of a socialpsychologist, or how every culture keeps time just a little bit differently*. New York: Basic Books.

Lim, Y. M., & Seers, A. (1993). Time dimensions of work: Relationships with perceived organizational performance. *Journal of Business and Psychology, 8,* 91–102.

Locke, E. A. (1994). Goal setting and productivity under capitalism and socialism. In Z. Zaleski (Ed.), *Psychology of future orientation* (pp. 157–170). Lublin, Poland: Towarzystwo Naukowe KUL.

Macan, T. H. (1994). Time management: Test of a process model. *Journal of Applied Psychology, 79,* 381–391.

March, J. G. (1991). Exploration and exploitation in organizational learning. *Organization Science, 2,* 71–87.

Marko, K. W., & Savickas, M. L. (1998). Effectiveness of a career time perspective intervention. *Journal of Vocational Behavior, 52,* 106–119.

McGrath, J. E., & Kelly, J. R. (1986). *Time and human interaction: Toward a social psychology of time.* New York: Guilford Press.

Mintzberg, H. (1978). Patterns in strategy formation. *Management Science, 24,* 934–948.

Mitchell, T. R., & James, L. R. (2001). Building better theory: Time and the specification of when things happen. *Academy of Management Review, 26,* 530–547.

Ouchi, W. G. (1981). *Theory Z: How American business can meet the Japanese challenge.* New York: Avon.

Schneider, B. (1987). The people make the place. *Personnel Psychology, 40,* 437–453.

Strathman, A., Gleicher, F., Boninger, D. S., & Edwards, C. S. (1994). The consideration of future consequences: Weighing immediate and distant outcomes of behavior. *Journal of Personality and Social Psychology, 66,* 742–752.

Usunier, J.-C. G., & Valette-Florence, P. (1994). Perceptual time patterns ('time-styles'): A psychometric scale. *Time & Society, 3,* 219–241.

Weber, M. (1958). *The Protestant ethic and the spirit of capitalism* (Talcott Parsons, Trans.). New York: Schribner's.

Weick, K. E. (1979). *The social psychology of organizing* (2nd ed.). Reading, MA: Addison-Wesley.

Zaheer, S., Albert, S., & Zaheer, A. (1999). Time scales and organizational theory. *Academy of Management Review, 24,* 725–741.

15

Environmental Problems as Social Dilemmas: The Temporal Dimension

Jeff Joireman
Washington State University

> The most serious long-term threat facing the world is the danger that human actions are producing irreversible, harmful changes to the environmental conditions that support life on Earth. If this problem is not overcome, there may be no viable world for our descendants to inhabit.
>
> —Oskamp (2000, p. 373)

Ironically, one of the primary problems with environmentally destructive behaviors is the fact that their consequences are fairly minimal when we consider the impact of a single individual at a given point in time. Indeed, the global temperature will not rise appreciably if I choose to take my car to work tomorrow. Our natural resources will not suddenly disappear if I choose to throw a recyclable can in the trash. And press on the earth's capacity to feed its inhabitants will not change after the birth of a new baby. And yet, over the long run, overreliance on cars, failure to recycle, overpopulation, and a variety of related behaviors have contributed to the serious environmental challenges we now face. In sum, many of our most pressing environmental problems can be viewed as the result of an insidious arrangement of conflicting short-term individual and long-term collective consequences that has gradually led us down a path that we might soon regret. Restated, many of our most challenging environmental problems can be framed as *social dilemmas* (cf. Joireman, Lasane, Bennett, Richards, & Solaimani, 2001; Karp, 1996; Steg, 2003), broadly defined as situations in which short-term individual and long-term collective interests are at odds (Komorita & Parks, 1994; Messick & Brewer, 1983).

Viewing proenvironmental behavior as a social dilemma suggests that when presented with a choice between more or less environmentally friendly behaviors, decision makers are faced with at least two basic underlying conflicts of interest: a *social conflict* (between individual and collective interests) and a *temporal conflict* (between immediate and delayed consequences of their actions). If true, understanding how decision makers resolve these conflicts of interest could yield valuable insights into the conditions under which people are willing to sacrifice for the environment. To that end, in this chapter I review theory and research on social dilemmas related to proenvironmental behavior that has either directly or indirectly addressed these issues. Given this book's focus on time, I pay particular attention to those theories and findings relevant to the temporal conflict underlying these proenvironmental dilemmas. However, because temporal concerns may be inferred from what are traditionally viewed as social concerns, I also briefly review work examining how social concerns impact cooperation in social dilemmas. Following my review, I outline several directions for future research.

SOCIAL DILEMMAS AND THE ENVIRONMENT

The use of natural resources has long been viewed as a social dilemma in which short-term individual and long-term collective interests are at odds. One of the best-known examples is Hardin's (1968) classic description of the "tragedy of the commons" in which citizens share access to a common grazing land. The dilemma arises because each person recognizes that he or she would be better off by placing one more cow on the commons, and yet, if all do this, the commons will eventually fail in its ability sustain the demand placed upon it by the individually rational citizens, and all will end up worse off than if all had exercised restraint. This type of situation, in which a behavior with immediate positive consequences for the self results in long-term negative consequences for the self and others, is commonly referred to as a *social (delayed) trap* (Platt, 1973; cf. Cross & Guyer, 1980). In the lab, social traps are modeled within experimentally created resource dilemmas in which participants manage a common pool of points, or more realistic resources like fish or trees. Typically, after "harvests" have been taken out of the common pool, the remaining resources are subject to some replenishment rate that determines the resources available on the next trial.

Other forms of proenvironmental behavior can also be viewed as social dilemmas. As an example, consider the decision to engage in proenvironmental political behavior. On the one hand, proenvironmental activism can be costly in terms of time, money, and effort. However, it can also produce long-term benefits for society. This type of situation, in which a behavior with immediate negative consequences for the self results in long-term positive consequences for the self and others, is commonly referred to as a *social (delayed) fence* (Platt, 1973; for additional metaphors used to describe social dilemmas, see Allison, Beggan, & Midgley, 1996). In the lab, social fences are frequently modeled within experimentally created public goods dilemmas in which members of a group must decide how much of some personal resource to give to a collective account in an effort to produce a public good (e.g., a bonus, environmental legislation that protects the environment and improves public health). One of the challenges of getting people to contribute to public goods is that people can enjoy the benefits of the public good without contributing to its provision by free-riding on the contributions of others (Olson, 1965).

In theory, social traps, social fences, and a host of conceptually related types of social dilemmas (e.g., the classic prisoner's dilemma, commons dilemmas, and public goods dilemmas) pose serious problems for the well-being of societies. The root of the problem in social dilemmas is that it is in each person's own best interest to behave noncooperatively regardless of what others do (noncooperation dominates cooperation), yet all would be better off if all had cooperated than if all had cooperated (mutual cooperation produces better outcomes than mutual noncooperation) (see Liebrand, 1983, for a somewhat less restrictive definition that does not require a dominating strategy). According to a rational actor model (Luce & Raffia, 1957; Olson, 1965), the first property makes it unlikely than any (individually rational) person will cooperate in a social dilemma. Despite its elegance, several decades of research call this assumption into question, as people often do cooperate in social dilemmas (for reviews, see Biel, 2000; Dawes & Messick, 2000; Kollock, 1998; Komorita & Parks, 1994; Van Lange & Messick, 1996). The key question is why.

Social Concerns

One reason people cooperate in social dilemmas is that they value the well-being of others. This concern may be driven by a person's disposition or features of the situation. For example, it is well established that individuals differ in the weight they attach to their own and others' well-being. These weights form the basis of an individual difference construct known as *social value orientation* (McClintock, 1978; Messick & McClintock, 1968; Van Lange, Otten, De Bruin, & Joireman, 1997). Given that all social dilemmas involve a social conflict between personal and collective interests, individual differences in social value orientation have understandably been the focus of much research in the field of social dilemmas. A review of the literature on social value orientation is beyond the scope of the present chapter. Suffice it to say that more than 30 years of research has clearly established that prosocials (those concern with joint gain or equality; Van Lange, 1999) are more likely than proselfs (those concerned with their own gain, or their relative advantage over others) to cooperate in a variety of experimentally created social dilemmas. Most directly relevant within the current context are studies showing that prosocials exercise more restraint in experimentally created commons dilemmas simulating the management of natural resources (e.g., Kramer, McClintock, & Messick, 1986; Liebrand, 1984; Parks, 1994; Roch & Samuelson, 1997). In line with these lab-based findings, several applied studies also suggest that prosocials are more concerned than proselfs with promoting the well-being of the environment. For example, prosocials often express stronger preferences for commuting by public transportation (e.g., Joireman, Van Lange, Kuhlman, Van Vugt, & Shelley, 1997; Van Lange, Van Vugt, Meertens, & Ruiter, 1998; Van Vugt, Meertens, & Van Lange, 1995), and prosocials have been shown to express stronger support for legislation aimed at reducing environmental problems (e.g., Cameron, Brown, & J. G. Chapman, 1998; see also, Joireman, Van Lange et al., 2001), and stronger intentions to engage in proenvironemental political action (Joireman, Lasane et al., 2001).

Concern with the well-being of others can also arise out of a sense of collective identification. For example, numerous studies reveal that cooperation in social dilemmas increases when people are led to experience a strong sense of collective identification (e.g., Brewer & Kramer, 1986; De Cremer & Van Vugt, 1999; Wit & Kerr, 2002). Parallel findings have been observed in field studies examining the effect of naturally occurring levels of community identification on people's willingness to conserve water (e.g., Van Vugt, 2001, Study 1) and involvement in political action (Klandermans, 2000). In sum, a good deal of research has demonstrated that people's concern and/or identification with others can encourage higher levels of cooperation in social dilemmas.

Temporal Concerns

In comparison with social concerns, temporal concerns (i.e., concerns with the delayed consequences of one's actions) have, until recently, received relatively little attention in the social dilemma literature. In one early exception, Kelley and Grzelak (1972) observed that people who chose to cooperate in a 20-trial n-person prisoner's dilemma believed that cooperation was the best strategy for maximizing their own individual outcomes. In interpreting this finding, Kelley and Grzelak comment that:

> In a sense, [the cooperative people who responded in this way] may be said to be responding "incorrectly" if they should take "own score" to mean their own score *in the short run*. From a long-run point of view, however, they are reporting more correctly than their colleagues. The implication of this result may be that these cooperative persons tend more to view their *own* score from a long time perspective and to see the implication of the game, that in the long run their own scores are advanced by the more frequent choice of the [cooperative] action. (p. 194)

Several additional studies in the 1970s implicated, either indirectly or directly, the importance of temporal concerns within social dilemmas. In one study, Kuhlman and Marshello (1975) had people, initially classified as cooperators, individualists, or competitors, play 30 trials of a two-person prisoner's dilemma game with one of three preprogrammed strategies: 100% cooperative, tit-for-tat, or 100% noncooperative. Whereas competitors showed relatively low levels of cooperation, regardless of their partner's strategy, cooperators and individualists were affected by their partner's strategy: Cooperators showed relatively high levels of cooperation with all but the 100% noncooperative partner, whereas individualists only showed a high level of cooperation with the tit-for-tat partner. The typical interpretation of the latter finding is that individualists learned over time that cooperation was in their own long-term best interest (for a fuller discussion of how temporal factors shape cooperative behavior, more broadly defined, see chaps. 11 and 12, this volume).

In another study, more directly related to the current chapter's focus on proenvironmental behavior, Stern (1976) examined people's choices in a multitrial experimentally created social dilemma framed as a choice between commuting by car or by carpool. Two key findings bearing directly on the importance of temporal concerns showed that people demonstrated more conservation behavior (joining a carpool) when consumption (use of gasoline, a function of the number of solo drivers) resulted in price increases and when people were given detailed information about the long-term consequences of their actions.

Since this time, studies directly examining the role of temporal factors in social dilemmas have been relatively sparse. In the early 1980s, Messick and McClelland (1983) drew (renewed) attention to the distinction between social and temporal traps, and highlighted the importance of the purely temporal trap by demonstrating that, although single individuals were more successful at managing a common resource than were groups of three or six, more than half of the individuals who were solely responsible for maintaining the pool for 50 trials failed to do so. Based on their data, Messick and McClelland concluded that "while the social trap is important, the temporal trap is not trivial even for individuals working alone" (p. 108).

In another study, Schroeder and Johnson (1982) suggested that people's willingness to restrain harvests within a social trap could be enhanced by heightening people's concern with the long-term consequences of their actions. In developing their argument, Schroeder and Johnson forwarded a model of a consumer's *consumption rate* (CR) within social traps that took into account the weight a person assigns to short-term (w_{st}) and long-term outcomes (w_{lt}), and the perceived value of the short-term (v_{st}) and long-term (v_{lt}) outcomes, as shown here (cf. Björkman, 1984):

$$CR = f[(w_{st}*v_{st}), (w_{lt}*v_{lt})]$$

To test their model, Schroeder and Johnson manipulated the long-term value of the resource by informing participants that, at the end of the game, the total number of points remaining in the resource pool would be multiplied by one of four weighting factors (.5, 1, 2, 4) before being divided up among the (three) group members. Consistent with their reasoning, participants in the last two conditions left significantly more points in the pool when compared with the group whose weighting factor was 1.0, indicating that increasing the value of the long-term consequences led people to become more conservative in their harvests.

Several years later, Kramer and colleagues (1986) reported that when common resources became more scarce as a result of overharvesting, prosocials (but not proselfs) responded by reducing their harvests, presumably in an effort to extend the life of the resource. Interestingly, on the last (series of) trial(s), prosocials increased their harvests, in this case presumably because they no longer perceived any long-term negative consequences associated with overharvesting.

In sum, aside from a few exceptions, early research paid relatively little direct attention to the importance of temporal concerns in social dilemmas. More recent theory and research, however, are beginning to address the importance of temporal concerns directly. For example, Dewitte and De Cremer (2001) have recently drawn parallels between problems of self-control (primarily viewed in terms of a conflict between short-term and long-term goals) and problems of cooperation in mixed-motive situations (primarily viewed in terms of a conflict between individual and collective interests), and have argued that factors that drive cooperation and self-control may share important parallels. Others have argued that researchers should begin to capitalize on the power of computer simulations to develop a dynamic systems model of cooperative behavior in resource dilemmas, as such simulations allow for an examination of how cooperation may develop over longer periods of time than is typically feasible within the laboratory (Mosler & Brucks, 2003). Finally, Hendrickx, Poortinga, and van der Kooij (2001) have offered a detailed discussion of the theoretical relevance of temporal factors (e.g., time horizon, discount rates, regeneration rates) within resource dilemmas, concluding that "if discount rates affect cooperation in [resource dilemmas], then all situational factors that have been shown to affect discount rates constitute potential determinants of [resource dilemma] behavior" (p. 144) (cf. G. B. Chapman, 2001; Guyse, Keller, & Eppel, 2002; Mannix, 1991; Vlek & Keren, 1992).

One category of temporal factors Hendrickx et al. (2001) identify as relevant in resource dilemmas (and proenvironmental behavior, by extension) are individual differences in the tendency to discount the value of future outcomes. In line with this reasoning, a number of recent studies have demonstrated that individuals who attach a high degree of importance to the delayed consequences of their actions tend to be more likely than those who emphasize immediate consequences to act in the interests of the environment (e.g., Nicolaij & Hendrickx, 2003; Svenson & Karlsson, 1989). One construct that has received increasing attention in the proenvironmental literature is an individual difference construct known as the *consideration of future consequences* (CFC) (Strathman, Gleicher, Boninger, & Edwards, 1994).

Individual Differences in the Consideration of Future Consequences

Individual differences in the consideration of future consequences (CFC) are defined in terms of the relative weight an individual assigns to the immediate versus delayed consequences of his or her actions (Strathman et al., 1994). On one end of the CFC continuum are individuals who attach a high degree of importance to the immediate consequences of behavior, and very little importance to the delayed consequences of their behavior, or to use Schroeder and Johnson's (1982) earlier notation (see previous discussion), those with high values of w_{st}, and low values of w_{lt}. On the other end of the CFC continuum are individuals who attach a high degree of importance to the future consequences of their behavior, and very little importance to the immediate consequences of behavior (i.e., those with low w_{st} and high w_{lt} values).

In their original article, Strathman et al. (1994) linked high levels of CFC with higher levels of a composite measure of proenvironmental behaviors (recycling, energy use, travel mode choice). More important, Strathman et al. demonstrated a theoretically meaningful interaction between CFC and the perceived consequences of an environmentally relevant decision, offshore drilling. In this study, Strathman and colleagues manipulated the expected time course of costs and benefits associated with offshore drilling (i.e., the v_{st} and v_{lt} values in Schroeder and Johnson's equation). Results revealed that individuals low in CFC had more favorable attitudes toward offshore drilling when the advantages were immediate (and the disadvantages were delayed), whereas individuals high in CFC had more favorable attitudes toward offshore drilling when the advantages were delayed (and the disadvantages were immediate). Additional analyses revealed that individuals low (vs. high) in CFC did not differ significantly in the number of positive thoughts they generated about drilling, but individuals high in CFC generated significantly more negative thoughts about drilling than those low in CFC.

Subsequent research has provided additional support for the link between CFC and proenvironmental beliefs and intentions. For example, individuals high in CFC report greater willingness to recycle (Ebreo & Vining, 2001; Lindsay & Strathman, 1997), stronger proenvironmental political intentions and behavior (Joireman, Lasane et al., 2001), and stronger preferences for public transportation, especially when they believe that commuting by car produces negative consequences for the environment (Joireman, Van Lange, & Van Vugt, 2004). An additional study has shown complex, but meaningful links between CFC and support for structural solutions to real-world transportation dilemmas (Joireman, Van Lange et al., 2001). Interestingly, individuals low in CFC were more likely than those high in CFC to support a structural solution to their transportation dilemma (i.e., to fund increases in public transportation) when they believed it would be personally beneficial and would reduce congestion (two more immediate consequences of the proposed transit plan).

Following Strathman et al.'s (1994) article, subsequent research has tended to evaluate the validity of the CFC scale by testing two theoretically relevant models, shown in Fig. 15.1.

According to the *mediation (or awareness) model*, individual differences in CFC influence the perceived consequences of an action, which in turn influence the outcome of interest. For example, an individual high in CFC may be less likely to drive because they believe (are aware) that driving has detrimental long-term consequences for people's health. According to the *moderation (or concern) model*, individual differences in CFC influence the sensitivity to immediate versus delayed consequences of an action. In this case, two individuals, one low, the other high in CFC may be equally convinced that commuting by car is bad for the environment, but individuals high in CFC are less likely to

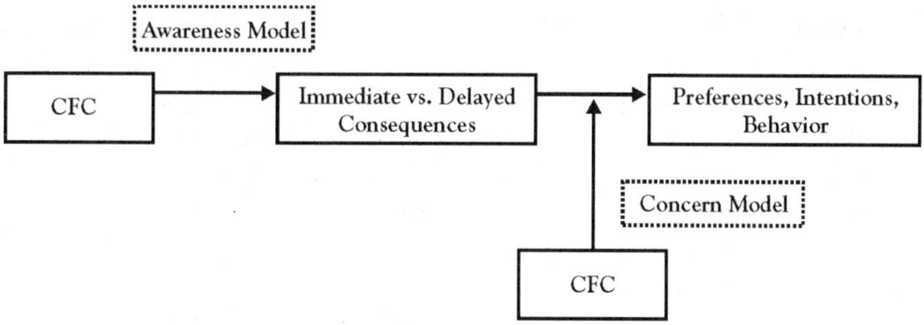

FIG. 15.1. Awareness and concern models of consideration of future consequences.

commute by car, because they are more concerned with the delayed consequences on the environment. Of course, in theory, it is also possible that both models apply at the same time, as when individuals high in CFC are both more aware of, and concerned with, the delayed environmental consequences of commuting by car (e.g., Joireman et al., 2004). These models are important, because they demonstrate the processes by which individual differences in CFC can operate. Moreover, as noted earlier, a growing number of studies have supported the mediation and/or moderation models, lending increasingly strong support for the CFC construct.

Additional Factors with Temporal Implications

A number of other factors more broadly related to temporal concerns have been found to affect cooperation in social dilemmas. Many of these factors revolve around two related questions bearing on anticipated outcomes in a social dilemma: First, what will happen if a person chooses to cooperate or defect in a social dilemma, and second, will an individual's actions have any real impact on the outcome in a dilemma? For example, cooperation can be increased through incentives for cooperation and punishments for defection (e.g., Parks, 2000). Cooperation is also enhanced when one trusts that others will cooperate (Parks, Henager, & Scamahorn, 1996; Pruitt & Kimmel, 1977; Yamagishi & Sato, 1986) and when a person believes that his or her decision matters; for example, when a person believes his or her decision is critical (van de Kragt, Orbell, & Dawes, 1983), or more generally, when a person has a high degree of perceived efficacy with respect to the outcome in a dilemma (e.g., Kerr, 1989, 1992). Also relevant in many environmental social dilemmas is the finding that cooperation in social dilemmas declines as the size of the resource becomes more uncertainty (e.g., Gärling, Gustafsson, & Biel, 1999; Gustafsson, Biel, & Gärling, 1999), in part because when resources are uncertain, people overestimate the size of the resource (the big-pool illusion; Messick & McClelland, 1983), and take more than they should if they are truly trying to maintain the resource.

Each of the factors just noted highlights the impact of temporal concerns on people's willingness to cooperate in their primary or first-order dilemma. Temporal concerns are also relevant when people try to find solutions to solve their primary dilemma. Two approaches to solving social dilemmas have typically been delineated (Messick & Brewer, 1983). So-called individual solutions attempt to influence individual-level decision making, without necessarily altering the incentive structure of the original social dilemma (e.g., via appeals to conscience). So-called *structural solutions,* on the other hand, represent collective action aimed at altering the decision-making authority (e.g., by electing a leader), or altering the incentive structure of the decision (e.g., by making the cooperative response more attractive). Structural solutions are particularly interesting within the context of the current chapter, because they often require people to make sacrifices that may not pay off for some time, as when voters faced with traffic congestion agree to pay additional taxes to fund improvement in public transportation that may require 10 or more years to complete (e.g., Joireman, Van Lange et al., 2001). In more general terms, structural solutions involve temporal concerns because they involve transition costs from the certain status quo to the less certain "solution."

One model that has been used to predict people's support for structural solutions is Samuelson's (1993; Samuelson & Messick, 1995) *multiattribute evaluation model* of structural change in social dilemmas. According to Samuelson's model, people should be more likely to support a structural solution to a social dilemma when they believe that the proposed solution will provide sufficient benefits along at least four dimensions, including efficiency, self-interest, fairness, and freedom. Samuelson's model further assumes that the importance of these dimensions will vary between people (e.g., prosocials should be especially sensitive to the perceived fairness of the solution, whereas proselfs should place more emphasis on self-interest).

Samuelson's model has received support in both the lab (e.g., Samuelson, 1993) and the field (e.g., Joireman, Van Lange et al., 2001; Van Vugt, 1997). Van Vugt, for example, found that British commuters' support for privatizing the country's rail system was negatively related to the plan's perceived transition costs, and positively related to its perceived personal and collective benefits. Van Vugt also found that the relative importance of both transition costs and personal benefits varied as a function of the individual's social value orientation. When evaluating the proposed privatization plan, proselfs placed more weight on personal concerns whereas prosocials placed more weight on the plan's transition costs. In a more recent study, my colleagues and I found additional support for Samuelson's model when we surveyed commuters about their support for a 10-year, $4 billion plan to increase public transportation in and around the city of Seattle (Joireman, Van Lange et al., 2001). In line with Samuelson's model, support for the plan was higher when people believed it would be personally beneficial, be fair, and reduce congestion and pollution. Additional results revealed that prosocials were especially sensitive to the plan's perceived fairness, whereas proselfs were especially sensitive to the plan's perceived personal benefits and its ability to reduce congestion. More directly relevant to the temporal focus of this chapter, we also found that those scoring low in CFC were especially sensitive to the plan's perceived personal benefit and ability to reduce congestion (two more immediate outcomes) whereas those high in CFC were somewhat more sensitive to the plan's ability to reduce pollution.

FUTURE DIRECTIONS

This review has highlighted the various ways in which temporal concerns have been examined within social dilemmas relevant to the environment. Though encouraging, much remains to be learned about how temporal concerns shape social dilemma behavior relevant to the environment. One especially promising direction for future research, it seems, would be to more fully investigate the role of individual differences in the consideration of future consequences. To date, the majority of work in this area has focused on the relationship between CFC and various proenvironmental behaviors such as recycling, commuting preferences, or broad indexes of proenvironmental behavior. Future research should begin to examine the role of CFC within social dilemmas that contain a temporal dimension. This research could help complement and extend the large body of research that has heretofore focused on the role of individual differences in social concerns (i.e., social value orientation).

Future research should also begin to more closely integrate research on social dilemmas related to the environment with more mainstream theory and research on proenvironmental behavior. Environmental issues are often cited as examples of social dilemmas, yet to date, there has been no systematic attempt to integrate the two bodies of theory and research. One contemporary theory of proenvironmental behavior that could begin to bridge that gap is Stern, Dietz, and Kalof's (1993) *extended norm activation model* (ENAM) of proenvironmental behavior. This model comes directly out of Schwartz's (1970) *norm activation model*. According to Schwartz's original model, moral norms influence behavior only when people believe that certain actions have consequences for another's well-being (i.e., awareness of consequences) and when people accept responsibility for producing those actions (i.e., ascription of responsibility). According to Stern et al.'s ENAM, individuals are motivated to act in a proenvironmental fashion based on the extent to which such actions have consequences for things they value (i.e., the self, others, and biosphere). The general specification of the model is:

$$M = V_{ego}AC_{ego} + V_{soc}AC_{soc} + V_{bio}AC_{bio}$$

where M is the motivation to act, *V* is the value an individual assigns to personal, social, and biospheric outcomes, and the three *AC* terms represent an individual's perception that the state of the environment has personal, social, and biospheric consequences, respectively.

Stern et al. (1993) provided evidence for the existence of each value orientation (*V*) by demonstrating a positive relationship between each *AC* term and proenvironmental political intentions (cf. Schultz, 2001). Subsequent research applying this model has focused on the role of universal values (e.g., Schultz & Zelezny, 1999; Stern, Dietz, Kalof, & Guagnano, 1995), providing some support for the claim that perceived consequences mediate the relationship between such values and proenvironmental political intentions (cf. Stern, Dietz, & Black, 1986).

Because the consequences represented in each *AC* term are frequently delayed, the ENAM represents a model of proenvironmental behavior that clearly recognizes the importance of temporal concerns. In addition, two of the value orientations correspond to the typical concerns studied within research on social dilemmas (e.g., concern with self, concern with others). As such, the ENAM provides an avenue for integrating research on proenvironmental behavior and social dilemmas related to the environment. Future efforts in this direction are likely to lead to long-term benefits in our understanding of how to encourage behaviors that protect the environment.

REFERENCES

Allison, S. T., Beggan, J. K., & Midgley, E. H. (1996). The quest for "similar instances" and "simultaneous possibilities": Metaphors in social dilemma research. *Journal of Personality and Social Psychology, 71,* 479–497.

Biel, A. (2000). Factors promoting cooperation in the laboratory, in common pool resource dilemmas, and in large-scale dilemmas. In M. Van Vugt, M. Snyder, T. Tyler, & A. Biel (Eds.), *Cooperation in modern society: Promoting the welfare of communities, states, and organizations* (pp. 25–41). New York: Routledge.

Björkman, M. (1984). Decision making, risk taking and psychological time: Review of empirical findings and psychological theory. *Scandanavian Journal of Psychology, 25,* 31–49.

Brewer, M. B., & Kramer, R. M. (1986). Choice behavior in social dilemmas: Effects of social identity, group size, and decision framing. *Journal of Personality and Social Psychology, 50,* 543–549.

Cameron, L. D., Brown, P. M., & Chapman, J. G. (1998). Social value orientation and decisions to take proenvironmental action. *Journal of Applied Social Psychology, 28,* 675–697.

Chapman, G. B. (2001). Time preferences for the very long run. *Acta Psychologica, 108,* 95–116.

Cross, J. G., & Guyer, M. J. (1980). *Social traps.* Ann Arbor: University of Michigan Press.

Dawes, R. M., & Messick, D. M. (2000). Social dilemmas. *International Journal of Psychology, 35,* 111–116.

Dewitte, S., & De Cremer, D. (2001). Self-control and cooperation: Different concepts, similar decisions? A question of the right perspective. *The Journal of Psychology, 135,* 133–153.

De Cremer, D., & Van Vugt, M. (1999). Social identification effects in social dilemmas: A transformation of motives. *European Journal of Social Psychology, 29,* 871–893.

Ebreo, A., & Vining, J. (2001). How similar are recycling and waste reduction? Future orientation and reasons for reducing waste as predictors of self-reported behavior. *Environment and Behavior, 33,* 424–448.

Gärling, T., Gustafsson, M., & Biel, A. (1999). Managing uncertain common resources. In M. Foddy, M. Smithson, M. Hogg, & S. Schneider (Eds.), *Resolving social dilemmas: Dynamic, structural, and intergroup aspects* (pp. 219–225). Philadelphia: Psychology Press.

Gustafsson, M., & Biel, A., & Gärling, T. (1999). Outcome-desirability bias in resource management problems. *Thinking & Reasoning, 5,* 327–337

Guyse, J. L., Keller, L. R., & Eppel, T. (2002). Valuing environmental outcomes: Preferences for constant or improving sequences. *Organizational Behavior and Human Decision Processes, 87,* 253–277.

Hardin, G. (1968). The tragedy of the commons. *Science, 162,* 1243–1248.

Hendrickx, L., Poortinga, W., & van der Kooij, R. (2001). Temporal factors in resource dilemmas. *Acta Psychologica, 108,* 137–154.

Joireman, J. A., Lasane, T. P., Bennett, J., Richards, D., & Solaimani, S. (2001). Integrating social value orientation and the consideration of future consequences within the extended norm activation model of proenvironmental behavior. *British Journal of Social Psychology, 40,* 133–155.

Joireman, J. A., Van Lange, P. A. M., Kuhlman, D. M., Van Vugt, M., & Shelley, G. P. (1997). An interdependence analysis of commuting decisions. *European Journal of Social Psychology, 27,* 441–463.

Joireman, J. A., Van Lange, P. A. M., & Van Vugt, M. (2004). Who cares about the environmental impact of cars? Those with an eye toward the future. *Environment and Behavior, 36,* 187–206.

Joireman, J. A., Van Lange, P. A. M., Van Vugt, M., Wood, A., Vander Leest, T., & Lambert, C. (2001). Structural solutions to social dilemmas: A field study on commuters' willingness to fund improvements in public transit. *Journal of Applied Social Psychology, 31,* 504–526.

Karp, D. G. (1996). Values and their effect on proenvironmental behavior. *Environment and Behavior, 28,* 111–133.

Kelley, H. H., & Grzelak, J. (1972). Conflict between individual and common interest in an n-person relationship. *Journal of Personality and Social Psychology, 21,* 190–197.

Kerr, N. L. (1989). Illusions of efficacy: The effects of group size on perceived efficacy in social dilemmas. *Journal of Experimental Social Psychology, 25,* 287–313.

Kerr, N. L. (1992). Efficacy as a causal and moderating variable in social dilemmas. In W. B. G. Liebrand, D. M. Messick, & H. A. M. Wilke (Eds.), *Social dilemmas: Theoretical issues and research findings* (pp. 59–80). Elmsford, NY: Pergamon Press.

Klandermans, B. (2000). Identity and protest: How group identification helps to overcome collective action dilemmas. In M. Van Vugt, M. Snyder, T. Tyler, & A. Biel (Eds.), *Cooperation in modern society: Promoting the welfare of communities, states, and organizations* (pp. 162–183). New York: Routledge.

Kollock, P. (1998). Social dilemmas: The anatomy of cooperation. *Annual Review of Sociology, 24,* 183–214.

Komorita, S. S., & Parks, C. D. (1994). *Social dilemmas.* Dubuque, IA: Brown & Benchmark.

Kramer, R. M., McClintock, C. G., & Messick, D. M. (1986). Social values and cooperative response to a simulated resource conservation crisis. *Journal of Personality, 54,* 576–592.

Kuhlman, D. M., & Marshello, A. F. J. (1975). Individual differences in game motivation as moderators of preprogrammed strategic effects in prisoner's dilemma. *Journal of Personality and Social Psychology, 32,* 922–931.

Liebrand, W. B. G. (1984). The effect of social motives, communication and group size on behavior in an *n*-person multi-stage mixed-motive game. *European Journal of Social Psychology, 14,* 239–264.

Lindsay, J. J., & Strathman, A. (1997). Predictors of recycling behavior: An application of a modified health belief model. *Journal of Applied Social Psychology, 27,* 1799–1823.

Luce, R. D., & Raffia, H. (1957). *Games and decisions.* New York: Wiley.

Mannix, E. A. (1991). Resource dilemmas and discount rates in decision making groups. *Journal of Experimental Social Psychology, 27,* 379–391.

McClintock, C. G. (1978). Social values: Their definition, measurement and development. *Journal of Research and Development in Education, 12,* 121–137.

Messick, D. M., & Brewer, M. B. (1983). Solving social dilemmas: A review. In L. Wheeler & P. Shaver (Eds.), *Review of personality and social psychology* (Vol. 4, pp. 11–44). Beverly Hills, CA: Sage.

Messick, D. M., & McClelland, C. L. (1983). Social traps and temporal traps. *Personality and Social Psychology Bulletin, 9,* 105–110.

Messick, D. M., & McClintock, C. G. (1968). Motivational basis of choice in experimental games. *Journal of Experimental Social Psychology, 4,* 1–25.

Mosler, H-J., & Brucks, W. M. (2003). Integrating commons dilemma findings in a general dynamic model of cooperative behavior in resource crises. *European Journal of Social Psychology, 33,* 119–133.

Nicolaij, S., & Hendrickx, L. (2003). The influence of temporal distance of negative consequences on the evaluation of environmental risks. In L. Hendrickx, W. Jager, & L. Steg (Eds.), *Human decision making and environmental perception: Understanding and assisting human decision making in real-life settings* (pp. 47–67). Groningen, the Netherlands: Regenboog Drukkerij.

Olson, M. (1965). *The logic of collective action: Public goods and the theory of groups.* Cambridge, MA: Harvard University Press.

Oskamp, S. (2000). Psychological contributions to achieving an ecologically sustainable future for humanity. *Journal of Social Issues, 56,* 373–390.

Parks, C. D. (1994). The predictive ability of social values in resource dilemmas and public goods games. *Personality and Social Psychology Bulletin, 20,* 431–438.

Parks, C. D. (2000). Testing various types of cooperation rewards in social dilemmas. *Group Processes and Intergroup Relations, 3,* 339–350

Parks, C. D., Henager, R. F., & Scamahorn, S. D. (1996). Trust and reactions to messages of intent in social dilemmas. *Journal of Conflict Resolution, 40,* 134–151.

Platt, J. (1973). Social traps. *American Psychologist, 28,* 641–651.

Pruitt, D. G., & Kimmel, M. J. (1977). Twenty years of experimental gaming: Critique, synthesis, and suggestions for the future. *Annual Review of Psychology, 28,* 362–392.

Roch, S. G., & Samuelson, C. D. (1997). Effects of environmental uncertainty and social value orientation in resource dilemmas. *Organizational Behavior and Human Decision Processes, 70,* 221–235.

Samuelson, C. D. (1993). A multiattribute evaluation approach to structural change in resource dilemmas. *Organizational Behavior and Human Decision Processes, 55,* 298–324.

Samuelson, C. D., & Messick, D. M. (1995). When do people want to change the rules for allocating shared resources? In D. Schroeder (Ed.), *Social dilemmas* (pp. 143–162). New York: Praeger.

Schroeder, D. A., & Johnson, D. E. (1982). Utilization of information in a social-trap situation. *Psychological Reports, 50,* 107–113.

Schultz, P. W. (2001). The structure of environmental concern: Concern for self, other people, and the biosphere. *Journal of Environmental Psychology, 21,* 327–339.

Schultz, P. W., & Zelezny, L. (1999). Values as predictors of environmental attitudes: Evidence for consistency across 14 countries. *Journal of Environmental Psychology, 19,* 255–265.

Schwartz, S. H. (1970). Moral decision making and behavior. In J. Macauley & L. Berkowitz (Eds.), *Altruism and helping behavior* (pp. 127–141). New York: Academic Press.

Steg, L. (2003). Motives and behavior in social dilemmas relevant to the environment. In L. Hendrickx, W. Jager, & L. Steg (Eds.), *Human decision making and environmental perception: Understanding and assisting human decision making in real-life settings* (pp. 83–102). Groningen, the Netherlands: Regenboog Drukkerij.

Stern, P. C. (1976). Effects of incentives and education on resource conservation decisions in a simulated commons dilemma. *Journal of Personality and Social Psychology, 34,* 1285–1292.

Stern, P. C., Dietz, T., & Black, J. S. (1986). Support for environmental protection: The role of moral norms. *Population and Environment, 8,* 204–222.

Stern, P. C., Dietz, T., & Kalof, L. (1993). Value orientations, gender, and environmental concern. *Environment and Behavior, 25,* 322–348.

Stern, P. C., Dietz, T., Kalof, L., & Guagnano, G. A. (1995). Values, beliefs, and proenvironmental action: Attitude formation toward emergent attitude objects. *Journal of Applied Social Psychology, 25,* 1611–1638.

Strathman, A. Gleicher, F., Boninger, D. S., & Edwards, C. S. (1994). The consideration of future consequences: Weighing immediate and distant outcomes of behavior. Journal *of Personality and Social Psychology, 66,* 742–752.

Svenson, O., & Karlsson, G. (1989). Decision-making, time horizons, and risk in the very long-term perspective. *Risk Analysis, 9,* 385–399.

van de Kragt, A. J. C., Orbell, J. M., & Dawes, R. M. (1983). The minimal contributing set as a solution to public goods dilemmas. *American Political Science Review, 77,* 112–122.

Van Lange, P. A. M. (1999). The pursuit of joint outcomes and equality in outcomes: An integrative model of social value orientation. *Journal of Personality and Social Psychology, 77,* 337–349.

Van Lange, P. A. M., & Messick, D. M. (1996). Psychological processes underlying cooperation in social dilemmas. In W. Gasparski, M. Mlicki, & B. Banathy (Eds.), *Social agency: Dilemmas and educational praxiology* (Vol. 4, pp. 93–112). New Brunswick, NJ: Transaction.

Van Lange, P. A. M., Otten, W., De Bruin, E. M. N., & Joireman, J. A. (1997). Development of prosocial, individualistic, and competitive orientations: Theory and preliminary evidence. *Journal of Personality and Social Psychology, 73,* 733–746.

Van Lange, P. A. M., Van Vugt, M., Meertens, R. M., & Ruiter, R. A. C. (1998). A social dilemma analysis of commuting preferences: The roles of social value orientation and trust. *Journal of Applied Social Psychology, 28,* 796–820.

Van Vugt, M. (1997). Concerns about the privatization of public goods: A social dilemma analysis. *Social Psychology Quarterly, 60,* 355–367.

Van Vugt, M. (2001). Community identification moderating the impact of financial incentives in a natural social dilemma: Water conservation. *Personality and Social Psychology Bulletin, 27,* 1440–1449.

Van Vugt, M., Meertens, R. M., & Van Lange, P. A. M. (1995). Car versus public transportation? The role of social value orientations in a real-life social dilemma. *Journal of Applied Social Psychology, 25,* 258–278.

Vlek, C., & Keren, G. (1992). Behavioral decision theory and environmental risk management: Assessment and resolution of four "survival" dilemmas. *Acta Psychologica, 80,* 249–278.

Wit, A. P., & Kerr, N. L. (2002). "Me versus just us versus us all": Categorization and cooperation in nested social dilemmas. *Journal of Personality and Social Psychology, 83,* 616–637.

Yamagishi, T., & Sato, K. (1986). Motivational bases of the public goods problem. *Journal of Personality and Social Psychology, 50,* 67–73.

16

Any Time Is Trinidad Time! Cultural Variations in the Value and Function of Time

James M. Jones
University of Delaware

William T. Brown
Norwalk Community College

In 1973, I spent a year in Trinidad and Tobago studying calypso humor (Jones & Liverpool, 1976). But the cultural expression that captured my attention and has persisted in my own research is the idea that, "Any Time is Trinidad time." What does that mean? My initial assumption was that in Trinidad, like my experiences in the United States, things operated on "colored people's time" (CPT). CPT I was familiar with. Events started when the principals arrived, not when the appointed hour was reached. They ended in similar fashion. If someone said she would met you at 7:00, you might inquire "Is that CPT time?" to know how to manage your own behavior. In essence, CPT is a variable, casual attitude about and value of time that has consequences for behavior.

What I discovered in Trinidad is that time perspective is a cultural attribute, a syndrome even. As a cultural attribute, Any Time is Trinidad Time (ATTT) is not just an attitude about time, it is a value of personal and cultural meaning. That is, *time* has no inherent value; it is not per se an asset. Rather, one's behavior is almost entirely determined by social relationships, personal intentions, preferences, and motives. Time is not a metric by which progress and accomplishment are calculated, but a "yoked control" that mirrors behavior, but exerts little effect on it. The anthropologist E. T. Hall (1983) referred to behavior in time as the *dance of life* that intimately connects feelings, emotions, actions, and verbal and nonverbal displays in a temporal *pas de deux*.

McGrath (1988) refers to a society that values time as a *temponomic* society. In this society, time has inherent value and is treated like an asset. What I discovered in Trinidad was a comparative cultural difference between this temponomic view of time, and a view that was indifferent to time. In this chapter, we call this temporal indifference a *temponostic* orientation. When time is inherently valued, it is treated like any other asset (time is money) and as a result, can be saved, invested, and should not be wasted. Furthermore, the values that arise in conjunction with this temponomic view become fundamental to the character of a society. Behavior that regards time as an asset is rewarded, and behaving indifferently and capriciously toward time is punished (the Puritans called their children to account for their time: "How have you spent your time today?"; cf. Sobel, 1987; see also the aphorism "An idle child is the Devil's playmate").

By contrast, when time has no inherent value, it has little effect on behavior. In temponomic society, time functions as an *independent* variable that often dictates and structures behavior. In a temponostic society, time functions as a *dependent* variable that is structured and given meaning primarily as a consequence of behavior. Temponomic society operates on "clock" time according to schedules and rules of engagement that structure the flow of daily living. Temponostic society operates on "event" time (Levine, 1997), by which the flow of daily living is determined by the transactions and events that occur. The temporal profile of the society is computed from the behavior patterns produced by events, not the other way around.

The purpose of this chapter is to explore the ramifications of cultural differences in temporal orientation. The ATTT temponostic perspective is contrasted with the temponomic perspective and the two are the fulcrum on which the present and future time orientations turn. It is possible as well to consider the past as an aspect of time that shows cultural as well as individual variation. At the individual level, past orientation is often associated with a variety of neurotic and adverse mental states (Holman & Silver, 1998; Jones, Banicky, Lasane, & Pomare, 2004), although aspects of a past orientation have also been linked with more positive behaviors (cf. Zimbardo & Boyd, 1999, and Boyd &

Zimbardo, chap. 5, this volume). But as a cultural syndrome, a strong past orientation may play a more constructive role in the management of psychological and behavioral patterns because of the associations with values of tradition, security, and conformity (cf., Schwartz, 1992).

The view we adopt suggests also that present orientation plays a pivotal role in the mechanisms that emerge to cope with oppressive circumstances. If one lives in a social condition where the contingencies between present behavior and future outcomes are diminished if not severed, then achieving desirable future outcomes may depend on the ability to repeatedly and consistently manage and control the immediate situation or context. This necessity promotes a present-time focus and it could be argued that in a present-oriented person or society the "present drives the future." In a future-oriented person or society, distal goals organize a person's present motivations and behavioral decisions; thus the "future drives the present" (Jones, 1988).

More specifically, we propose that cultural orientation to and value of time can importantly influence behaviors at the individual level, and that these variations are linked to different strategies and preferences for achieving health and well-being, socializing children, and structuring family dynamics. These basic ideas are illustrated and argued from the perspective of African Americans specifically, and persons from African descent more generally. The first section discusses culture principles of time, different ways in which time is conceived and influences behavior, and the values that arise. The second section explores the idea that time is a functional instrumentality in individual and collective lives. Whereas means–ends patterns promote the temponomic agenda, living in the now defines the temponostic strategy. The third section argues that temponomic and temponostic perspectives can and do coexist and taken together promote a comprehensive temporal perspective.

CULTURAL PRINCIPLES OF TIME

Hall (1983) puts time at the center of culture by declaring that, "Because time is a core system of all cultures, and because culture plays such a prominent role in the understanding of time as a cultural system, it is virtually impossible to separate time from culture at some levels" (p. 4). There has been much theorizing and empirical research regarding the origins, structure and consequences of time perspective as an individual-difference construct (cf. Zimbardo & Boyd, 1999). Culture operates at the collective level like personality functions at the individual level. In this section, we briefly discuss the fundamental nature of time as a cultural construction. The overriding point is that time perspective is a cultural value as well as an organizing principle for relationships, norms, and expectations.

There are a myriad of definitions of culture but we prefer the Kroeber and Kluckhohn (1952) definition, which provides a broader perspective on it:

> Culture consists of patterned ways of thinking, feeling and creating, acquired and transmitted mainly by symbols, constituting the distinctive achievements of human groups, including their embodiments in artifacts; the essential core of culture consists of traditional (i.e., historically derived and selected) ideas and especially their attached values. Culture systems may ... be considered products of action, [or] as conditioning elements of future actions. (p. 86)

By these criteria, culture is psychological, symbolic, historical, value based, and dynamic. By dynamic we mean it is both a template for behavior and the nature of meaning as well as a consequence of behavior. That is, we create culture continuously. Our orientation to and value of time is constantly shaping our culture-meaning systems while also responding to them.

Let us illustrate by an example from Jones' Trinidad sojourn. When is it time to leave a party? The night before I was to leave Trinidad after a 1-year stay, I was at a party given by a local journalist. I had to be at the airport by 11:00 a.m. the next morning and still had some packing to do and final preparations to make. I had stopped wearing a watch so I inquired about the time. I was told it was 1:00 a.m. and was amazed. I immediately said, "I have to go. I have to prepare for my trip tomorrow." I was chastised and rebuked for letting the clock dictate my behavior: "What is dat? You drinking, dancing and havin' a good time and now you look at the clock and say you have to go. Come nah mon, stay!" I knew enough to know that my "reasons" were not valued in this cultural milieu. Value lay in the now, not in preparing for the next thing. I stayed and still made my flight the next morning!

Hofstede's Four Dimensions of Culture. One of the earliest comprehensive empirical studies of culture was the classic work of Hofstede (2001) in which he identified four dimensions of culture: *power distance*—substantial power inequalities exist between those high and low in authority and are commonly accepted and valued; *individualism*—the degree to which ties among individuals are loose, and people pursue goals and acquire identities that are self-contained, valued, and contrasted with *collectivism*—which is the value of inclusion and connectedness of individuals to their primary groups; *masculinity* is clearly demarcated from femininity in terms of both role expectations and individual characteristics. Masculinity is assertive and competitive and femininity is nurturing and modest. These qualities are separable from the genders of the people who enact them; and *uncertainty avoidance*—the degree to which one feels uncomfortable in unstructured situations. This tendency leads to the value placed on order, rules, and structure and a belief in absolute truth, whereas *uncertainty acceptance* is associated with flexibility, relativism, and tolerance.

The Fifth Dimension—Confucian Dynamism. Hofstede and Bond (1988) report a study in which traditional Chinese values of Confucianism were assessed across 22 countries and compared with the results from the original Hofstede study. What emerged is a construct labeled *Confucian dynamism,* which was subsequently relabeled as orientation toward the present or the future. High-scoring countries valued persistence, hierarchical ordering of relationships, thrift, and sense of shame. Low-scoring countries valued personal steadiness and stability, protecting one's face, respect for tradition, and reciprocity with others. Eastern countries tended to be low scorers whereas Western countries tended to be high scorers. The high-scoring end was associated with a future orientation, whereas the low-scoring end was associated with greater value placed on the present and the past.

There are two things to note about this fifth dimension. Confucian dynamism is not one polarity *versus* the other, but there are strong Confucian elements in both. Thus the critical element here is the need to find the "middle way" (cf. Nisbett, 2002). The second finding is that both polarities were positively related to economic change as measured by change in gross national product (GNP) from 1965 to 1985. That is, both high- and low-scoring nations on Confucian dynamics were ranked among those with strong economic growth. Thus with respect to time, if culture and time are intertwined, and if orientation to the past, the present, or the future can occur in combination not only as "pure forms," then cultures as well as individuals might importantly be analyzed regarding their "bitemporality" or biculturalilty. The dichotomy we posed at the beginning may be too simple. Rather than juxtaposing temponomic and temponostic approaches, we need to consider the conditions under which either or both are most likely to occur and with what consequence.

Cultural Variations in Time Perspectives. The basic distinction made about cultures comes down to a contrast between an orientation toward the future (temponomic) and a preference for living in the present (temponostic). Hall (1983) distinguishes these perspectives with the concepts of *monochronic time* (M-time) and *polychromic time* (P-time). M-time is characterized by doing one thing at a time, following schedules, and considering time to be tangible. P-time is characterized by "doing many things at once." Punctuality or even time-based appointments are regarded lightly, and time is intangible. P-time is social and thus is based on "transactions." M-time is arbitrary, imposed, and ultimately learned as a consequence of cultural socialization. But, Hall argues, M- and P-time are not mutually exclusive and often interact. M-time is business time, work time, official. P-time is often play time, or social time. We work (M-time) and we play (P-time), and people and cultures have a capacity to do both. But culture is "patterning" of behavior, and the dominant pattern defines the cultural character. So when I purchased a rug in Morocco, it was a long process involving green tea, music, chatter, and of course bargaining. It was a P-time social occasion as a means of producing an M-time business transaction. Even in the United States, business deals may get brokered on the golf course.

Brislin and Kim (2003) discussed these dimensions of time in relationship to international business dealings across cultural boundaries. Of the many dimensions of culture that might be relevant, Brislin and Kim delineate time as crucial. They propose 10 ways in which intercultural transactions may be complicated by differences in time perspective and values: (a) clock versus event time; (b) punctuality; (c) task versus social time as workday activities; (d) M-time versus P-time; (e) relative importance of work versus leisure time; (f) fast versus slow pace of life—fast pace related to both higher economic productivity and coronary heart disease; (g) long periods of silence—may be uncomfortable to clock time, temponomic cultures, but valuable opportunities to show respect and calm in event time societies; (h) past, present, and future orientation—variations indicate the relative importance of tradition, short-term versus long-term goals, efficiency, the value of time; (i) symbolic nature of time—time is money, control a sign of power, important issues get more time; (j) time efficiency—faster is better in some cultures, not in others.

But what about cultures for whom the future is not so prominent. Mbiti (1970) argues that in many African cultures, language does not recognize the future, only the present (*sasa*) and the past (*zamani*). By this reckoning, time moves "backward." That is, because the future is only a "hypothetical construct," it has no tangible place in human affairs. Ancestors are culturally crucial because their life and spirit is responsible for the present. What lies ahead of us is the role of ancestor. Once an ancestor, we belong to the past and a *zamani* existence. The life chain is a cycle from present to past with only a fleeting moment in which a short-term future may come into fuzzy view.

This *sasa*-to-*zamani* axis can be considered a cyclical view of time. It is contrasted with the more familiar linear view in which the past precedes the present which precedes the future. In this linear world, events are discrete, time and space are bound together, and Newtonian physics is gospel. To illustrate, Cottle and Klineberg (1974) developed the "lines test," in which a respondent first places a hash mark on a straight line to indicate the point of his or her birth. Then a mark is placed at the beginning of the present, the beginning of the future, and one's death. The net of this exercise is five temporal regions: the "historical" past (prior to one's birth), the personal past, the present, the personal future, and the "historical" future (time after one's death). This procedure demonstrates the presumed linearity of life and the flow of time from past to future.

In the cyclical world, space and time are only casually related, events overlap, things occur simultaneously, and Einsteinian relativity is a core operating system. Cottle and Klineberg (1974) also developed a "circles test," which simply asked respondents to draw three circles representing the past, the present, and the future. In this way, issues of overlapping temporal regions, the size of these regions, and so forth could be calculated. The circles test admits a more cyclical view of time (for a review of additional measures of time orientation, see chaps. 2 and 5, this volume).

Summary. Time is a core cultural concept; it's cultural significance achieves a level of value that gives rise to characteristic beliefs and behaviors. Time is generalized and abstract as well as concrete in situations.

TIME AS VALUE—TEMPONOMICS

Future orientation is a value that follows from the temponomic perspective. Hofstede, (2001) based his cultural studies on Rokeach's (1973) idea that a value is "an enduring belief that a specific mode of conduct or end-state of existence is personally and socially preferable to alternative modes of conduct or end-states of existence" (pp. 159–160).

One value arises from the attitude or belief in the contingent probabilities associated with present actions and future outcomes (Jones, 1988). Specifically, believing that a specific act in the present (e.g., studying hard) increases the likelihood of the occurrence of some future goal state (becoming a medical doctor, having a rewarding career), attaches value to those contingent beliefs, and provides behavioral control. In a "just" world, one could argue, people control their futures and the outcomes that follow.

Another way is to attach value to the behaviors that lead to attainment of desirable future goals. There is ample experimental evidence for the value-based belief in setting goals, and realizing them. For example, DeVolder and Lens (1982) defined "perceived instrumentality" as the perceived contingent relationship between present study behaviors and future outcomes. They found among Belgian male adolescents a positive relationship between perceived instrumentality and academic performance.

Individual Differences in Time Orientation

In our own work, we have found similar evidence of the link between characteristics of future orientation (temponomics) and academic achievement. To test these time orientation effects, Jones et al. (2004) developed a temporal orientation scale (TOS). The TOS is a 15-item self-report scale with 5 items each assessing one's past orientation (I think about the past a lot), present orientation (The joy in my life comes from what I am doing now, not from what I will be doing later), and future orientation (I am able to resist temptation when there is work to be done).

Past Orientation. A past-oriented person thinks about the past a lot, and seems to want to relive those experiences. The reasons for reliving them appear to be twofold: One is to change them, which implies that the experiences were not as positive as the person would have liked, and the other is to return to what may perhaps have been a happier time. Past-orientation scores were positively related to neuroticism, and negatively related to conscientiousness in the NEO Personality Inventory (Costa & McCrae, 1986), and positively associated with depression (Beck Depression Inventory [BDI]), rumination (Lyubormirsky & Nolen-Hoeksema, 1995), and pessimism (Scheier, Carver, & Bridges, 1994).

Present Orientation. A present temporal orientation suggests a person who truly lives in the present by not dwelling on the past and not looking ahead to the future. The present focus is on enjoyment, which is an important aspect of the present-oriented person's behavioral motivation. The present-oriented person is focused on the now (living day-to-day), not on the future or the past. Present-orientation scores were positively related to extraversion (NEO), optimism, and impulsivity (Zuckerman, Kuhlman, Joireman, Teta, & Kraft, 1993), and negatively related to goal orientation (Malouff et al., 1990), and concern with future consequences (Strathman, Gleicher, Boninger, & Edwards, 1994).

Future Orientation. A future-orientation is characterized by planful action, delay of gratification, self-discipline, perseverance, punctuality, and the use of means–ends behavioral chains to meet long-term goals. Future-orientation scores were significantly related to conscientiousness and extraversion in the NEO, optimism, goal orientation, and concern with future consequences, and negatively related to impulsivity (Zuckerman Kuhlman Personality Questionnaire, ZKPQ) and depression (BDI).

Using the TOS to assess temporal orientation, we have explored the association of temporal orientation and several markers of academic achievement. For example, college students who set academic goals, internalize those goals as part of their self-concept, and believe in their personal efficacy in reaching those goals, show higher academic performance than those who either do not formulate academic goals, or do not internalize them when they do (Lasane & Jones, 1999).

It is important to know that these linkages of future orientation and academic achievement involve not only the ability to formulate goals but also internalizing the value of avoiding or suppressing behaviors that are believed to interfere with reaching desired goal. Lasane and Jones (2000) showed that the relative value placed on social relationships in part determined whether a person would make behavioral decisions that undermined their already established academic goals. Present-oriented college students were significantly more likely than future-oriented to choose a socially desirable alternative that conflicted with an academically relevant one. Present orientation, it seems, is characterized by the value of interpersonal transactions and events, not schedules or goals.

Brown and Jones (2004) investigated the relationship between the future temporal orientation (FTO) and academic performance of African American high school students. We proposed a positive relationship between FTO and academic performance but predicted that it would be mediated by students' perceptions of the usefulness of an education and the value they placed on academic work. We hypothesized further that the relationships between FTO and these mediators would be moderated by students' perceptions of school and societal inequity. Consistent with general findings (cf. Raynor & Entin 1983; Van Calster, Lens, & Nuttin, 1987) future-oriented African American students tended to also have higher grade point averages (GPAs). Those students with strong orientations toward the future also tended to see education as more useful for reaching later life success, and indicated higher perceived intrinsic value for academic work and activities. In addition, perceptions of social equality moderated the relationship between educational usefulness and the intrinsic value of academic activity, such that those students who perceived that their schools were unfair valued academic work less and did more poorly but only if they were low in future orientation. So future orientation leads to positive academic performance because of its positive association with educational values and academic activity, although this relationship can be diminished by the perception of unfair bias against you or your cultural or ethnic group.

The findings at the individual level of personality support the general argument that temponomics and temponostics (M-time and P-time) reflect fundamentally different value systems and guide behaviors that correspond to them.

In a temponomic world, goals are valued and behaviors conform to the parameters of goal attainment. In a temponostic world, individual autonomy and social relationships are valued and behavior conforms to these parameters.

CONTROLLING TIME—TEMPONOSTICS

"Time flies when you're having fun." Why exactly does time "fly"? To answer that we must consider the alternative in which time slows down—"the watched pot never boils." In the latter case, if we attend too closely to the details of an event as it unfolds in time, we are unconsciously setting up a stopwatch and ticking down the seconds. The *implicit* time of an event is tracked as we pay close attention to it. But when we are having fun, what we are not doing is "paying attention." We are engaging in one moment, not monitoring a succession of moments. Every cognitive fixation that is linked to a temporal unit allows us to trace its temporal path. In a temponostic world, time is not attended to; it flies and we have fun!

So a temponostic society or person is one in which time is not closely attended to; it has no inherent value; it is a silent partner in life. So where does order come from? It comes from the requirements of cultural patterns and values, social relationships, and personal intentions. ATTT is a cultural pattern or prescription. It is also a value, and it describes how things get done. It is not the absence of order, but we might define the qualities that distinguish a temponostic from a temponomic order. Order is inductive, bottom-up, derivative of the flow of behavior. This is contrasted to the deductive, top-down, and generative way in which time dictates order in the temponomic mode. In the former, the individual has substantial freedom and control to do what they want when they want. In the latter, the individual is constrained by the order, by the schedule, and by structure imposed by time.

Temporal Aspects of Behavioral Control

Let's look more closely at ways in which temponostic perspectives enhance personal control. The context is the state of being enslaved, marginalized, and discriminated against—the African American and African diasporic case. In its essence, slavery created a necessity to live in the now, to manage and control what was happening or could happen to you at any moment in a capricious and unpredictable world. Being present oriented was a necessity. But to change your lot, you also had to look ahead to the future (e.g., how could one ever make the trip on the Underground Railroad from deep in the South to the North and to Canada?). One has to be acutely sensitive to the nuances of the present while keeping one's "eyes on the prize."

We turn now to a model of psychological and cultural adaptation predicated on contexts in which connections to the past and projections into the future are tenuous, and the present circumstance is threatening to one's physical and psychological well-being. Time and several variables associated with time combine to create a model of coping and adaptation labeled TRIOS.

The TRIOS Model: Being in Time

Jones (2003) has argued that present-time perspective is connected to a more general worldview of African origins consisting of five major characteristics referred to as TRIOS. These five qualities together—time, rhythm, improvisation, orality, and spirituality (or TRIOS)—describe a coordinated worldview that turns on the basic idea of present time as a context for behaving, creating, and being. In addition to the perspective on time (mentioned earlier in the first section), African culture also carried a strong oral tradition, and valued and utilized rhythm in rituals, celebrations, and communication. Furthermore, improvisation was a quality that surfaced in music and oral performances and became a crucial aspect of adaptation to the vagaries of the controlling system of oppression. Finally, spirituality connected one to a higher order that slowed down and expanded time, connecting one to ancestors. Nature and the universe thus transcended the capricious and oppressive circumstances, relieving one of having to internalize control in a world that did not make control easy. The future, in this world, follows from and is driven by the present and the success one has in controlling it.

ATTT encompasses all of these TRIOS elements, although they are coordinated by their focus on the present. In a temponostic society, what matters is the life being lived, not the life that is being planned. Rhythm is integral to ATTT. The rhythm of the language, cadences, inflections, pronunciations, all punctuate the flow of behavior. Hall (1983) captured the centrality of rhythm well when he noted that "individuals are dominated in their behavior by complex hierarchies of interlocking rhythms ... every facet of human behavior is involved in the rhythmic process.... Rhythm is, of course, the very essence of time" (p. 140). Rhythm is captured in the cadences of the calypso, the colors of the Carnival costumes, the jumping up on Jouvier morning preceding Ash Wednesday. Improvisation occurs in time as well, because its essence is the online success of creatively solving unanticipated challenges or the spontaneous expression of a thought, a desire, or means of attaining a goal. It is contrasted with the planning approach, which tries to foresee and even create the contexts in which a given behavior will unfold. Improvisation is a value expressed as preference for being-in-the-now, for creating one's life moment by moment in characteristic ways that define one's individuality. *Orality* is the composite of oral influences that also occur in time, and become the basis for the creation of meaning, the forging of social bonds, the transition of core values and beliefs across generations. The "story" helps to create the life; the "song" provides the motivational impetus to live it. Movements, like the Civil Rights Movement, draw their energy and power from orality. The songs ("We shall overcome"), the speeches ("I have a dream"), and the stories (of struggle and moral purpose, heroic acts and noble deeds) define intention, value, and possibility, and provide an anchor to the cultural representation of collective experience. Spirituality expands time to the universality of our common experience now and forever, before and after. *Spirituality* contextualizes the present in a timeless world in which the linearity of M-time and its rules and abstractions have little weight. In the spirit world, time does not move, it just is.

So TRIOS is a cultural worldview that is organized around the momentary capacities of the present to contain the meaning of life, and the vast possibilities of timelessness to embrace the past, present, and future simultaneously.

TRIOS and Context. Context describes all of the available mental representations, emotional experiences, and situational cues and challenges in a moment in time. In Lewin's (1943) field theory, the *principle of contemporaneity* proposes that "any behavior or any other change in a psychological field depends only on the psychological field at that time" (p. 294). The psychological field at a given time includes the psychological past and the psychological future. The psychological past is an "origin" of the psychological present, and the psychological future is a "consequence" of it. So context here describes more than the physical aspects of a situation; it also defines the psychological representations of the situation and what preceded it and what may follow from it.

By this reasoning, situations may vary as a function of the relative density of the psychological past or psychological future brought to bear. The more influence exerted by the psychological future, that is the more goal-related influences, the more constrained the present becomes. The more the psychological past influences the construal of the present, the more constrained one may be by past events and their representations. The greatest flexibility and control one may have in a situation is one where the psychological past and psychological future have minimal roles. That is the meaning of living or being in the now. It is this sense that TRIOS is proposed to capture.

TRIOS and Meaning. Encounters in the moment derive meaning from the relevance of ongoing behaviors and their interpretation. Thus being present oriented is not simply a matter of the relative tendency to focus attention on the present and the resulting individual differences in behavioral tendencies to be sociable, risk takers, and so on. It also means that values and goals are pursued in the moment, the present context. The situation contains all that is needed to define and control outcomes.

For example, from a contextual viewpoint, language is not dependent on semantic meaning so much as on its paralinguistic features, inflections, body language, facial cues, and so on. Conventional meanings of words are replaced by colloquial or neologistic meanings that privilege the speaker over the audience. In Trinidad, "mamaguy" describes verbal utterances whose meaning is opposite to its semantic content. "Your hair looks very nice today," means just the opposite when a person is "mamaguying" you. More commonly, we are culturally aware, now, that "bad" can mean "good," as can "stupid" or "dope." Understanding is not just cognitive ("I understand") or perceptual ("I see what you're saying," or "I hear you") but emotional ("I feel you"). By strategic use of inflection, a simple affirmation ("uh-huh" with a rising inflection and head nod) can become a negation ("uh-huh" with a falling inflection and head shake). Alternative linguistic conventions in this contextual arsenal include the diminutive alternative (home equals "crib"), the graphically illustrated action (to leave is to "bounce"; to show appreciation is to "love"; to be an exemplar of the group or geographical area is to "represent").

Language provides a compelling argument for power "in" the situation. In each of these cases, the language captures the interpersonal, intragroup, and the intrapsychic meaning of things, and links the speaker and the audience in a union fortified against the outsider who, absent cultural understanding, is marginalized and stripped of power to harm. What an utterance means or an actor intends is defined by the parameters in the context itself.

Hall (1983) made a similar point in his distinction between *context-rich* and *context-poor* communications. Context-rich communications are semantically sparse and thus their meaning is derived by locating the utterances in a rich web of cultural nuance and meaning. Context-poor communications, on the other hand, are of necessity semantically dense and rely on the literal meaning of words that are both explicit and durable over time and place.

The community of perceivers who know the culture symbols get it, and outsiders don't. Thus one gains a measure of control when meaning is context-dependent. Conversely, imposed meanings that are instantiated and defined in a hostile culture impose external controls and reduce one's flexibility at self-definition. It is reasonable then to perceive this reliance on context as a means of gaining personal control in the situation and ultimately, control of one's self-worth.

TRIOS Functions in Time: The TRIOS Scale. Finally, we present some data that provide evidence that TRIOSity is more broadly characteristic of African Americans and Africans than Whites, Asians, and Latinos. Jones (2003) reports the development of a scale to assess the degree to which one endorses the dimensions of TRIOS. The dimensions are highly related and using a cumulative composite score is generally warranted.

Time is represented broadly as a present-time orientation—living in the now (e.g., Preparing for what might happen in the future is often a waste of time; It's better to live the present moment to the fullest than to plan for the future; When I try to envision the future, I draw a blank).

Rhythm reflects the synchrony and timing of behavior in relation to the environment and with others (e.g., I often feel anxiety when I am late for a scheduled event; If I feel someone is attacking me, I sometimes struggle not knowing what to do; I often feel that my experiences are not "real" until I tell someone about them). The three items that form this factor were actually written to tap—time, improvisation, and orality, respectively. Although they do not reflect the original ideas of rhythm, they do seem to reflect an asynchrony in relationships between a person and his or her surroundings.

Improvisation is a reflection of the belief that one can successfully overcome unforeseen obstacles, can achieve in spite of external barriers to success, and that one's manner of accomplishing this is heavily based on personal qualities that are self-defining (e.g., When a situation arises, I usually know two to three different ways to handle it; When things do not go as planned, it is easy for me to devise another plan right on the spot; I can figure my way out of almost any situation; When something disrupts my goals, I often figure out how to achieve them anyway). Improvisation is captured by creative and effective problem solving in a challenging context. Uncertainty of expectations is countered by the belief that one will handle whatever arises. In this belief resides a feeling of control.

Orality was conceived principally as the oral expression of meaning through words and song in a social context. Orality conveys meanings handed down over time through stories, but also establishes social bonds through the privileged meanings, styles of speech, and preferences for in-group relations (e.g., It is important to be yourself at all times; It is important for me to be comfortable in a situation in order to be successful; In my social group, laughter often holds us together; It is important for me to maintain harmony in my group). Orality presupposes a high-context social environment and is characterized by a generalized sensitivity to interpersonal relationships in a social context. Relations with friends and in-group harmony reflect the use of orality to maintain social boundaries and promote in-group cohesion. The socially constructed self is reflected by the need for a personal social identity defined by personal properties and invariance across settings. Furthermore, feeling comfort in the social context may be a precondition for psychological well-being.

Spirituality is defined by a belief in a higher power as a functional element of one's daily life (e.g., Belief in God or a greater power helps me deal with the circumstances of my life; In most every aspect of my life, I am strengthened by my spiritual beliefs; I believe that the world is full of powerful and unknowable forces; There are reasons beyond our understanding for everything that happens). There are two aspects to this measure of spirituality: first, that everyday functionality of spirituality such that spiritual beliefs and priorities help one cope with everyday life; and second, that spirituality reflects powerful and unknown forces that intervene to determine life events.

TRIOS: Initial Results. The TRIOS Scale was administered to 1,415 respondents of whom two thirds were women and one third men, a little more than 40% were White, 21% were Black, 19% were Latin, and 11% were Asian (Jones, 2003). The age range was 14 to 62, with the average age for each group between 20 and 21 years. African Americans scored significantly higher than all other groups on the composite of TRIOS, and in addition, they scored significantly higher than or as high as any of the other racial/ethnic groups on each of the individual TRIOS dimensions. Latinos/as scored higher than Whites and Asian Americans on TRIOS, and Asians and Whites did not differ.

We sought to further test the implications of a high level of TRIOSity by exploring the moderating role of TRIOS in the relationship between self-reported stress and psychological well-being (Jones, 2004). If TRIOS has the sort of control and adaptive functions suggested earlier, then, we reasoned, high TRIOS level will buffer the relationship between stress and well-being so that at high levels of stress, psychological well-being will be less severely reduced.

College students at a public predominantly White university, a private historically Black university, a community college serving predominantly African Americans, and a university in the African country of Ghana completed the TRIOS scale and measures of stress and well-being. Stress was measured in two ways: a college students hassle scale (ICSRLE: Kohn, Lafreniere, & Gurevich, 1990) and the Perceived Stress Scale (PSS; Cohen, Kamarck, & Mermelstein, 1983). The ICSRLE and the PSS each assesses the degree to which a person has experienced a variety of "hassles" or "stressors" during the past month. Psychological well-being was also measured over the previous month in two ways: Depression was assessed using the CES-D Scale (Radloff, 1977), which asks participants to indicate how often they have experienced certain feelings such as fear, happiness, crying fits; sadness, and positive and negative affect were assessed using the Positive and Negative Affect Scale (PANAS; Watson, Clark, &

Tellegen, 1988), which asks respondents to indicate to what extent they have felt 10 positive and 10 negative states such as hostile, excited, scared, strong, and so on.

Results showed that Africans and African Americans were higher in TRIOS than Whites, and not different from each other. These race main-effect differences in endorsement of the content of TRIOS suggests stable racial differences in worldview. To determine if TRIOS level played the same or a different role across racial groups in the relationship between self-reported stress and psychological well-being, we conducted multiple regression analyses with well-being as the dependent variable, and stress, TRIOS, and the stress × TRIOS interaction as independent variables. We conducted these regression analyses on the entire sample, and separately for Africans, African Americans, and Whites.

Results for the entire sample showed that stress was negatively associated with and TRIOS was a positive predictor of psychological well-being. Furthermore, stress was negatively associated with TRIOS. Analyses showed that TRIOS level moderated the relationship between stress and well-being such that stress and well-being were significantly negatively related when TRIOS was low, but were unrelated or slightly positive when TRIOS was high. This relationship held for Africans but did not hold for either African Americans or Whites, although it was in the same direction. Thus for Africans, TRIOS level moderated the relationship between stress and well-being. The same pattern was found for improvisation and spirituality for Africans but not for African Americans and Whites.

TRIOS: Summary. TRIOS is connected to time through the direct assessment of time as a dimension in the model, and through the functional association and centrality of time to the other four dimensions. We argue that TRIOS is a worldview that centers on beliefs about the meaning and functionality of time orientation and the values that are associated with it. We refer to it as a temponostic worldview in which being in the now, deriving the meaning of things, communicating, building social bonds, and enjoying the creative flow of personal expression and collective activity is contrasted with the rule-based, prescriptive temponomic worldview. It seems to better capture the worldview of African Americans and Africans than other groups, and plays a more functional role in everyday affairs for Africans than for American Whites or Blacks.

CONCLUSION

Discussion of temporal orientation or perspective seems to presume that the main temporal regions of past, present, and future are separate and function independently. Individuals as well as cultures are characterized by the single dimension on which they earn the highest scores. But as Lewin (1943) theorized, the past and the future are or can be integral to the feelings, motivations, thoughts, and behaviors that occur in any given situation. To the extent that the future and the past are relatively less influential in the present, then a person or culture may be characterized by the attributes associated with present-time perspective. Similarly, the more forcefully future considerations enter in the calculus of present-time experience, then a future orientation may dominate. Those who are continually influenced by the past to the relative disregard of the future may be "stuck in the past."

Equi-temporality

We tend to think temporal regions are related in a linear fashion with the past preceding the present and followed by the future. However, as Lewin argued, from a psychological view all temporal regions are accessible at any moment in time, thus there is the possibility of overlap and mutual influence among temporal regions. This mutuality and overlapping possibilities are better suited to a circular or cyclical relationship. This circularity is also found in cultural perspectives on time as we noted in the *sasa–zamani* axis of the flow of time in Africa.

Temporal regions may be "primed" by cultural values, episodic situations, or goal-oriented motivations. Goals may be distal, which make a future orientation more likely, or proximal, which makes a present orientation more functional.

In the TOS data, both present- and future-oriented participants were optimistic, and both showed areas of ability and achievement (PTO had higher SAT scores whereas FTO had higher GPAs). It is probable that the flexibility of adopting and utilizing multiple temporal perspectives and orientations as a matter of living is an effective way to organize one's life. We believe that a case can been made for the equi-temporality principle—that flexibly incorporating multiple temporalities and deploying them strategically in the service of situational demands and personal life values will yield the best results. Although we developed the argument that African culture promotes a present and past time orientation, we have found that African Americans are more bitemporal. (Boyd & Zimbardo, chap. 5, this volume, make a similar argument.)

So the conclusion should not be that temponomic cultures are better than temponostic ones. What we have learned is that one's individual or collective circumstances dictate which orientations to time have value and thus the capacity to organize and motivate behavior at an individual as well as a collective level. It was Jones' experience with ATTT that brought out the equi-temporality principle and allowed for adaptation and adjustment in culture-relative ways.

REFERENCES

Brislin, R. W., & Kim, E. S. (2003). Cultural diversity in people's understanding and uses of time. *Applied Psychology: An International Review, 52*(3), 363–382.

Brown, W. T., & Jones, J. M. (2004). The substance of things hoped for: A study of the future orientation, minority status perceptions, academic engagement, and academic performance of Black high school students. *Journal of Black Psychology, 30*(2), 248–273.

Cohen, S., Kamarck, T., & Mermelstein, R. (1983). A global measure of perceived stress. *Journal of Health and Social behavior, 24*(4), 385–396.

Costa, P.-T., & McCrae, R. R. (1986). Normal personality assessment in clinical practice: The NEO Personality Inventory. Special Section: Assessing personality characteristics in clinical settings. *Psychological Assessment, 4*(1), 5–13.

Cottle, T. J., & Klineberg, S. L. (1974). *The present of things future: Explorations of time in human experience*. New York: The Free Press.

DeVolder, M. L., & Lens, W. (1982). Academic achievement and future timer perspective as a cognitive motivational concept. *Journal of Personality and Social Psychology, 42*, 566–571.

Hall, E. T. (1983). The dance of life: The other dimension of time. Garden City, NY: Anchor Press/Doubleday.

Hofstede, G. (2001). *Culture's consequences* (2nd ed.). Newbury Park, CA: Sage.

Hofstede, G., & Bond, M. H. (1988). The Confucius connection: From cultural roots to economic growth. *Organizational Dynamics, 16*(4), 4–21.

Holman, E. A., & Silver, R. C. (1998). Getting "stuck" in the past: Temporal orientation and coping with trauma. *Journal of Personality and Social Psychology, 74*(5), 1146–1163.

Jones, J. M. (1988). Cultural differences in temporal perspectives. In J. E. McGrath (Ed.), *The social psychology of time* (pp. 21–38). Newbury Park, CA: Sage.

Jones, J. M. (1994). An exploration of temporality in human behavior. In R. Schank & E. Langer (Eds.), *Beliefs, reasoning, and decision making: Psycho-logic in honor of Bob Abelson* (pp. 389–411). Hillsdale, NJ: Lawrence Erlbaum Associates.

Jones, J. M. (2003). TRIOS: A psychological theory of African legacy in American culture. *Journal of Social Issues, 59*, 217–241.

Jones, J. M. (2004). TRIOS: A model for coping with the universal context of racism. In G. Philogene (Ed.), *Kenneth B. Clark: Essays in honor of a social activist and scholar* (pp. 161–190). Washington, DC: American Psychological Association.

Jones, J. M., Banicky, L., Lasane, T. P., & Pomare, M. (2004). *A temporal orientation scale: Focusing attention on the past, present and future*. Unpublished manuscript, University of Delaware, Newark.

Jones, J. M., & Liverpool, H. (1976). Calypso humour in Trinidad. In A. Chapman & H. Foot (Eds.), *Humour and laughter: Theory, research and applications* (pp. 259–286). London: Wiley.

Kohn, P. M., Lafreniere, K., & Gurevich, M. (1990). The inventory of college students' recent life experiences: A decontaminated hassles scale for a special population. *Journal of Behavior Medicine, 13*, 619–630.

Kroeber, A. L., & Kluckhohn, C. (1952). *Culture: A critical review of concepts and definitions*. New York: Random House.

Lasane, T. P., & Jones, J. M. (1999). Temporal orientation and academic goal-setting: The mediating properties of a motivational self. *Journal of Social Behavior and Personality, 14*(1), 31–44.

Lasane, T. P., & Jones, J. M. (2000). When socially induced temporal myopia interferes with academic goal-setting. *Journal of Social Behavior and Personality, 15*(5), 75–86.

Levine, R. E. (1997). A *geography of time*. New York: Basic Books.

Lewin, K. W. (1943). Defining the "field at a given time." *Psychological Review, 50*, 292–310.

Lyubormirsky, S., & Nolen-Hoeksema, S. (1995). Effects of self-focused rumination on negative thinking and interpersonal problem solving. *Journal of Personality and Social Psychology, 69,* 176–190.

Mbiti, J. S. (1970). *African religions and philosophy.* Garden City, NY: Anchor/Doubleday.

McGrath, J. E. (Ed.). (1988). *The social psychology of time.* Newbury Park, CA: Sage.

Nisbett, R. E. (2002). *The geography of thought.* New York: The Free Press.

Radloff, L. S. (1977). The CES–D Scale: A self-report depression scale for research in the general population. *Applied Psychological Measurement, 11*(3), 385–401.

Raynor, J. O., & Entin, E. E. (1983). The function of future orientation as a determinant of human behavior in step-path theory of action. *International Journal of Psychology, 18,* 463–487.

Rokeach, M. J. (1973). *The nature of human values.* New York: The Free Press.

Scheier, M. F., Carver, C. S., & Bridges, M. W. (1994). Distinguishing optimism from neuroticism (and trait anxiety, self-mastery, and self-esteem): A reevaluation of the Life Orientation Test. *Journal of Personality and Social Psychology, 67*(6), 1063–1078.

Schwartz, S. H. (1992). Universals in the content and structure of values: Theoretical advances and empirical test in 20 countries. In M. P. Zanna (Ed.), *Advances in experimental social psychology* (Vol. 25, pp. 1–67). San Diego: Academic Press.

Sobel, M. (1987). *The world they made together: Black and White values in eighteenth-century Virginia.* Princeton, NJ: Princeton University Press.

Strathman, A., Gleicher, F., Boninger, D. S., & Edwards, S. C. (1994). The consideration of future consequences: Weighing immediate and distant outcomes of behavior. *Journal of Personality and Social Psychology, 66,* 742–752.

Van Calster, K., Lens, W., & Nuttin, J. R. (1987). Affective attitude toward the personal future: Impact on motivation in high school boys. *American Journal of Psychology, 100*(1), 1–13.

Watson, D., Clark, L. A., & Tellegen, A. (1988). Development and validation of brief measures of positive and negative affect. *Journal of Personality and Social Psychology, 54,* 1063–1070.

Zimbardo, P. G., & Boyd, J. N. (1999). Putting time in perspective: A valid, reliable individual-differences metric. *Journal of Personality and Social Psychology, 77,* 1271–1288.

Zuckerman, M., Kuhlman, D. M., Joireman, J., Teta, P., & Kraft, M. (1993). A comparison of three structural models for personality: The Big Three, the Big Five, and the Alternative Five. *Journal of Personality and Social Psychology, 65,* 757–768.

IV

Conclusion

17

Further Study of Behavior in the Context of Time

Jeff Joireman
Washington State University

Alan Strathman
University of Missouri–Columbia

The topic of time has long been of interest to psychologists. To date, thousands of studies have, in some fashion or another, addressed temporal issues such as how people perceive and experience time, which temporal domain people emphasize, and whether people will discount the value of future outcomes or delay gratification. In fact, we know a good deal about these basic temporal issues. What we need to better understand is the extent to which such temporal processes impact behavior across a wide range of settings. In this book, we took a domain-driven approach to this question by identifying broad categories of behavior that past theory and research have suggested involve temporal concerns. The contributors to this book have clearly demonstrated the many different domains in which time plays a role in behavior. The range of perspectives and applications our authors have highlighted suggest that the future is bright for research aimed at understanding behavior in the context of time. In this final chapter, we outline several directions for future research based on the chapters in this book.

INDIVIDUAL DIFFERENCES IN TIME ORIENTATION

One topic that has received a fair amount of attention in past research is time orientation. Unfortunately, as Lasane and O'Donnell have discussed in their chapter, research on time orientation has often been plagued by problems with measurement. Within the past 10 years, however, two scales have emerged that reliably assess the extent to which people orient toward the past, present, or future (i.e., the Zimbardo Time Perspective Inventory [ZTPI]; Zimbardo & Boyd, 1999), and the extent to which people emphasize the immediate versus delayed consequences of their actions (i.e., the Consideration of Future Consequences [CFC] Scale; Strathman, Gleicher, Boninger, & Edwards, 1994). Both scales have shown promise in predicting a range of behaviors with significant personal and social ramifications (e.g., risk taking, health behavior, safe sex, aggression, and proenvironmental behavior), and both scales have been shown to overlap to some degree: Those scoring high in CFC score high on the future and low on the present-hedonistic time orientation scales from the ZTPI. Little is known, however, about the discriminant validity of these scales. For example, we do not know whether either scale can predict behavior over and above the other. We also do not know a great deal about the underlying mechanisms that can explain why these various constructs impact behavior, or the conditions under which their impact is more or less important. Some research employing the CFC construct has begun to address these questions, but much remains to be learned about the operation of these constructs, over and above their simple relationships with behavior. Thus, future research should begin to compare the relative impact of the two scales, their underlying mechanisms, and the boundary conditions that determine when each construct does and does not impact behavior. When combined, studies addressing these issues will also help to highlight the relative advantages of each scale.

As Boyd and Zimbardo have illustrated in this volume, one strength of the five-dimensional ZTPI is its ability to provide time orientation profiles. Using this approach, Boyd and Zimbardo have argued that the best time orientation profile may be one characterized by a combination of high future orientation and high present-hedonistic orientation (i.e., the ability to work toward future goals while still allowing one to stop and enjoy life in the moment). Future research exploiting the ZTPI's ability to provide such time orientation profiles will surely advance our understanding of the impact of time orientation on behavior.

Future research should also attempt to merge insights from work on time orientation and work on temporal discounting. As the chapters by Boyd and Zimbardo and by Finke clearly demonstrate, the emphasis people place on immediate versus delayed outcomes has important implications for their well-being. One the whole, future orientation, and low levels of temporal discounting, are associated with a cluster of more positive life outcomes. Surprisingly, however, these related lines of research on time orientation and temporal discounting have developed in relative isolation from the other. Thus, one direction for future research would be to merge insights from the research on temporal discounting with those on time orientation. This research could help to identify a basic mechanism that underlies the impact of time orientation, and to locate temporal discounting within a broader personality framework. These dual lines of research could, in turn, also draw insights from Nurmi's work on the development of future orientations across the life span. Integrating these three lines of research promises to aid in our understanding of how people come to value and act upon their concern with immediate and distant outcomes.

INTRAINDIVIDUAL PROCESSES

Several of our contributors broke new ground in highlighting how time and temporal concerns shape the way we think and feel. Future research in this area could tackle a number of interesting questions. Routledge and Ardnt, for example, have provided a strong argument for why people find it necessary to track time. Their theorizing and initial research suggest that tracking time helps us deal with the fear of our ultimate demise. This mortality-driven concern with time, in turn, appears to have important implications for how we judge others, especially those people who remind us of our own mortality (e.g., the elderly), judgments that in turn are likely to play a role in how people behave toward this population. Future research exploring these possibilities, and the more general domain of future anxiety as outlined by Zaleski, could yield important insights into how to promote positive attitudes and interactions with the elderly.

Sanna and colleagues' imagination, goals, and affect (IGoA) model of mental simulations about the past, present, and future also offers researchers a comprehensive and integrative framework within which to examine a range of interesting questions revolving. One of the more interesting applications, in our view, is in the area of personality. It would be particularly interesting to examine convergence between the different types of mental simulation specified in the IGoA model and the various time orientation profiles discussed by Boyd and Zimbardo in an effort to better understand the nature of the differences between people with different profiles. The IGoA model could also be used to help elucidate the way people set, work toward, and reflect back on goals, as Sheldon and Vansteenkiste discuss in their chapter on goals as a form of time travel.

Brown and Albarracín discuss how time is inherent in the process of attitude change. Their presentation of models emphasizing the role and importance of prior attitudes affords an interesting view of the process of attitude change. These models may also serve to guide future research on a topic of long-standing interest to social psychologists.

INTERPERSONAL RELATIONS

Another topic ripe for future research is the role of temporal concerns within the context of interpersonal relations. Social interdependence theory (Kelley & Thibaut, 1978) has long posited that temporal concerns help to shape the way people approach interdependent interactions, and research is beginning to bear this out, as the chapters by Rumble, and Parks and Posey, illustrate. Recent work on aggression, couched within the general aggression model (GAM; C. A. Anderson & Bushman, 2002), also suggests that temporal concerns may play an important role in determining how people respond in settings of interpersonal conflict, as Wood and K. A. Anderson have highlighted in their chapter on time and aggressive behavior. Though interdependence theory and the GAM were developed with somewhat different goals in mind, they both converge in suggesting that temporal concerns may influence interpersonal behavior. Social interdependence theory suggests that this occurs when people transform a given situation into an effective situation that is more closely related to behavior, based on social and temporal concerns. Within the GAM, temporal concerns are likely to operate at two locations: first, as person and situation inputs that influence people's thoughts, feelings, and actions, and second, at the appraisal and decision stage of the model. Future research should evaluate these assumptions by examining how temporal concerns shape people's *transformation* of the outcomes in interdependent relationships (cf. Joireman, Kuhlman, Van Lange, Doi, & Shelley, 2003), and how various person and situation variables relevant to temporal concerns (e.g., future orientation and future interaction with a potential target of aggression) serve as inputs into people's thoughts, feelings, and decision-making processes within conflict settings. This research will help to address the range of interpersonal settings across which temporal concerns play a role, as well as the precise mechanisms that can account for the effects of those temporal concerns.

Future research on the role of temporal concerns in interpersonal relations could also be advanced by drawing on the theory and research discussed in Sanna et al.'s chapter on how people think about the past, present, and future, as well as Routledge and Arndt's chapter on time and terror management theory. For example, the former model could help us understand how we respond following a negative social encounter. Quite likely, we will be unhappy, and will consider how things might have been different. We may look back and consider how things might have been different, and then turn around and use those mental simulations to avoid confrontations in the future (cf. Boninger, Gleicher, & Strathman, 1994). Thus, the types of mental simulations that Sanna and colleagues discuss are likely to explain variance in the way people respond in conflict settings. The chapter by Routledge and Arndt can also help us to understand how concern with our ultimate demise may impact our functioning in interpersonal relations.

BEHAVIOR IN GROUPS

Time not only impacts the individual and the dyad, but also society at large. Several of our chapters dealt with the relevance of temporal factors for group- and societal-level problems. As Joireman has noted, temporal concerns play an important theoretical role in people's willingness to cooperate in social dilemmas that model environmental problems, and there is some more direct evidence that willingness to engage in proenvironmental behavior is more likely when people are concerned about the future. Nevertheless, much remains to be learned in this domain. Especially interesting, it seems, would be a parallel line of studies in the lab (using the social-dilemma paradigm) and the field that address how features of the situation (e.g., persuasive messages emphasizing the types of outcomes people value; immediate or delayed) combine with individual differences in concern with immediate versus delayed outcomes to impact proenvironmental behavior.

As Bluedorn has argued, temporal concerns (future focus and depth) are also likely to play a key role in the success of an organization. Yet, as Bluedorn notes, little research has actually examined these questions. How do future focus and depth develop in organizations? Are future focus and depth always associated with better outcomes, or should an organization tailor their level of future depth to the circumstances at hand? Some organizational goals require taking a long-term perspective and thinking well into the future, whereas other goals require the company to be nimble and to adjust to rapidly changing circumstances. Future research examining this match could help to identify strategies for companies wishing to maximize their outcomes.

Finally, as Jones and Brown discuss, any comprehensive treatment of the role of time in behavior must recognize that there are cross-cultural differences in time orientation. These differences raise many interesting questions. Given the increasingly interdependent nature of our world, it is likely that understanding those differences in time orientation can pave the way to smoother cross-cultural interactions, which could carry ramifications in a wide range of settings.

Thus, future research should begin to address the impact of these cross-cultural differences on social interactions.

REFERENCES

Anderson, C. A., & Bushman, B. J. (2002). Human aggression. *Annual Review of Psychology, 53,* 27–51.

Boninger, D. S., Gleicher, F., & Strathman, A. (1994). Counterfactual thinking: From what might have been to what may be. *Journal of Personality and Social Psychology, 67,* 297–307.

Joireman, J. A., Kuhlman, D. M., Van Lange, P. A. M., Doi, T., & Shelley, G. P. (2003). Perceived rationality, morality, and power of social choice as a function of interdependence structure and social value orientation. *European Journal of Social Psychology, 33,* 413–437.

Kelley, H. H., & Thibaut, J. W. (1978). *Interpersonal relations: A theory of interdependence.* New York: Wiley.

Strathman, A., Gleicher, F., Boninger, D. S., & Edwards, C. S. (1994). CFC: Weighing immediate and distant outcomes of behavior. *Journal of Personality and Social Psychology, 66,* 742–752.

Zimbardo, P. G., & Boyd, J. N. (1999). Putting time in perspective: A valid, reliable individual differences metric. *Journal of Personality and Social Psychology, 77,* 1271–1288.

Author Index

Numbers in *italics* indicate pages with complete bibliographic information.

C

D

L

M

N

O

P

R

T

Subject Index

R

S

T

U